PROGRESS IN

Nucleic Acid Research and Molecular Biology

Volume 24

PROGRESS IN
Nucleic Acid Research and Molecular Biology

edited by

WALDO E. COHN

Biology Division
Oak Ridge National Laboratory
Oak Ridge, Tennessee

Volume 24

1980

ACADEMIC PRESS

A Subsidiary of Harcourt Brace Jovanovich, Publishers

New York   London   Toronto   Sydney   San Francisco

COPYRIGHT © 1980, BY ACADEMIC PRESS, INC.
ALL RIGHTS RESERVED.
NO PART OF THIS PUBLICATION MAY BE REPRODUCED OR
TRANSMITTED IN ANY FORM OR BY ANY MEANS, ELECTRONIC
OR MECHANICAL, INCLUDING PHOTOCOPY, RECORDING, OR ANY
INFORMATION STORAGE AND RETRIEVAL SYSTEM, WITHOUT
PERMISSION IN WRITING FROM THE PUBLISHER.

ACADEMIC PRESS, INC.
111 Fifth Avenue, New York, New York 10003

United Kingdom Edition published by
ACADEMIC PRESS, INC. (LONDON) LTD.
24/28 Oval Road, London NW1 7DX

LIBRARY OF CONGRESS CATALOG CARD NUMBER: 63–15847

ISBN 0–12–540024–1

PRINTED IN THE UNITED STATES OF AMERICA

80 81 82 83   9 8 7 6 5 4 3 2 1

Contents

LIST OF CONTRIBUTORS vii
ABBREVIATIONS AND SYMBOLS ix
SOME ARTICLES PLANNED FOR FUTURE VOLUMES xiii

Structure of Transcribing Chromatin

Diane Mathis, Pierre Oudet, and Pierre Chambon

I. Introduction 2
II. Ultrastructure of Transcriptionally Active Chromatin 3
III. Analysis of the Structure of "Active" Chromatin by Nuclease Digestion Reveals the Presence of an Altered Chromatin Subunit 11
IV. Proteins Associated with "Active" Chromatin 20
V. Conclusions, Conjectures, and Prospects 42
References 49

Ligand-Induced Conformational Changes in Ribonucleic Acids

Hans Günter Gassen

I. Introduction 57
II. The Role of the Ribosomal Protein S1 in Adjusting the Conformation of Either 16 S RNA or mRNA 58
III. Induced Conformational Changes in tRNA 71
References 82

Replicative DNA Polymerases and Mechanisms at a Replication Fork

Robert K. Fujimura and Shishir K. Das

I. General Properties of Replicative DNA Polymerases 87
II. Overview of Chain Elongation at a Replication Fork 90
III. Mechanisms of DNA Polymerase Action at a Replication Fork 92
IV. Concluding Remarks on DNA Polymerase and the Replication Complex 102
References 104

Antibodies Specific for Modified Nucleosides: An Immunochemical Approach for the Isolation and Characterization of Nucleic Acids

Theodore W. Munns and M. Kathryn Liszewski

I.	Introduction	110
II.	Nucleic Acid Methylation	114
III.	Immunochemical Procedures	124
IV.	Immunochemical Isolation of Oligonucleotides and Nucleic Acids Possessing Specific Modified Constituents	134
V.	Immunochemical Approaches for Assessing the Function of Modified Constituents	142
VI.	Immunoelectron Microscopy: The Mapping of Specific Determinants within Nucleic Acids	148
VII.	Conclusion	156
	References	158

DNA Structure and Gene Regulation

R. D. Wells, T. C. Goodman, W. Hillen, G. T. Horn, R. D. Klein, J. E. Larson, U. R. Müller, S. K. Neuendorf, N. Panayotatos, and S. M. Stirdivant

I.	Introduction	168
II.	DNA Structure	172
III.	Preparation of Large Amounts of DNA Restriction Fragments	200
IV.	Transcription Recognition Sites	203
V.	Recognition at the Origins of RNA-Primed DNA Replication	218
VI.	Effect of Protein Binding on DNA Conformations	224
VII.	Structure in Single-Stranded Viral DNAs	238
VIII.	DNA Secondary Structures in Intercistronic Regions	244
IX.	Conclusions and Prospects for the Future	253
	References	255
	Note Added in Proof	267

Index ... 269

Contents of Previous Volumes ... 273

List of Contributors

Numbers in parentheses indicate the pages on which the authors' contributions begin.

PIERRE CHAMBON (1), *Laboratoire de Génétique Moléculaire des Eucaryotes du CNRS, U. 184 de l'INSERM de Biologie Moléculaire et de Génie Génétique, Institut de Chimie Biologique, Faculté de Médecine, 67085 Strasbourg Cedex, France*

SHISHIR K. DAS (87), *Biology Division, Oak Ridge National Laboratory, and The University of Tennessee–Oak Ridge Graduate School of Biomedical Sciences, Oak Ridge, Tennessee 37830*

ROBERT K. FUJIMURA (87), *Biology Division, Oak Ridge National Laboratory, and The University of Tennessee–Oak Ridge Graduate School of Biomedical Sciences, Oak Ridge, Tennessee 37830*

HANS GÜNTER GASSEN (57), *Fachgebiet Biochemie der Technischen Hochschule, Darmstadt, D-6100 Darmstadt, Federal Republic of Germany*

T. C. GOODMAN (167), *University of Wisconsin, Department of Biochemistry, College of Agricultural and Life Sciences, Madison, Wisconsin 53706*

W. HILLEN* (167), *University of Wisconsin, Department of Biochemistry, College of Agricultural and Life Sciences, Madison, Wisconsin 53706*

G. T. HORN† (167), *University of Wisconsin, Department of Biochemistry, College of Agricultural and Life Sciences, Madison, Wisconsin 53706*

R. D. KLEIN (167), *University of Wisconsin, Department of Biochemistry, College of Agricultural and Life Sciences, Madison, Wisconsin 53706*

J. E. LARSON (167), *University of Wisconsin, Department of Biochemistry, College of Agricultural and Life Sciences, Madison, Wisconsin 53706*

M. KATHRYN LISZEWSKI (109), *Washington University School of Medicine, Rheumatology Division, St. Louis, Missouri 63110*

DIANE MATHIS (1), *Laboratoire de Génétique Moléculaire des Eucaryotes du CNRS, U. 184 de l'INSERM de Biologie*

---

*Present address: Institut für Organische Chemie und Biochemie, Technische Hochschule, Darmstadt, D-6100 Darmstadt, Federal Republic of Germany.

†Present address: Department of Structural Biology, Stanford University, Stanford, California 94305.

Moléculaire et de Génie Génétique, Institut de Chimie Biologique, Faculté de Médecine, 67085 Strasbourg Cedex, France

U. R. MÜLLER (167), *Department of Microbiology, School of Medicine, East Carolina University, Greenville, North Carolina 27834*

THEODORE W. MUNNS (109), *Washington University School of Medicine, Rheumatology Division, St. Louis, Missouri 63110*

S. K. NEUENDORF (167), *University of Wisconsin, Department of Biochemistry, College of Agricultural and Life Sciences, Madison, Wisconsin 53706*

PIERRE OUDET (1), *Laboratoire de Génétique Moléculaire des Eucaryotes du CNRS, U. 184 de l'INSERM de Biologie Moléculaire et de Génie Génétique, Institut de Chimie Biologique, Faculté de Médecine, 67085 Strasbourg Cedex, France*

N. PANAYOTATOS (167), *University of Wisconsin, Department of Biochemistry, College of Agricultural and Life Sciences, Madison, Wisconsin 53706*

S. M. STIRDIVANT (167), *University of Wisconsin, Department of Biochemistry, College of Agricultural and Life Sciences, Madison, Wisconsin 53706*

R. D. WELLS (167), *University of Wisconsin, Department of Biochemistry, College of Agricultural and Life Sciences, Madison, Wisconsin 53706*

Abbreviations and Symbols

All contributors to this Series are asked to use the terminology (abbreviations and symbols) recommended by the IUPAC-IUB Commission on Biochemical Nomenclature (CBN) and approved by IUPAC and IUB, and the Editor endeavors to assure conformity. These Recommendations have been published in many journals (*1, 2*) and compendia (*3*) in four languages and are available in reprint form from the Office of Biochemical Nomenclature (OBN), as stated in each publication, and are therefore considered to be generally known. Those used in nucleic acid work, originally set out in section 5 of the first Recommendations (*1*) and subsequently revised and expanded (*2, 3*), are given in condensed form (I–V) below for the convenience of the reader. Authors may use them without definition, when necessary.

I. Bases, Nucleosides, Mononucleotides

1. *Bases* (in tables, figures, equations, or chromatograms) are symbolized by Ade, Gua, Hyp, Xan, Cyt, Thy, Oro, Ura; Pur = any purine, Pyr = any pyrimidine, Base = any base. The prefixes S–, H_2, F–, Br, Me, etc., may be used for modifications of these.

2. *Ribonucleosides* (in tables, figures, equations, or chromatograms) are symbolized, in the same order, by Ado, Guo, Ino, Xao, Cyd, Thd, Ord, Urd (Ψrd), Puo, Pyd, Nuc. Modifications may be expressed as indicated in (1) above. Sugar residues may be specified by the prefixes r (optional), d (=deoxyribo), a, x, l, etc., to these, or by two three-letter symbols, as in Ara-Cyt (for aCyd) or dRib-Ade (for dAdo).

3. *Mono-, di-, and triphosphates of nucleosides* (5') are designated by NMP, NDP, NTP. The N (for "nucleoside") may be replaced by any one of the nucleoside symbols given in II-1 below. 2'-, 3'-, and 5'- are used as prefixes when necessary. The prefix d signifies "deoxy." [Alternatively, nucleotides may be expressed by attaching P to the symbols in (2) above. Thus: P-Ado = AMP; Ado-P = 3'-AMP] cNMP = cyclic 3':5'-NMP; Bt_2cAMP = dibytyryl cAMP, etc.

II. Oligonucleotides and Polynucleotides

1. Ribonucleoside Residues

(a) Common: A, G, I, X, C, T, O, U, Ψ, R, Y, N (in the order of I-2 above).

(b) Base-modified: sI or M for thioinosine = 6-mercaptopurine ribonucleoside; sU or S for thiouridine; brU or B for 5-bromouridine; hU or D for 5,6-dihydrouridine; i for isopentenyl; f for formyl. Other modifications are similarly indicated by appropriate *lower-case* prefixes (in contrast to I-1 above) (*2, 3*).

(c) Sugar-modified: prefixes are d, a, x, or l as in I-2 above; alternatively, by *italics* or boldface type (with definition) unless the entire chain is specified by an appropriate prefix. The 2'-*O*-methyl group is indicated by *suffix* m (e.g., -Am- for 2'-*O*-methyladenosine, but -mA- for 6-methyladenosine).

(d) Locants and multipliers, when necessary, are indicated by superscripts and subscripts, respectively, e.g., -m_2^6A- = 6-dimethyladenosine; -s^4U- or -^{4}S- = 4-thiouridine; -ac^4Cm- = 2'-*O*-methyl-4-acetylcytidine.

(e) When space is limited, as in two-dimensional arrays or in aligning homologous sequences, the prefixes may be placed *over the capital letter*, the suffixes *over the phosphodiester symbol*.

2. Phosphoric Acid Residues [left side = 5', right side = 3' (or 2')]

(a) Terminal: p; e.g., pppN... is a polynucleotide with a 5'-triphosphate at one end; Ap is adenosine 3'-phosphate; C > p is cytidine 2':3'-cyclic phosphate (*1, 2, 3*); p < A is adenosine 3':5'-cyclic phosphate.

(b) Internal: hyphen (for known sequence), comma (for unknown sequence); unknown sequences are enclosed in parentheses. E.g., pA-G-A-C(C_2,A,U)A-U-G-C $>$ p is a sequence with a (5′) phosphate at one end, a 2′:3′-cyclic phosphate at the other, and a tetranucleotide of unknown sequence in the middle. (**Only codon triplets should be written without some punctuation separating the residues.**)

3. Polarity, or Direction of Chain

The symbol for the phosphodiester group (whether hyphen or comma or parentheses, as in 2b) represents a 3′-5′ link (i.e., a 5′ . . . 3′ chain) unless otherwise indicated by appropriate numbers. "Reverse polarity" (a chain proceeding from a 3′ terminus at left to a 5′ terminus at right) may be shown by numerals or by right-to-left arrows. Polarity in any direction, as in a two-dimensional array, may be shown by appropriate rotation of the (capital) letters so that 5′ is at left, 3′ at right when the letter is viewed right-side-up.

4. Synthetic Polymers

The complete name or the appropriate group of symbols (see II-1 above) of the repeating unit, **enclosed in parentheses if complex or a symbol,** is either (a) preceded by "poly," or (b) followed by a subscript "n" or appropriate number. **No space follows "poly"** (2, 5).

The conventions of II-2b are used to specify known or unknown (random) sequence, e.g.,

polyadenylate = poly(A) or A_n, a simple homopolymer;

poly(3 adenylate, 2 cytidylate) = poly(A_3C_2) or $(A_3,C_2)_n$, an *irregular* copolymer of A and C in 3:2 proportions;

poly(deoxyadenylate-deoxythymidylate) = poly[d(A-T)] or poly(dA-dT) or $(dA-dT)_n$ or $d(A-T)_n$, an *alternating* copolymer of dA and dT;

poly(adenylate,guanylate,cytidylate,uridylate) = poly(A,G,C,U) or $(A,G,C,U)_n$, a random assortment of A, G, C, and U residues, proportions unspecified.

The prefix copoly or oligo may replace poly, if desired. The subscript "n" may be replaced by numerals indicating actual size, e.g., $A_n \cdot dT_{12-18}$.

III. Association of Polynucleotide Chains

1. *Associated* (e.g., H-bonded) chains, or bases within chains, are indicated by a *center dot* (not a hyphen or a plus sign) separating the *complete* names or symbols, e.g.:

poly(A) · poly(U)   or $A_n \cdot U_m$
poly(A) · 2 poly(U)   or $A_n \cdot 2U_m$
poly(dA-dC) · poly(dG-dT)   or $(dA-dC)_n \cdot (dG-dT)_m$.

2. *Nonassociated* chains are separated by the plus sign, e.g.:

2[poly(A) · poly(U)] → poly(A) · 2 poly(U) + poly(A)
or $2[A_n \cdot U_m] \rightarrow A_n \cdot 2U_m + A_n$.

3. Unspecified or unknown association is expressed by a comma (again meaning "unknown") between the completely specified chains.

Note: In all cases, each chain is completely specified in one or the other of the two systems described in II-4 above.

IV. Natural Nucleic Acids

RNA	ribonucleic acid or ribonucleate
DNA	deoxyribonucleic acid or deoxyribonucleate
mRNA; rRNA; nRNA	messenger RNA; ribosomal RNA; nuclear RNA
hnRNA	heterogeneous nuclear RNA
D-RNA; cRNA	"DNA-like" RNA; complementary RNA

mtDNA	mitochondrial DNA
tRNA	transfer (or acceptor or amino-acid-accepting) RNA; replaces sRNA, which is not to be used for any purpose
aminoacyl-tRNA	"charged" tRNA (i.e., tRNA's carrying aminoacyl residues); may be abbreviated to AA-tRNA
alanine tRNA or tRNA^{Ala}, etc.	tRNA normally capable of accepting alanine, to form alanyl-tRNA, etc.
alanyl-tRNA or alanyl-tRNA^{Ala}	The same, with alanyl residue covalently attached. [*Note:* fMet = formylmethionyl; hence tRNA^{fMet}, identical with tRNA$_f^{Met}$]

Isoacceptors are indicated by appropriate subscripts, i.e., tRNA$_1^{Ala}$, tRNA$_2^{Ala}$, etc.

V. Miscellaneous Abbreviations

P$_i$, PP$_i$	inorganic orthophosphate, pyrophosphate
RNase, DNase	ribonuclease, deoxyribonuclease
t_m (not T_m)	melting temperature (°C)

Others listed in Table II of Reference 1 may also be used without definition. No others, with or without definition, are used unless, in the opinion of the editor, they increase the ease of reading.

Enzymes

In naming enzymes, the 1978 recommendations of the IUB Commission on Biochemical Nomenclature (4) are followed as far as possible. At first mention, each enzyme is described *either* by its systematic name *or* by the equation for the reaction catalyzed *or* by the recommended trivial name, followed by its EC number in parentheses. Thereafter, a trivial name may be used. Enzyme names are not to be abbreviated except when the substrate has an approved abbreviation (e.g., ATPase, but not LDH, is acceptable).

REFERENCES*

1. *JBC* **241**, 527 (1966); *Bchem* **5**, 1445 (1966); *BJ* **101**, 1 (1966); *ABB* **115**, 1 (1966), **129**, 1 (1969); and elsewhere.†
2. *EJB* **15**, 203 (1970); *JBC* **245**, 5171 (1970); *JMB* **55**, 299 (1971); and elsewhere.†
3. "Handbook of Biochemistry" (G. Fasman, ed.), 3rd ed. Chemical Rubber Co., Cleveland, Ohio, 1970, 1975, Nucleic Acids, Vols. I and II, pp. 3–59.
4. "Enzyme Nomenclature" [recommendations (1978) of the Nomenclature Committee of the IUB]. Academic Press, New York, 1979.
5. "Nomenclature of Synthetic Polypeptides," *JBC* **247**, 323 (1972); *Biopolymers* **11**, 321 (1972); and elsewhere.†

Abbreviations of Journal Titles

Journals	Abbreviations used
Annu. Rev. Biochem.	ARB
Arch. Biochem. Biophys.	ABB
Biochem. Biophys. Res. Commun.	BBRC

*Contractions for names of journals follow.

†Reprints of all CBN Recommendations are available from the Office of Biochemical Nomenclature (W. E. Cohn, Director), Biology Division, Oak Ridge National Laboratory, Box Y, Oak Ridge, Tennessee 37830, USA.

Biochemistry	Bchem
Biochem. J.	BJ
Biochim. Biophys. Acta	BBA
Cold Spring Harbor Symp. Quant. Biol.	CSHSQB
Eur. J. Biochem.	EJB
Fed. Proc.	FP
Hoppe-Seyler's Z. physiol. Chem.	ZpChem
J. Amer. Chem. Soc.	JACS
J. Bacteriol.	J. Bact.
J. Biol. Chem.	JBC
J. Chem. Soc.	JCS
J. Mol. Biol.	JMB
Nature, New Biology	Nature NB
Nucleic Acid Research	NARes
Proc. Nat. Acad. Sci. U.S.	PNAS
Proc. Soc. Exp. Biol. Med.	PSEBM
Progr. Nucl. Acid Res. Mol. Biol.	This Series

Some Articles Planned for Future Volumes

Splicing of Viral mRNAs
  Y. Aloni

Metabolism and Function of Cyclic Nucleotides
  W. Y. Cheung

Structure, Replication, and Transcription of the SV40 Genome
  G. C. Das and S. K. Niyogi

Accuracy of Protein Synthesis: A Reexamination of Specificity in Codon–Anticodon Interaction
  H. Grosjean and R. Buckingham

Mechanisms of DNA Replication and Mutagenesis in Ultraviolet-Irradiated Bacteria and Mammalian Cells
  J. D. Hall and D. W. Mount

The Regulation of Initiation of Mammalian Protein Synthesis
  R. Jagus, W. F. Anderson, and B. Safer

Mechanism of Interferon Action
  G. Sen

The Regulatory Function of the 3'-Region of mRNA and Viral RNA Translation
  U. Littauer and H. Soreq

Participation of Aminoacyl-tRNA Synthetases and tRNAs in Regulatory Processes
  G. Nass

Queuine
  S. Nishimura

DNA Methylation and Its Possible Biological Roles
  A. Razin and J. Friedman

Structure of Transcribing Chromatin

DIANE MATHIS,
PIERRE OUDET,
AND PIERRE CHAMBON

Laboratoire de Génétique Moléculaire des Eucaryotes du CNRS
U. 184 de l'INSERM de Biologie Moléculaire et de Génie Génétique
Institut de Chimie Biologique, Faculté de Médecine
Strasbourg, France

I. Introduction	2
II. Ultrastructure of Transcriptionally Active Chromatin	3
A. Ribosomal Gene Chromatin	3
B. Nonribosomal Gene Chromatin	5
C. Does Electron Microscopy of Spread Preparations Reflect the *in Vivo* Situation?	7
III. Analysis of the Structure of "Active" Chromatin by Nuclease Digestion Reveals the Presence of an Altered Chromatin Subunit	11
A. Actively Transcribed DNA Sequences Are Found in Repeating Subunits	11
B. Transcribing Chromatin Is Preferentially Digested by DNase I	12
C. Micrococcal Nuclease Analysis Confirms That Active Genes Reside within an Altered Chromatin Subunit	16
D. DNase II May Also Recognize Some Distinctive Feature of Transcriptionally Active Chromatin	17
E. Differential Nuclease Sensitivity May Be Useful in Isolating Template-Active Chromatin Fractions	19
IV. Proteins Associated with "Active" Chromatin	20
A. The Protein Complement of "Active" Chromatin as Deduced from Nuclease Studies	21
B. Protein Content of Isolated "Native" Chromatin	38
V. Conclusions, Conjectures, and Prospects	42
A. The Basic Structure of "Active" Chromatin: A Unique, Repeating, Altered Nucleosomal Structure?	42
B. Is Histone H1 Present in "Active" Chromatin?	44
C. Factors Involved in the Nuclease Hypersensitivity of "Active" Chromatin	45
D. Generation of the "Active" Chromatin Conformation. Modulation of Gene Activity via Changes in Chromatin Structure	46
References	49

I. Introduction

Regulation at the transcriptional level is one of the mechanisms involved in the control of gene expression in eukaryotic cells (for reviews, see Refs. *1* and *2*). How the structure of chromatin is related to this regulation and whether chromatin conformational changes are a prerequisite for transcription or are secondary to the transcriptional events have been long-standing problems (for review of early work on relationship between euchromatin and active chromatin, see Refs. *3* and *4*).

The approach to these questions has been radically changed by the very rapid progress made in the field of chromatin structure following the discovery of the nucleosome, the repeating subunit of the bulk of nuclear chromatin (for reviews, see Refs. *1, 5–7, 7a*). Supporting the model originally proposed by R. D. Kornberg (*8*), electron microscopic and biochemical studies have established that the bulk of chromatin is organized as nucleosomes, corresponding to the first level in a stepwise compaction mechanism that leads to packaging of DNA at high concentrations in the interphase nucleus or metaphase chromosome. Nucleosomes are particles about 11–13 nm in diameter, which appear to be in close apposition in micrographs of the native compact state of chromatin. They contain a histone core (an octamer made up of two each of the four histones H2A, H2B, H3, and H4) associated with a well defined length of DNA (the DNA repeat length, generated at early times of micrococcal nuclease digestion). A molecule of H1 or H1-like histone is also associated with each nucleosome. The apparent packing ratio of DNA compared with its extended B-form length is about 5. Nucleosome "cores" (core particles) are derived from nucleosomes by further micrococcal nuclease digestion. They contain the histone octamer core (but not H1) and a constant amount of DNA (145 base-pairs), regardless of the amount of DNA originally contained in the nucleosome. The remaining nucleosomal DNA, variable in length, is termed the "linker." The length of the linker varies from about 15 base-pairs to about 100 base-pairs depending upon the organism and tissue from which the nucleosomes are isolated. In fact, there is evidence supporting heterogeneity of the linker length even within a single cell. Thus, the DNA content of nucleosomes can vary between 160 and 240 base-pairs.

The first level of chromatin organization is, then, a nucleosomal chain (the basic 100 Å "thin" fiber) consisting of a linear array of close-packed particles. Higher orders of structure are generated by folding of this nucleosomal chain. The second level of organization is a thread about 25 to 30 nm in diameter (the "thick" fiber), which might be generated by the helical coiling of the nucleosomal chain or the apposition of clusters of nucleosomes (the superbeads) (for Refs., see *9–11*).

What happens to nucleosomes during transcription? Are nucleosomes present in transcribing chromatin? If so, is their structure modified, and by what mechanism? All of these and related problems have been extensively studied during the past four years using both electron microscopic and biochemical approaches. In particular, the recent progress in preparing specific DNA probes has been extremely useful in analyzing the chromatin structure of defined genes. This review is an attempt to summarize the recent studies in order to draw a contemporary picture of the structure of transcribing chromatin.

II. Ultrastructure of Transcriptionally Active Chromatin

To date, most workers have used the "spreading technique" developed by Miller *et al.* (12, 13) to visualize ribosomal or nonribosomal transcription units. With this technique, however, the 250 Å fiber is converted to the 100 Å fiber, and the individual nucleosomal particles of the bulk ("inactive") chromatin are not always in close apposition, as is the case in bulk native chromatin, but exhibit various degrees of nucleosome spacing, most likely related to various extents of nucleosomal DNA unraveling. Instead of the apparent DNA packing ratio of about 5, characteristic of a nucleosomal chain composed of a linear array of close-packed nucleosomes, the Miller technique results in packing ratios of 1.6 to 2.5 (*11, 14–17*). As we will see (Section II,C), these variations in the apparent DNA packing ratio of chromatin that has been dispersed and spread under low ionic strength conditions introduce some difficulties in interpreting the exact significance of the lower DNA packing ratios characteristic of transcriptionally active chromatin.

A. Ribosomal Gene Chromatin

The Miller technique has revealed an absence or marked reduction in the number of nucleosomes in transcriptionally active ribosomal chromatin. This observation is true not only for fully "active" pre-rRNA genes, but also for pre-rRNA matrix units of reduced transcriptional activity and for the apparent spacer regions interspersed between matrix units.

Fully active transcription units are visualized as matrix units (for definitions, see Ref. *12*) showing close spacing of lateral ribonucleoprotein fibrils anchored to the chromatin DNA axis by RNA polymerase A(I) molecules (12–15 nm in diameter) (see Fig. 1a and b) (*14–27*). There is no evidence for particles free of lateral ribonucleoprotein fibrils, and, therefore, nucleosomes appear to be absent from these very actively transcribed ribosomal genes. In accord with this absence, it appears that the DNA of such chroma-

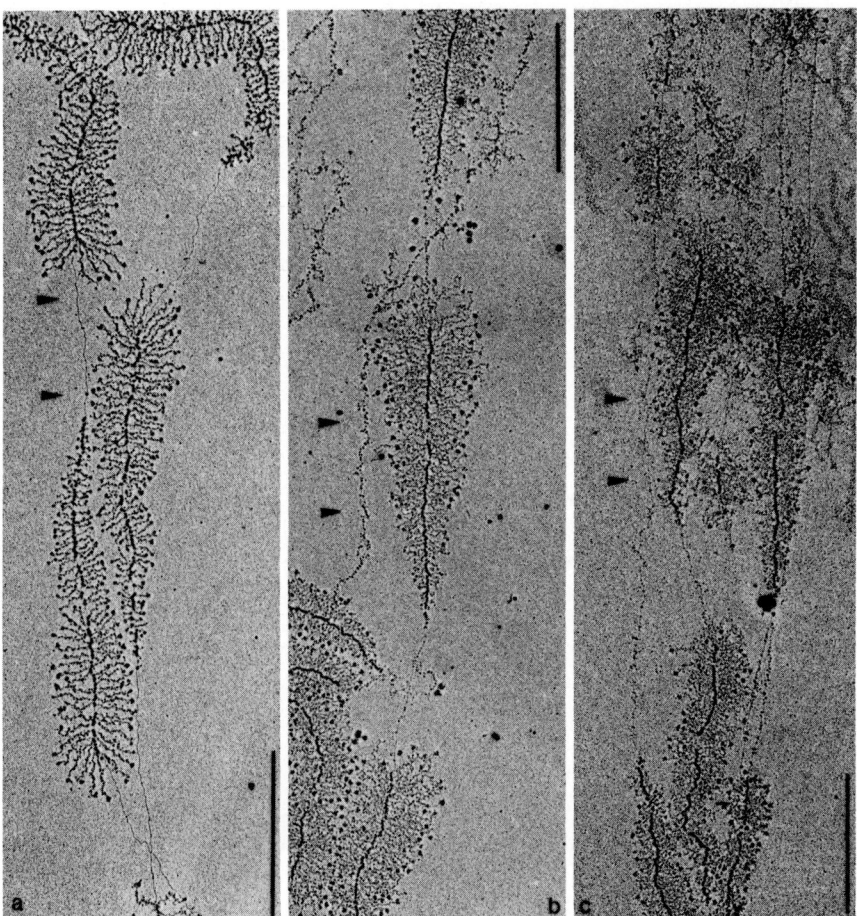

FIG. 1. "Active" ribosomal DNA transcription pattern visualized by the Miller technique (from M. Trendelenburg and R. G. McKinnell, personal communication). (a) "Active" ribosomal chromatin from a *Xenopus laevis* oocyte. Note the nonbeaded appearance of the spacer region, as indicated by arrowheads. (b) Ribosomal transcription units from a full grown oocyte of *Rana pipiens*. The beaded configuration of "inactive" chromatin fibers is indicated by arrowheads. (c) The same material as in (b) was exposed to high concentration (0.1%) of Sarkosyl. The ribosomal transcription unit did not change morphology, but a loss of the nucleosomal configuration in "inactive" chromatin is observed, as indicated by arrowheads. The bar indicates 0.5 μm.

tin segments is not contracted more than about 1.1-fold relative to B-form DNA (*14–16, 25, 28–30*). That the rDNA of the ribosomal transcription unit is not significantly compacted by the presence of nucleosomes is also supported by the preservation of its main structural features after treatment of the chromatin with the detergent Sarkosyl (Fig. 1c), which can dissociate most of the chromatin proteins, including histones, from the DNA (*19, 23, 31*).

Ribosomal transcription units of reduced transcriptional activity, with reduced densities of transcriptional complexes, also exhibit a nonbeaded morphology (15, 23, 24, 31). This apparent absence of nucleosome-sized particles is in agreement with length measurements that indicate that the rDNA is apparently in an extended form. However, the rDNA does appear to be associated with basic proteins, and, when stained with phosphotungstic acid, has a width of 73 ± 17 Å (15, 27).

In some cases, either prior to transcription or shortly after the cessation of transcription, it is possible to identify pre-rRNA genes apparently not associated with lateral ribonucleoprotein fibrils but, nevertheless, showing a nonbeaded appearance ("smooth" or "rho" chromatin) and little or no DNA compaction (14, 15, 19, 23, 24, 31). In contrast, when it has been possible to identify transcriptionally inactive nucleolar chromatin, its morphology and the contraction of the rDNA were indistinguishable from that of the nucleosome-beaded bulk chromatin (see Fig. 1b) (11, 15, 23, 24, 31). The transition of rDNA-containing chromatin from the beaded, condensed state to the nonbeaded, unfolded state, and vice versa, has been observed in stages preceding or following transcription (15, 31). It appears, therefore, that the extended state is not necessarily a consequence of the transcription process, and also that the change from the compact nucleosomal to the extended state is not directly associated with the transcription event.

There is some disagreement on whether or not nucleosomes are present in the apparently nontranscribed spacer regions interspersed between fully "active" matrix units. Several authors have described the presence of nucleosome-sized particles in spacer regions and have concluded that they correspond to nucleosomes (14, 15, 17, 27, 28). However, the apparent contraction ratio of the rDNA spacer chromatin appears to be relatively low, about 1.2 (17). Moreover, Franke and his collaborators have presented several lines of evidence supporting the conclusion that those few nucleosome-sized particles that can be seen in the spacer regions are not of nucleosomal nature. At least some of them could represent molecules of RNA polymerase A (19, 22-24, 32). However, the spacer regions between "active" ribosomal genes in *Drosophila* chromatin are complexed with histones, since they react specifically with anti-histone antibodies (17). Franke and collaborators have also studied the spacers of nonmaximally transcribed ribosomal chromatin. The spacer regions adjacent to some matrix units of reduced transcriptional activity show few or no nucleosome-sized particles (24, 31), whereas the spacer regions of "inactive" nucleolar chromatin are in the beaded configuration (see 11, 31).

B. Nonribosomal Gene Chromatin

The morphology of very actively transcribed nonribosomal genes has been studied in lampbrush chromosome loops (24, 31, 33), in *Drosophila melanogaster* embryo chromatin (*16, 34*), and in the putative transcriptional unit of the silk fibroin gene (*35*). In no case is there evidence for nucleosome-sized particles located in between the RNA polymerase B molecules, and the transcribed DNA is apparently extended. For example, a DNA contraction ratio of about 1.1 has been calculated for the silk fibroin transcription unit (*16, 35*).

In contrast, nucleosome-sized particles are present between the RNA polymerase B(II) molecules of nonribosomal transcription units of low lateral fibril density, indicative of a markedly reduced transcriptional activity (*14-16, 20, 24, 25, 34, 36-39*). The nucleosomal nature of these particles is suggested by their disappearance when lampbrush chromosome loops are treated with Sarkosyl (*31*). In addition, immunological studies show that histones H2B and H3 could be associated with transcriptionally active *Drosophila melanogaster* embryo genes (*16*). In no case has it been possible to determine directly the contraction ratio of these nonmaximally transcribed DNAs, because only undefined genes have been examined (*16, 25*).

However, the recent isolation and spreading of chromatin from the "active" transcription units in Balbiani rings of *Chironomus tentans* provides an example of a defined, nonmaximally transcribed gene (*40*). Whereas the "inactive" bulk chromatin fiber exhibits a uniform beaded conformation (28 beads per micrometer), the "active" transcription units are sparsely and irregularly beaded (4 or 5 beads per micrometer) (Fig. 2A). The DNA is more extended in the transcription units than in "inactive" chromatin, since a DNA contraction ratio of 1.6 can be calculated for the transcription units, whereas that of the bulk "inactive" chromatin is about 1.9. Taken together with previous calculations [DNA packing ratios of 1.6 to 1.9 for transcription units with a relatively low fiber frequency (*14, 39, 41*)], this recent observation (*40*) is in agreement with the proposal (*16*) that, in contrast with the ribosomal transcription units, the degree of chromatin extension in "active" nonribosomal units might be directly proportional to the density of RNA polymerase molecules on the chromatin. In this respect, it is of interest that the *in situ* length of a loop in a Balbiani ring puff might be related not only to the size of the transcribed gene, but also to the density of transcribing RNA polymerase along the "active" unit (*40*). The above proposal is also supported by the demonstration that the number of nucleosomes and contraction ratio of the DNA of transcribing SV40 "minichromosomes" bearing one growing RNA chain is indistinguishable from that of the bulk "inactive" viral chromatin (*42*). Therefore, it appears that after the passage of a polymerase the nucleosomes can be re-formed faster in nonribosomal than in ribosomal (see above) transcription units of moderate or low activity. Alternatively, these differences might be induced during the electron microscopic treatment and may not accurately reflect the *in vivo* situation.

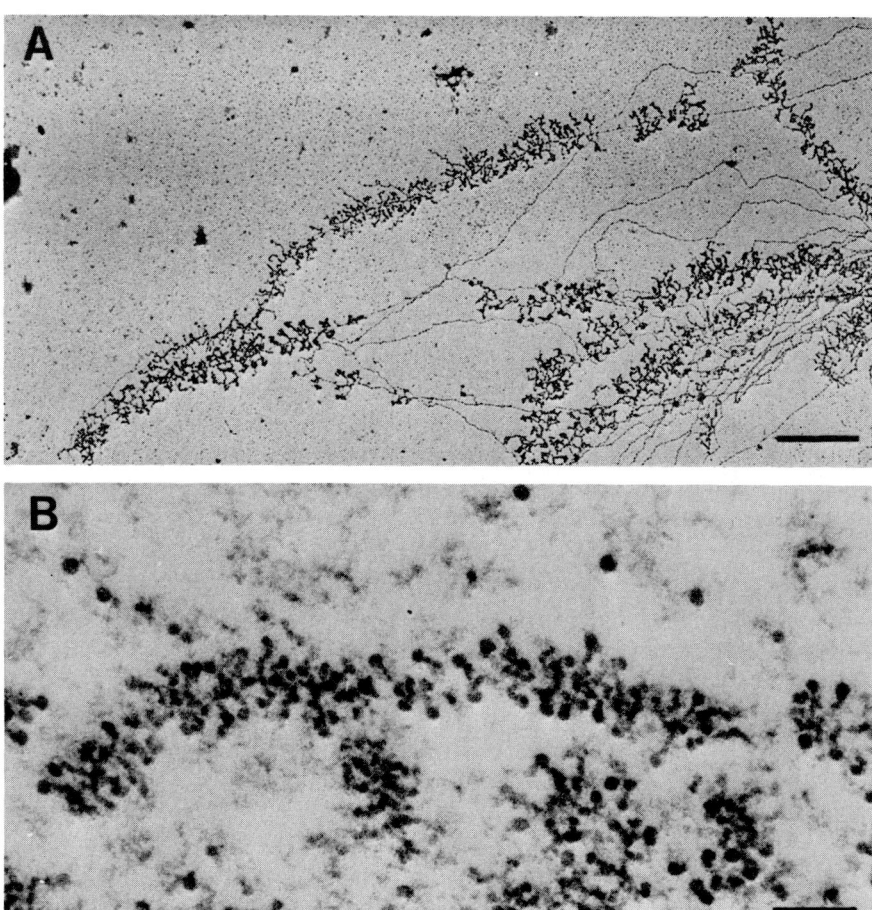

FIG. 2. Electron micrographs (from M. M. Lamb and B. Daneholt, personal communication) of active 75 S RNA transcription units in Balbiani rings of *Chironomus* chromosomes as spread by the Miller method (A) and as observed *in situ* by conventional electron microscopy (B). The bar in panel A corresponds to 1 μm, and that in B to 0.2 μm.

C. Does Electron Microscopy of Spread Preparations Reflect the *in Vivo* Situation?

All the electron microscopic data summarized so far strongly suggest that transcriptionally active and inactive genes have different chromatin structures when examined by electron microscopy using the Miller method. Both highly "active" ribosomal and nonribosomal transcription units appear devoid of nucleosomes, and their DNA is only slightly contracted (about 1.1-fold relative to B-form DNA). However, since the Miller method involves chromatin dispersal in a medium of very low ionic strength and the use of detergents, one cannot rule out that such a treatment alters the structure of the transcribed chromatin (17, 24). This situation is, in fact, very likely, since the compaction ratio of the bulk "inactive" chromatin "thin" fiber is only about 2 under these low ionic strength conditions (11, 14–17), whereas it should be about 5 for the native, unstretched, "inactive" chromatin "thin" fiber (see 1, 5–7a). One can therefore question the real significance of a decreased contraction ratio in transcribing chromatin. Does it mean that the DNA is not compacted *in vivo* relative to its extended B-form or rather that it is compacted, but in altered nucleosomal structures that unfold very readily when the chromatin is stretched under low-salt conditions?

One extreme view would be that all the chromatin DNA is similarly compacted *in vivo* whether it is transcribed or not, but that the actively transcribed chromatin regions are characterized by an increased lability to spreading under low-ionic-strength conditions. Depending on the transcriptional complex frequencies and whether the transcribed gene is a ribosomal or a nonribosomal gene, there might be different degrees of lability, accounting for the observed chromatin structural differences. Such an extreme view is tenable *a priori*, since there is light and electron microscopic immunological evidence that histones could be associated with actively transcribing chromatin (16, 43, 44). That the DNA contraction ratio measured on preparations obtained with the Miller method does not correspond to the *in vivo* situation is supported by recent observations (Lamb and Daneholt, personal communication) that the contraction ratio of the active transcription units in Balbiani rings is about 2 to 3 when measured in sectioned chromosome IV (see Fig. 2B), whereas it is only about 1.6 in Miller spreads. Knowing that the diameter of the animal RNA polymerase B molecule is about 150 Å (Oudet, unpublished results), it is interesting that this 2- to 3-fold contraction is precisely that expected from the number of RNA polymerase molecules per transcription unit, if one assumes that all the DNA of the transcription unit is packed in nucleosomes with the exception of about 50 to 100 base-pairs around each transcriptional complex. In any case, these observations of Lamb and Daneholt indicate that the DNA of an actively transcribed nonribosomal gene is less contracted than that of the bulk "inactive" chromatin, but more contracted than was anticipated from measurements on preparations obtained by the Miller method. Thus, it seems likely that, in between the transcriptional events, the bulk of the infrequently or moderately transcribed nonribosomal chromatin is organized in nucleosome-like structures where the DNA is as compacted as in the nucleosomes of "inactive" chromatin. On the other hand, the observations on sectioned chromosome IV indicate that it is very unlikely that the nucleosomal chain is folded in a higher-order structure when chromatin is transcribed.

From the above considerations, one can understand why fully "active" ribosomal or nonribosomal transcription units have the same extended organization when visualized with the Miller spreading method. It is more difficult to understand why moderately "active" or even "inactive" ribosomal genes that will soon be, or have recently been, transcribed appear fully extended, whereas moderately or weakly "active" nonribosomal genes are significantly compacted under the same spreading conditions. There are two possibilities: (a) the moderately "active" (or nearly "active") ribosomal chromatin is actually fully extended *in vivo*; or (b) it is compacted *in vivo* in the same way as the moderately "active" nonribosomal chromatin, but is characterized by an increased lability to spreading under low ionic strength conditions. There is no way at the present time to distinguish between these two possibilities, but in any case these differences indicate that there should be qualitative differences in the protein complement of the two types of "active" chromatin. Similar explanations could be invoked to account for the low contraction ratio of the apparently nontranscribed spacer DNA of spread "active" ribosomal units and for the apparent discrepancies among various authors (see above) concerning the presence or the absence of nucleosomes in these spacer regions. That the structure of spacer chromatin is altered in some way relative to bulk "inactive" chromatin is clearly demonstrated by the finding (45) that, after injection into germinal vesicles of *Xenopus laevis*, plasmids containing both spacer and pre-rRNA gene regions exhibit nucleosome-like particles only in the nontranscribed plasmid DNA region, but not in the transcribed gene region and the adjacent apparently nontranscribed spacer region (see Fig. 3). It has been suggested (17) that this altered structure might reflect a potential for transcription of the spacer DNA, which normally would not be efficiently transcribed only because of RNA chain termination at the 3' end of the pre-rRNA gene. Alternatively, the nontranscribed spacer could contain sequences at which RNA polymerase initiates only rarely (45–45b).

To conclude, it appears from all of these electron microscopic studies that the structures of ribosomal and nonribosomal transcribing chromatin are different. What is questionable is whether these differences result *in vivo* in different contraction ratios for ribosomal and nonribosomal chromatins exhibiting the same degree of transcriptional activity. It is, however, clear that the overall contraction ratio of DNA from ribosomal or nonribosomal transcribing chromatin is lower than that of the DNA from "inactive" chromatin. This reduced compaction could be achieved by nucleosome disaggregation (dissociation of the histone core from the DNA) or by nucleosome unfolding (opening). As shown in Sections III,B–D, the second alternative is supported by biochemical studies on transcribing chromatin.

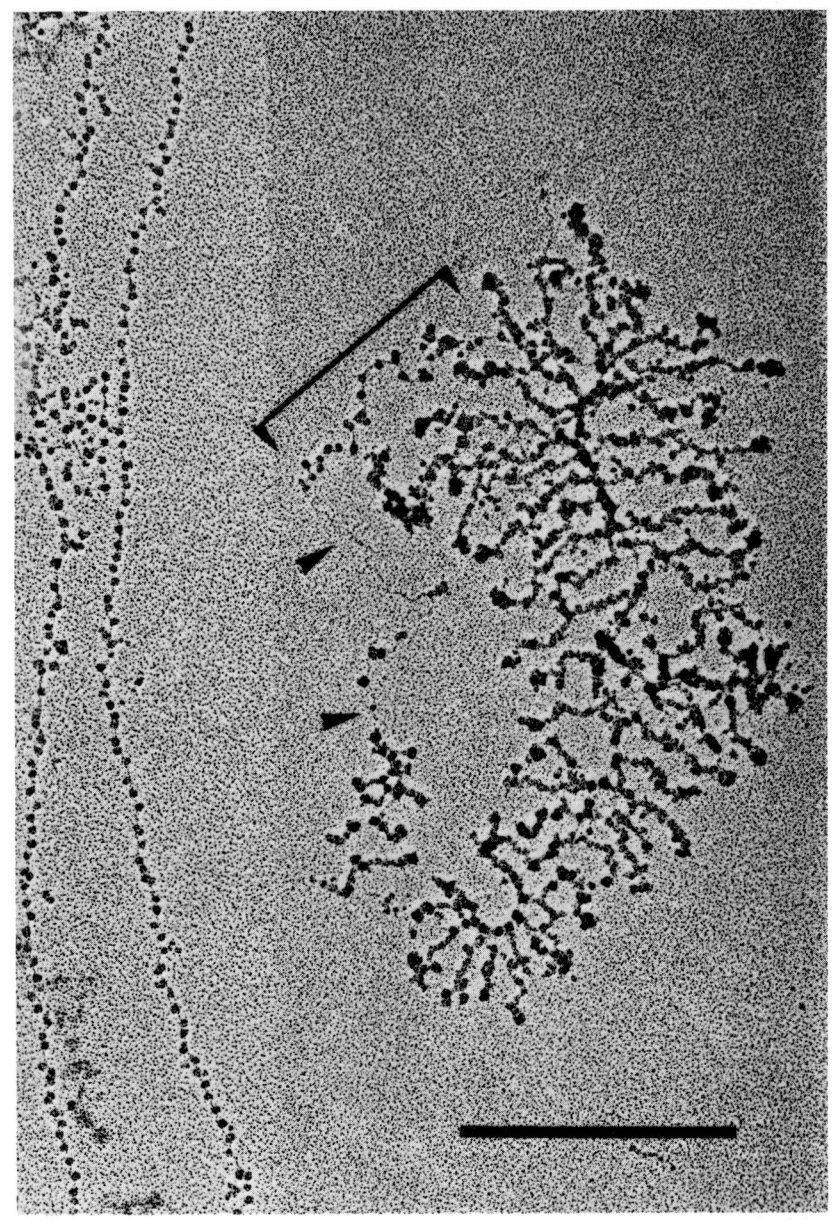

Fig. 3. Miller-spread preparation of an oocyte germinal vescicle injected with ribosomal DNA plasmid (from M. Trendelenburg, J. B. Gurdon, and W. W. Franke, personal communication). A *Xenopus laevis* oocyte was injected with ribosomal DNA-containing plasmid (as described in 45). The morphology of the transcribed region and the unbeaded structure of the "spacer" DNA (arrows) are very similar to those observed in the "active" rDNA transcription units of Fig. 1a. Note that the plasmid sequences (delineated by the pointer bar) have a beaded structure identical to the "inactive" cellular chromatin seen on the left site. The bar at lower right indicates 0.5 μm.

III. Analysis of the Structure of "Active" Chromatin by Nuclease Digestion Reveals the Presence of an Altered Chromatin Subunit

Nuclease digestion studies have produced evidence that chromatin has a repeating subunit structure, and have provided a means to dissect the internal organization of the nucleosome (for reviews, see 1, 5, 7). Mild digestion of chromatin with various enzymes (e.g., micrococcal nuclease, DNase II) reveals a periodicity of susceptible sites at approximately 200-base-pair intervals, thus delineating the basic biochemical subunit. More extensive digestion by micrococcal nuclease can degrade DNA within the subunits, resulting in relatively resistant "core" particles containing about 145 base-pairs of DNA complexed with the histone octamer. When the core is cleaved internally (e.g., with DNase I, DNase II, or extensive micrococcal nuclease digestion), a pattern of fragments based on cuts at about 10 base intervals emerges. Results from the electron microscopic studies of transcribing chromatin immediately raise the questions as to whether actively transcribed DNA sequences are also contained within repeating chromatin subunits, and whether these subunits are of similar architecture to those characteristic of bulk chromatin.

A. Actively Transcribed DNA Sequences Are Found in Repeating Subunits

The first evidence that "active" chromatin exhibits a repeating subunit structure came from studies that analyzed the DNA sequence content of 11 S nucleoprotein particles generated by micrococcal nuclease or DNase II digestion. It could be demonstrated that discrete structures sedimenting with the chromatin nucleoprotein monomers contained globin gene sequences in cells expressing this gene (46, 47), ovalbumin gene sequences in the hen oviduct (48, 50), rat liver DNA sequences coding for polysomal RNA-poly(A) (46), sequences specifying cytoplasmic RNA-poly(A) in human leukocytes (51), and 28 S, 18 S, and 5 S ribosomal genes in *Xenopus* embryonic and erythroid cells (52). It was generally concluded from these studies that actively transcribed DNA is complexed with protein to form a compact particulate structure. Whether the structure is in fact compact, does contain histone, and could possibly be a nucleosome are still questionable on the basis of these experiments. In addition, one can ask whether the observed particles actually exist *in vivo*, or result from the folding of a more open structure during nucleoprotein particle isolation.

A more rigorous criterion for the existence of nucleosomes is the detection of an approximate 200-base-pair repeat in the DNA digestion products generated by micrococcal nuclease. It is well documented that micrococcal nuclease reveals a periodicity of approximately 200 base-pairs in the chromatin containing the ribosomal cistrons (28, 53–56). However, since the ribosomal genes are multiple-copy, in none of these studies can it be stated with certainty that "active" cistrons are contained within the repeat. Likewise, the detection of sequences complementary to mouse liver nuclear RNA-poly(A) within the bulk chromatin repeat (56), is not irrefutable evidence that "active" chromatin exhibits a periodic structure. Nuclear RNA-poly(A) could contain stable RNA synthesized previously, or RNA transcribed from multiple-copy sequences, only some of which are "active" at any given time. At present, the ovalbumin gene in hen oviduct is the only example of transcribed single-copy gene sequences localized within an approximately 200-base-pair repeat (57).

Recent evidence from Wu et al. (58) indicates that, for some genes, the periodic chromatin structure can be disrupted after the onset of transcription. They observed that for certain heat-shock loci there was a progressive blurring of the micrococcal nuclease digestion pattern with increasing time after the heat-shock stimulus. This process was reversible, since after recovery from heat-shock the digestion pattern of the previously "active" sequences was as sharp as the bulk chromatin repeat pattern. Stadler et al. (59) provided another example of an "active" gene, the rRNA-coding sequences of *Physarum*, that does not exhibit the micrococcal repeat pattern typical of bulk chromatin. They quantitated the RNA sequences in the DNA band and interband regions of nuclear micrococcal nuclease digests and could demonstrate that the band/interband distinction was not as sharp for the ribosomal genes as for bulk chromatin. However, they did not rule out the possibility that the extrachromosomal ribosomal chromatin exhibits a "normal" periodic structure, but of a phase different from that of bulk chromatin.

In any case, the existence of a periodic structure similar to that of bulk chromatin does not prove that "active" chromatin is associated with the four core histones. It is conceivable that proteins other than histones can complex with DNA to form a regularly repeated structure. The observation that the ovalbumin sequences in the hen oviduct can be recovered in the metastable 145-base-pair core DNA fragment after extensive micrococcal nuclease digestion (Bellard et al., personal communication) implies that, at least for this gene, the repeat pattern results from a structure very much like the nucleosome. However, it should be kept in mind that the association of DNA with nonhistone proteins, like "DNA gyrase," could also lead to the protection of a 140-base-pair fragment (60).

B. Transcribing Chromatin Is Preferentially Digested by DNase I

The results discussed above indicate that the presence of DNA sequences within repeating chromatin subunits is not in itself sufficient to prevent transcription. The question remains whether actively transcribed sequences reside within subunits identical to bulk nucleosomes. DNase I probing of chromatin structure provides convincing evidence that, indeed, "active" chromatin has an altered, more accessible ("open") configuration. Weintraub and Groudine (61) showed that the globin gene sequences are hypersensitive to digestion by DNase I in nuclei of cells actively expressing these genes. Only those genes coding for adult-specific globin were digested preferentially in 18-day chick embryo red blood cells (which contain only adult-type globin chains), while only embryonic sequences were degraded more rapidly in 5-day chick embryo red blood cells (which contain only embryo-type globin chains). Neither set of sequences was hypersensitive to DNase I in fibroblast or brain cells; nor were the genes coding for ovalbumin mRNA degraded rapidly in red blood cell nuclei. Curiously, the adult-specific globin sequences are not preferentially digested in adult duck reticulocyte nuclei (62), which should have been as capable of globin RNA synthesis as the mature embryo red blood cells.

A structure hypersensitive to DNase I has since been detected for several other "active" genes: the ovalbumin gene in hen oviduct (50, 57, 58), the ribosomal genes in various organisms (63–65), several integrated viral genomes (66–72), and the induced heat-shock loci in Drosophila tissue culture cells (58). In addition, chromatin sequences complementary to the total population of nuclear RNA (61), nuclear RNA-poly(A) (77), or cytoplasmic RNA-poly(A) (74) are hypersensitive to DNase I digestion. Recently (75), the DNase I hypersensitivity of transcribed genes in nuclei has been exploited (75) to label preferentially the "active" chromatin by "nick-translation" of nuclear DNA. Applying this procedure to oviduct nuclei, over 85% of the label was introduced into sequences defined as actively transcribing by hybridization to total cellular RNA.

The enhanced DNase I digestibility of transcribed DNA sequences in chromatin can be envisaged as an intrinsic property of individual "active" subunits or as a property generated by the interaction between subunits not hypersensitive when isolated. For the globin gene in chick red blood cells and for integrated avian virus genomes, isolated monomer subunits retain the DNase I-hypersensitive structure, implying that interactions between subunits is not a factor in the enhanced accessibility of these genes (61, 70). However, there appears to be a loss of DNase I hypersensitivity for the ovalbumin gene in isolated subunit monomers from hen oviduct (48). The discrepancy in these results might be attributed to differences in the procedures for isolating subunit monomers if the property of DNase I sensitivity is a labile one (see Section IV,A,3,a).

The hypersensitivity of transcribed genes to DNase I provokes one to question whether "active" chromatin exhibits the ten-nucleotide repeat of protected DNA fragments typical of DNase I digests of bulk chromatin. The rapidly digested, putatively "active" rDNA-containing chromatin in *Tetrahymena* lacks the 10-base repeat (*64*). Whether this property is a special feature of ribosomal gene chromatin is not known because no nonribosomal gene hypersensitive to DNase I has been studied. Globin gene sequences within the 10-base ladder of duck reticulocyte digests have been detected, but this gene was not preferentially degraded by DNase I in the experiments (*62*).

To determine just how precisely an "open" chromatin subunit conformation reflects the transcription process, attempts have been made to map DNase I sensitivity along a specific stretch of DNA. In a comparison of the susceptibility of integrated adenovirus genes that code for mRNA with that of adjacent adenovirus sequences not expressed as mRNA in transformed hamster cells, it was observed that the property of preferential DNase I sensitivity is specified in a rather precise manner: all sequences coding for mRNA were sensitive, and the sensitivity extended only 3 to 4 nucleosomal DNA lengths beyond the 5' end and 2 to 15 nucleosomal DNA lengths beyond the 3' end of the stable transcript (*66*). It has also been concluded, though to a lower degree of resolution, that there is a fairly close correspondence between sequences transcribed into RNA and DNase I-hypersensitive sequences for the integrated adenovirus and SV40 genomes (*67–69*). On the other hand (Bellard *et al.*, personal communication), DNase I hypersensitivity extends at least 15 and 30 nucleosomal DNA lengths beyond the regions coding for the 5' and the 3' ends, respectively, of the ovalbumin pre-mRNA (*76*).

Since for some genes DNase I hypersensitivity can be rather precisely localized to messenger-coding sequences along a DNA segment, one is led to question whether the transcriptional machinery itself is responsible for the enhanced digestibility. That this is not the case has been demonstrated by several comparisons of the DNase I sensitivity of genes (apparently) transcribed at widely variant rates.

1. The accessibility to DNase I of a complex subset of genes rarely represented in the mRNA population of hen oviduct cells is indistinguishable from that of the ovalbumin gene, which is likely to be transcribed at a rate at least 10 times greater (*77*).

2. While the mRNA population of *Drosophila* cells changes after heat shock (the normally puffed loci being greatly repressed), the sequences coding for the preshock mRNA population are as sensitive to DNase I before and after heat-shock treatment (*78*).

3. In certain mouse erythroleukemia cell lines, the level of expression of the globin gene as mRNA differs by as much as 45-fold in the uninduced

versus Me$_2$SO-induced state; in these cells, the globin gene sequences are hypersensitive to DNase I digestion of nuclei both before and after induction (79).

4. When the potent antiestrogen tamoxifen is administered to estrogen-treated chicks, the level of ovalbumin production falls drastically; however, the ovalbumin genes become equally sensitive to DNase I digestion of chromatin in estrogen-stimulated or tamoxifen-treated chicks (see 80).

5. The adult globin genes in mature erythrocytes of 18-day-old chicken embryos (61) and the sheep γ-hemoglobin genes in adult bone marrow (81) are not detectably expressed as mRNA but do exhibit an enhanced DNase I sensitivity. However, DNase I hypersensitivity of the adult-specific globin sequences in adult duck reticulocytes is not observed (62).

6. Chicken fibroblast cells of the phenotype gs$^+$chf$^+$ contain about 100 copies per cell of RNA specified by the endogenous RAV-O viral genome, whereas those of the phenotype gs$^-$chf$^-$ do not detectably express the viral genome. In both cell types, the RAV-O sequences exhibit an elevated DNase I sensitivity (70).

7. In *Physarum*, the rRNA genes are transcribed maximally in late G$_2$ phase and are essentially inactive during mitosis. The DNase I sensitivity of rRNA-coding sequences is very similar in nuclei from cells synchronized at these two stages (65). In yeast, DNase I digests the nuclear DNA from growing cells more rapidly than it does the DNA from cells in the transcriptionally quiescent stationary phase. However, in both cases, transcribed sequences are degraded at the same rate as total DNA, implying that in yeast the entire genome participates in the conversion from an "active" to an "inactive" state (81a).

The conclusion from these studies is that the maintenance of an "active" chromatin conformation around specific DNA sequences is not merely a reflection of the RNA polymerase distribution: the initiation or acceleration of RNA synthesis depends on additional factors. Stated in other terms, an "open" chromatin subunit conformation around a specific subset of DNA sequences may only signify a potential, i.e., a cell's commitment to a particular differentiated state. The "open" conformation may be established before the onset of substantial RNA synthesis (79) or may be retained after synthesis is essentially stopped (61, 80, 81). However, one point of caution should be noted: all these studies assayed stable RNA or even the protein product as a measure of transcription, and it is possible that this assay does not reflect true transcription rates, since rapid transcript processing could have occurred.

In summary, DNase I analysis has shown that "active" genes reside within chromatin subunits of altered conformation. For some genes, the DNase I hypersensitivity can be rather precisely localized to mRNA coding regions along a stretch of DNA. This "open" configuration is probably not the result of an alteration in the interaction between nucleosomes and cannot be attributed to the mere presence of the transcription machinery. In fact, in some cases, it might only reflect a potential for transcription.

C. Micrococcal Nuclease Analysis Confirms That Active Genes Reside within an Altered Chromatin Subunit

It has often been stated that, while DNase I exhibits a preference for "active" DNA sequences in chromatin, micrococcal nuclease does not attack these genes preferentially. This conclusion was based on several reports that "active" gene sequences are not degraded (i.e., rendered nonhybridizable) faster than bulk DNA, when chromatin is digested by the latter enzyme (e.g., *48, 50, 57, 61*). However, it has now been clearly demonstrated that micrococcal nuclease can recognize some feature of "active" chromatin, which results in the preferential excision of "active" monomer and lower oligomer subunits. For example, titration of the content of ovalbumin and globin gene sequences in the multimeric series of DNA digestion products resulting from a very mild micrococcal nuclease treatment of hen oviduct nuclei produced a clear enrichment of ovalbumin gene sequences in the monomer and small oligomer fragments; conversely, the globin gene sequences were relatively impoverished in these fractions. Similar digestion of hen erythrocyte nuclei resulted in no preferential cleavage of either the ovalbumin or globin genes (*57*).

The finding that chromatin subunit monomers containing ovalbumin gene sequences are preferentially excised in cells actively expressing this gene has been confirmed (*82*). The accelerated excision of "active" sequences by micrococcal nuclease has been demonstrated for other non-ribosomal genes, for the *Drosophila* heat-shock loci (*58*), and for sequences coding for cytoplasmic RNA-poly(A) in trout testis (*83*).

The "active" ribosomal cistrons of certain organisms appear to be especially sensitive to micrococcal nuclease. At early stages of digestion, the *Physarum* ribosomal genes are preferentially recovered in two nucleoprotein monomer fractions (differing in sedimentation value, but both containing DNA fragments of about 140 base-pairs) (*84, 85*). In addition, at all stages of digestion, rRNA-coding sequences are preferentially converted to nonhybridizable material. This latter observation is consistent with several reports that the "active" ribosomal cistrons are preferentially rendered nonhybridizable when micrococcal nuclease digests *Xenopus* nuclei (*86–88*). The higher the rate of transcription of these genes, the more pronounced their enhanced micrococcal nuclease sensitivity.

STRUCTURE OF TRANSCRIBING CHROMATIN 17

These findings contrast with many claims (48, 50, 57, 61) that micrococcal nuclease does not rapidly degrade (to acid-soluble or nonhybridizable material) actively transcribed DNA within nuclei. The accelerated degradation may be a peculiarity of the ribosomal genes. However, there is both an example of a nonribosomal gene that is preferentially degraded (the integrated murine leukemia proviral genome—see 71) and examples of ribosomal cistrons that are not (53–55, 59, 89). The experiments claiming to demonstrate a preferential degradation of "active" genes to nonhybridizable fragments should be scrutinized on two accounts. First, most involve hybridization of very small DNA fragments (< 200 base-pairs), and the quantitation of hybridization of these small fragments is likely to be subject to error. Second, in none of these studies is there an attempt to assay endogenous nuclease activity; an endogenous DNase I-like activity would, for example, preferentially degrade "active" DNA sequences. This problem has been discussed at length (89).

In summary, micrococcal nuclease probing provides evidence that transcribing genes exhibit a periodic chromatin structure, but has also demonstrated that "active" sequences reside within chromatin subunits that differ in configuration from bulk nucleosomes. The sensitivity to micrococcal nuclease of "active" genes in chromatin has not been so well characterized as their sensitivity to DNase I. For example, it is not known how closely increased micrococcal nuclease susceptibility corresponds to the transcriptional unit in a linear sense; nor is it known whether sensitivity to this enzyme reflects transcriptional ability or potential. That is, it has not been determined whether the chromatin structural alteration responsible for enhanced micrococcal nuclease sensitivity is identical to that promoting DNase I hypersensitivity.

D. DNase II May Also Recognize Some Distinctive Feature of Transcriptionally Active Chromatin

Evidence that "active" chromatin exhibits an altered structure also derives from the chromatin fractionation studies of Gottesfeld *et al.* (for review, see 90). Their method of separating bulk chromatin and transcriptionally active chromatin depends on two factors: first, "active" chromatin seems to be more susceptible to DNase II digestion than "inactive" chromatin; and second, chromatin with associated RNA chains is soluble in 2 mM Mg^{2+} whereas "inactive" chromatin is not. The isolation of a template-active chromatin fraction has been claimed on the basis of the following criteria.

1. The amount of DNA recovered in the template-active fraction varies according to the transcriptional activity of the cell type under study (91).
2. According to hybridization studies, the DNA in the "active" fraction is

a specific subset of the genome (92, 93). This specificity applies both to the unique-copy and the middle repetitive classes of sequences. In addition, template-active fractions isolated from different tissues (e.g., rat brain versus rat liver) contain different subsets of sequences.

3. The DNA in the template-active fraction is enriched in sequences coding for total cellular RNA and cytoplasmic mRNA (92, 94). On the basis of the amount of DNA that hybridized with a vast excess of cellular RNA, it was estimated that about 60% of the DNA in the "active" fraction codes for RNA. That the remaining 40% of the DNA is also a specific subset of the genome, not a random contamination from bulk chromatin, could be argued on the basis of DNA renaturation kinetics (92).

4. Nascent RNA chains and RNA polymerase activity copurify with the DNase II-released, Mg^{2+}-soluble material (95–98).

A more rigorous test of the specificity of the DNase II/Mg^{2+} fractionation method comes from studies that determine the concentration of a specific gene in the putatively "active" fraction isolated from cells in which the gene is differentially expressed. Experiments comparing the concentration of globin sequences in the DNase II-released, Mg^{2+}-soluble fraction from uninduced versus induced mouse erythroleukemia cells have yielded conflicting results. There is no enrichment for the globin gene in the "active" fraction from uninduced Friend cells, but there is a sixfold enrichment (compared with the DNase II-resistant "inactive" fraction) in Me_2SO-induced cells (94). Others (99) observed no enrichment of globin sequences in the DNase II-released, Mg^{2+}-soluble fractions of either induced or uninduced Friend cells. Still different results indicate an approximately sevenfold enrichment (compared with whole cell DNA) for the globin sequences in the template-active fractions from both induced and uninduced cells (100). Whether these conflicting results derive from subtle differences in fractionation technique (e.g., method of preparing nuclei (100a) or extent of DNase II digestion) or whether they are related to differences in Friend cell clones is not known. For example, different clones express the globin gene to variant degrees in the uninduced state (see discussion in 79). In this regard, it is somewhat surprising that no enrichment of globin sequences in the "active" fraction from uninduced cells of clone M2 was observed (94). Uninduced cells of this clone exhibit only a threefold impoverishment of nuclear globin RNA when compared with induced cells (101). Conflicting results have also been published concerning the degree of enrichment for globin sequences in the DNase II-released, Mg^{2+}-soluble fraction from chick reticulocyte chromatin: no enrichment for these sequences (102) and a three- to fivefold enrichment (the parallel fraction from chick liver chromatin was not enriched) (103). The difference between these two sets of results is likely to be due to the more extensive DNase II digestion employed in the former case (102).

On the assumption that template-active sequences are localized preferentially in the DNase II-released, Mg^{2+}-soluble chromatin fraction, Gottesfeld and Butler (*104*) have characterized the structure of the resultant chromatin fragments. The presumably "active" chromatin exhibits a DNase II DNA fragment repeat (199 base-pairs) indistinguishable from that of bulk chromatin (198 pairs). However, the nucleoprotein monomers and oligomers are of higher sedimentation values than their bulk chromatin counterparts, e.g., 14 S versus 11 S for the monomer. When treated with RNase, the 14 S structures convert to 11 S, suggesting that the differences in sedimentation values can be attributed to associated RNA. It is interesting to note that the putatively "active" 14 S monomers are DNase I-hypersensitive, implying that DNase I and DNase II recognize the same regions of template-active chromatin. Even though hypersensitive to DNase I, the DNase II-released, Mg^{2+}-soluble material exhibits a typical 10-nucleotide repeat after DNase I digestion. This finding contrasts with a report that the DNase I-hypersensitive, putatively "active" rDNA-containing chromatin in *Tetrahymena* does not exhibit a 10-nucleotide repeat (see Section III,B).

In summary, the evidence that DNase II specifically recognizes actively transcribed DNA sequences in chromatin is suggestive, but not conclusive. Results from studies analyzing the structure of the putatively "active" chromatin fraction derived from the DNase II/Mg^{2+} procedure are consistent with results from micrococcal nuclease probing. That is, transcribed genes exhibit a periodic chromatin organization, but "active" chromatin is in an altered, more acessible conformation.

E. Differential Nuclease Sensitivity May Be Useful in Isolating Template-Active Chromatin Fractions

Over the past 10 years, several groups have directed their efforts toward the isolation of an "active" chromatin fraction consisting of a mixed population of transcribed genes. The approach has generally been to shear chromatin randomly, and then to separate the template-active and -inactive fractions on the basis of some physical property, such as sedimentation rate or t_m. This approach has not resulted in any separation scheme that can satisfy the criterion of an enrichment for specific transcribed DNA sequences within the putative template-active chromatin fraction (for reviews, see *3, 89, 90, 105–109*).

The preferential nuclease susceptibility of transcribed DNA sequences in chromatin could be useful in devising "active" chromatin fractionation schemes. The hypersensitivity of transcribed genes to DNase I is probably not of value in this respect, because the "active" sequences are rapidly degraded to acid-soluble material. On the other hand, in most cases micrococcal nuclease preferentially excises "active" DNA sequences as monomer or short oligomer nucleoprotein particles and does not preferentially degrade the "active" DNA to acid solubility. Thus, a monomer fraction was isolated from hen oviduct nuclei that is enriched 5- to 6-fold for ovalbumin gene sequences when compared with unfractionated chromatin and 25- to 36-fold when compared with an "inactive" chromatin fraction (82). This procedure seems to be based on chromatin transcriptional activity because neither the ovalbumin gene in hen liver nor the globin gene in hen oviduct were localized in the template-active fraction. Similarly, two monomer fractions from trout testis nuclei enriched about 7-fold for sequences complementary to cytoplasmic RNA-poly(A) have been isolated (83). Tata and Baker (110) also employed mild micrococcal nuclease digestion in their very gentle method to subfractionate rat liver nuclei. They obtained an "active" fraction containing 10% of the nuclear DNA as aggregates of 6 to 30 "nucleosomes" with 85% of the template-engaged RNA polymerase II (B). This finding suggests that "active" chromatin fragments have been isolated, but the concentration of transcribed DNA sequences within this fraction has not yet been determined.

As discussed above, the DNase II/Mg^{2+} fractionation procedure (90) yielded a chromatin fraction defined as template active by several criteria. However, whether this fraction is enriched in specific "active" DNA sequences is still questionable, and furthermore, it is not known to what extent DNase II preferentially degrades transcribed DNA sequences to acid-soluble material.

The use of nucleases to release preferentially actively transcribed genes will, no doubt, be of great value in devising "active" chromatin fractionation procedures. The possibility of specifically labeling "active" chromatin by "nick-translation" should also accelerate the development of new fractionation schemes (75). It should be kept in mind that a major disadvantage of the nuclease digestion approach is that "active" genes are generally fragmented in order to be excised from bulk chromatin and to be maintained in a soluble state. Moreover, the population of nuclease-released "active" genes may contain sequences differing widely in degree of transcription. Since the strategies for regulation of these genes might be vastly different, it could prove difficult to draw a generalized picture of "active" chromatin structure. Thus, there is a limit to the questions that can be answered using this approach, but it may be possible to determine, in gross terms, whether "active" and "inactive" fractions differ in protein content or nucleosome structure.

IV. Proteins Associated with "Active" Chromatin

Nuclease digestion analysis has led to the conclusion that transcribed genes are complexed with proteins to form a periodic structure similar to, but more accessible than, that of the nucleosome-packaged bulk chromatin. Since biochemical and electron microscopic evidence indicate that the presence of the transcriptional machinery is not necessary to maintain chromatin in an "active" configuration (see Sections II,A and II,B) much recent effort has been directed at elucidating those factors that might be responsible for this altered, more "open" conformation. One hypothesis has been that "active" chromatin regions exhibit an atypical histone composition: a depletion of some or all of the core histones (*111*, *112*), an absence of H1 (*3*), or the presence of specific primary sequence histone variants (*113*). Another popular view is that template-active chromatin exhibits a typical histone complement, but that postsynthetic histone modifications effect an alteration in "active" chromatin structure that facilitates transcription (*114*). Finally, nonhistone proteins have often been suggested as factors that could modulate chromatin transcription (*115*, *116*). Two general approaches have been favored in attempts to assess the role of these various factors. First, several studies have analyzed the proteins released together with actively transcribed DNA sequences after mild nuclease digestions. Second, there have been attempts to characterize the proteins associated with specific genes isolated in chromatin form.

A. The Protein Complement of "Active" Chromatin as Deduced from Nuclease Studies

Nuclease probes now provide a means by which to study chromatin constituents that appear to be associated with transcribed genes. Any component that cosolubilizes with "active" sequences after DNase I digestion, coisolates with "active" monomers after very limited micrococcal nuclease digestion, or cofractionates with "active" sequences using the DNase II/ Mg^{2+} procedure (*90*) could be involved in the maintenance or generation of an "active" chromatin conformation. This approach has provided information on whether transcribing chromatin exhibits an atypical histone composition, contains histones subject to extensive postsynthetic modification, or is characterized by a specific complement of high-mobility group or other nonhistone proteins.

1. THE HISTONE COMPLEMENT: CORE HISTONES, H1, PRIMARY SEQUENCE VARIANTS

As discussed in Section III,A, the fact that "active" chromatin exhibits a periodic structure implies, but does not prove, that the four core histones are associated with transcribed genes. More direct evidence that the core histones are components of transcribing chromatin has come from the immunoelectronmicroscopic detection of H2A and H3 associated with transcriptionally active *Drosophila* embryo genes (*16*, *17*). Unfortunately, this approach provides no information about histone stoichiometry.

Further evidence for an association of the core histones with "active" chromatin derives from analysis of the proteins released with transcribed DNA sequences at early stages of digestion by DNase I, micrococcal nuclease, or DNase II. A solubilization of all four core histones could be demonstrated at early stages of DNase I digestion (*117, 118*). There was no obvious depletion of any core histone, but there was no way in these studies to determine an overall histone/DNA ratio for the solubilized material. The putatively "active" chromatin fraction released by micrococcal nuclease digestion has likewise been observed to contain the four core histones (*83–85, 110, 119, 120*). Isolated monomers enriched in sequences coding for trout testis RNA-poly(A) were reported to contain equimolar amounts of each core histone (*120*); in contrast, monomers enriched in *Physarum* rRNA-coding sequences appeared to have reduced amounts of the *Physarum* equivalents of H3 and H4 (*84, 85*). The supposedly template-active chromatin fraction resulting from the DNase II/Mg^{2+} fractionation procedure (*90*) seems also to contain all of the core histones (*104, 121, 122*). The histone/DNA ratio of this fraction was reported to be similar to that of bulk chromatin (*104*), and the stoichiometry of the individual core histones did not differ substantially from that of bulk chromatin (*121*).

On the basis of these results, it can probably be stated that "active" chromatin contains the core histones. What is not definitively established is the exact stoichiometry of the individual histones, as well as the overall core histone/DNA ratio. A depletion in any or all of the core histones will, in fact, be difficult to demonstrate convincingly. Artifactual losses due to proteolysis or incomplete extraction will have to be stringently eliminated.

Several early chromatin fractionation studies reported that "active" chromatin is depleted in H1 (for review, see *3*). However, the validity of this conclusion has been subject to some doubt, since H1 is the histone most prone to proteolysis (*123*) as well as that most easily dissociated from chromatin (*124, 125*). Results from more recent nuclease digestion studies have not been able to resolve the question of whether H1 is, indeed, associated with "active" chromatin. A substantial solubilization of H1 during early stages of DNase I digestion of nuclei has been observed (*117, 118*), suggesting that this histone could be associated with transcribing regions. On the other hand, putatively "active" chromatin fractions preferentially released by micrococcal nuclease or DNase II digestion have been reported both to contain H1 (*83, 104, 121*) and to be H1-depleted (*110, 122*). Further complicating is the finding (*126*) that DNase I preferentially releases chromatin subunits deficient in H1, while micrococcal nuclease exhibits no such specificity. Again, proteolysis could explain an H1 impoverishment; furthermore, the nuclease digestion, itself, could dislodge H1 from its *in vivo* binding sites, and artifactual migration to new sites could ensue (*127*).

Although difficult to answer, the question of whether H1 is associated with transcribing chromatin is an important one. The absence of H1 might explain certain characteristics of "active" chromatin: an apparent lack of higher-order folding (see Section II,C), some features of the preferential nuclease sensitivity (e.g., the rapid excision of "active" monomers by micrococcal nuclease; cf. *128, 128a*), and an increased solubility in 0.1 to 0.15 M NaCl (*91, 92, 95, 98, 120*). It is also relevant to consider that, from a careful quantitation of histone stoichiometry in mouse cell nuclei, it was calculated that there is less than one copy of H1 per nucleosome (*129*); hence, H1 could be absent in transcribing chromatin.

Primary sequence variants of histones H1, H2A, H2B, and H3 have been identified with the aid of polyacrylamide gel electrophoresis, especially in the presence of Triton X-100 (*113, 130, 131*). Both species-specific and tissue-specific patterns of histone subfractions have been described. More interesting, developmental stage-specific variants have been observed during sea urchin embryogenesis. These findings suggest that primary sequence histone variants could play a role in the coarse control of gene expression; however, any evidence for histone heterogeneity being involved in even this coarse control is at present circumstantial. As yet, no study has exploited the nuclease probes described above to determine whether specific primary sequences histone variants are preferentially associated with transcribed genes. All the major mouse cell primary sequence variants seem to be represented equally in the spectrum of nucleoprotein particles generated by micrococcal nuclease (*132*); unfortunately, the rather extensive digestion employed (to 18% DNA acid solubility) precludes the possibility of drawing any conclusions about the histones associated with "active" chromatin. More recently, a difference in the ratio of two H1 variants in monomer versus larger nucleoprotein particles was observed (*129*), but again, owing to the advanced stage of digestion, it is not possible to assess the significance of this result in terms of the protein complement of "active" chromatin.

A different type of histone variant is represented by protein A24, originally identified as a component of rat liver nucleoli (for review, see *133*). Protein A24 is a branched protein, composed of H2A and ubiquitin (*134*), and appears to be a component of rat liver chromatin, present on about 10–20% of the nucleosomes (*135*). This stoichiometry has been confirmed in mouse cells, and it is calculated that H2A is not precisely equimolar with the other core histones (*129*). That this protein is, in fact, an integral component of certain nucleosome cores was demonstrated by its presence in purified, 0.55 M NaCl-washed core particles and by its integration into *in vitro*-assembled nucleohistone complexes in a manner analogous to H2A (*136*). It has also been reported that ubiquitin, the nonhistone protein moiety of A24, is found in association with chromatin, possibly as a component of the nucleosome linker region (*137*).

In order to determine whether protein A24 or free ubiquitin might be preferentially associated with transcribed DNA sequences, Busch et al. (122) fractionated rat liver chromatin by the DNase II/Mg^{2+} method (90); free ubiquitin partitioned in the DNase II-released, Mg^{2+}-soluble (putatively "active") fraction while A24 was preferentially associated with supposedly "inactive" fractions. Unfortunately, the actual partitioning of transcribed DNA sequences was not quantitated.

In summary, the histone complement of template-active chromatin remains ill-defined. The core histone/DNA ratio, the stoichiometry of the core histones, the presence of H1, and the contribution of primary sequence histone variants are all in question.

2. Postsynthetic Histone Modifications: Acetylation

Postsynthetic histone modification by phosphorylation, acetylation, methylation, and ribosylation by poly(ADP-ribose) is a well characterized feature of chromatin metabolism (for recent reviews, see 114, 133, 138, 139). It has been hypothesized that these transient modifications provide chromatin with the flexibility dictated by the processes of transcription and replication. Since the evidence is sketchy for the involvement of histone phosphorylation, methylation, and ribosylation by poly(ADP-ribose) in the regulation of gene expression, these modifications will not be discussed. On the other hand, acetylation of histone has repeatedly been invoked as the mechanism by which inert chromatin is activated for RNA transcription (e.g., 140). This hypothesis is attractive for two reasons. First, a striking correlation between the enhancement of histone acetylation and the enhancement of RNA synthesis has been detailed in a number of systems (114, 138). Second, since acetylation normally is found in the basic NH_2-terminal "arms" of core histones, it is easy to envision how this modification could alter DNA · histone binding to allow RNA polymerase read-through. A model has been devised whereby the acetylation–deacetylation reaction provides the mechanism to transduce laterally across an "active" gene an increased accessibility to RNA polymerase (141).

Until recently, it has been extremely difficult to study histone acetylation because of the normally low levels of this modification in most types of chromatin. Some attempts have been made to study chemically acetylated histones or chromatin (142–145), but these experiments must be interpreted with some caution because chemical acetylation has been shown to modify histone fractions that are not modified in vivo, often at sites not normally acetylated (146, 146a).

At the 1977 Cold Spring Harbor Symposium on Chromatin, a method to greatly enhance levels of histone acetylation in vivo was described (147). For instance, by treating growing cells with millimolar concentrations of sodium butyrate, 80% of histone H4 was acetylated to some degree. Depending on the cell type, all four core histones could be modified (148, 149). Most important, the effect of butyrate treatment is to inhibit deacetylation without perceptibly altering the acetylation reaction (148–155), implying that the butyrate-induced, highly acetylated molecules have been modified at their proper *in vivo* sites. Consequently, it has been possible to use nuclease probes on the chromatin from butyrate-treated cells to find out whether high levels of histone acetylation could be involved in the nuclease hypersensitivity of transcribing chromatin.

a. DNase I Studies. The first question that several laboratories attempted to answer was whether highly acetylated chromatin from butyrate-treated cells exhibits the heightened DNase I sensitivity characteristic of actively transcribed DNA sequences. In most experiments, the DNA within nuclei of butyrate-treated cells is degraded to acid-soluble material faster than that within nuclei of control, nontreated cells. Under conditions of digestion similar to those of Weintraub and Groudine (61), the increase in initial rate of digestion has been variously estimated as undetectable (117), 1.5- to 2-fold, (156), 3-fold (154, 157), or 5- to 10-fold (158). These values are to be compared with an approximately 5- to 7-fold enhanced rate of conversion of template-active DNA sequences to nonhybridizable material (77).

An attempt has also been made to determine whether acetylated chromatin is more sensitive to DNase I digestion in a system not chemically induced. There are polyoma strains that differ drastically in their histone acetylation levels (159). When "minichromosomes" from these stains are digested with DNase I, the highly acetylated viral chromatin is degraded about three times more rapidly (initial rate) than the relatively nonacetylated viral chromatin (Schaffhausen, personal communication). It should be noted that the acetylation levels compared were similar to those in butyrate-treated versus control cells, and that the "minichromosomes" were prepared in a 0.4 M NaCl buffer, which would remove many noncore histone proteins, including the polyoma histone H1.

More direct evidence that acetylated histones are associated with DNA sequences preferentially susceptible to DNase I was obtained by analyzing those proteins solubilized at early times of digestion. There is a depletion of di- and triacetylated H4 in the chromatin fraction resistant to DNase I after 10% solubilization of total DNA from nuclei of butyrate-treated HeLa cells (tetra-acetylated H4 was not resolved) (154). More dramatically, there is a 10-fold enrichment in the ratio of tetra-acetylated H4 to nonacetylated H4 when soluble versus resistant fractions of nuclei from butyrate-treated HTC cells digested to 5% DNA acid-solubility are compared (117). In addition, acetylated H3 and H2b are preferentially released with the solubilized DNA.

That acetylated histones are preferentially released during early stages of

DNase I digestion in naturally occurring systems has also been demonstrated. The specific activities of [^{3}H]acetyl-labeled H3 and H4 is higher in the DNase I-released than in the DNase I-resistant fraction after digestion of avian erythrocyte nuclei to 11–12% DNA acid-solubility (*154*). Similarly, there is a preferential association of acetylated histones with the smallest nucleoprotein particles generated by mild DNase I digestion of HTC cell nuclei (*126, 160*).

The DNase I hypersensitivity of acetylated chromatin could be due to an increased DNA accessibility within individual chromatin subunits or could result from a more rapid digestion of the DNA linking subunits. There is evidence that both processes do occur (*126, 160*). When the nucleoprotein particles produced by mild DNase I digestion of HTC cell nuclei are electrophoresed in the absence of urea, it is possible to observe core, nucleosome and oligonucleosome particles, but no subcore particles. Acetylated histones are preferentially associated with the smallest nucleoprotein particles, suggesting that the linker regions adjacent to chromatin subunits enriched in acetylated histones are DNase I-hypersensitive. When similar digests are electrophoresed in the presence of urea, subcore particles enriched in acetylated histones may be detected, indicating that an elevated acetylation level can also be correlated with an increased DNA accessibility within subunits.

Since histone acetylation could be a feature of chromatin subunits hypersensitive to DNase I without being actively involved in promoting the sensitivity, it was of interest to determine whether, indeed, the increased accessibility to DNase I resides in the core histones and their interaction with DNA. Simpson (*158*) compared the DNase I digestion rates of isolated "cores" from butyrate-treated and control cells and found no difference in kinetics, which implies that the acetylation of core histones is not sufficient to dictate DNase I hypersensitivity. However, this finding should be interpreted with caution on two accounts. First, Simpson compared the DNase I digestion kinetics of acetylated and control core particles in different reaction mixtures (*158*). The hazards of this approach have been discussed at some length (*117, 157*), emphasizing that histone acetylation could influence the K_m of the enzyme reaction without altering its V_{max}. Thus, at saturating levels of substrate, no difference in overall kinetics would result. Second, as already discussed, the conflicting results (*48, 61*) on whether the DNase I sensitivity of active sequences can be recovered in isolated monomers suggest that DNase I hypersensitivity may be labile, sensitive to the conditions of monomer isolation. Relevant to this point is the finding that an enhanced DNase I susceptibility is characteristic of *in vitro*-assembled complexes of SV40 form I DNA and the isolated core histones from butyrate-treated cells (*156*). This result makes unlikely the interpretation that acetylated core particles are rendered less sensitive to DNase I by the loss of some component during the core particle isolation. Simpson's interpretation of his results (158) was that the acetylation of the amino terminals of H3 and H4 alters their interactions with adjacent core particles, and it is only when such core particle interactions are expressed, as in native chromatin, that acetylation can be seen to affect DNase I digestibility. This interpretation is not totally inconsistent with the observed DNase I hypersensitivity of *in vitro*-assembled complexes containing butyrate-treated cells, and is consistent with the suggestion (126) that core histone acetylation may promote the digestibility of adjacent linker regions.

b. Micrococcal Nuclease Digestion Studies. Actively transcribed sequences can be preferentially released as monomer-sized DNA fragments after very limited digestion of nuclei with micrococcal nuclease (Section III,C). It was of obvious interest to determine whether acetylated monomers are likewise released by this enzyme. The highly modified H4 subspecies in butyrate-treated HeLa cells are associated with those isolated monomers that are rapidly excised (158). Furthermore, when the 5' terminal-labeled monomers from control and treated cells are digested with micrococcal nuclease, the label associated with the acetylated monomers is solubilized 2- to 3-fold faster than that of the relatively unacetylated monomers (158).

A recent report by Levy-Wilson and co-workers (161) provides some evidence that highly acetylated histones are also preferentially released by micrococcal nuclease in a naturally occurring system. When trout testis chromatin is mildly digested and pelleted, the resulting supernatant is enriched in acetylated H4 subspecies. It should be noted that this fraction does not actually contain chromatin subunit monomers, defined by material sedimenting at about 11 S. Thus, it may not be valid to correlate these results with the finding that "active" chromatin monomers are preferentially excised by micrococcal nuclease. Although it was assumed by the authors that the soluble fraction was enriched in transcribed sequences, the degree of enrichment was not actually quantitated.

Two reports contradict the finding that acetylated histones are preferentially associated with micrococcal nuclease-susceptible chromatin regions. Davie and Candido (162) indicate that micrococcal nuclease treatment of trout testis nuclei produced monomers not significantly enriched in their proportion of acetylated histone subspecies. The apparent contradiction with Levy-Wilson *et al.* (161) may possibly be attributed to differences in stage of digestion. Davie and Candido (162) were studying quite advanced stages of digestion, when it was likely that the monomer particle was no longer enriched in transcribed DNA sequences. Similarly, Nelson *et al.* (126) have reported that the linker regions adjacent to chromatin subunits enriched in acetylated histones are not preferentially attacked by micrococcal nuclease.

Again, the stage of digestion (10% DNA acid-solubility) may have been too advanced to detect an enrichment for modified histones in the nucleoprotein monomers.

c. DNase II Digestion Studies. Davie and Candido (*121*) attempted to determine whether acetylated histones in trout testis nuclei partition in the same manner as active sequences during the Gottesfeld fractionation procedure (*90*). Their results indicate that the DNase II-released/Mg^{2+}-soluble material does contain an unusually high level of acetylated H4 (but, interestingly, not H3) subspecies. When this Mg^{2+}-soluble material was treated with RNase, the highly acetylated H4 was recovered with the Mg^{2+}-insoluble material. These results are suggestive of an association of acetylated histones with "active" sequences, but there are problems apparent on closer examination of the data. For example, in certain fractions there seems to be a depletion in the amount of total H4, and one is led to question whether selective losses of specific subspecies could have occurred. Also, the putative mono-, di-, and triacetylated species of H4 do not migrate in a regular fashion; that is, the distance between the mono- and diacetylated species is different from that between the mono- and nonacetylated fractions. It is possible, then, that modifications other than acetylation were being analyzed. Finally, there is no quantitation of the actual enrichment for actively transcribed DNA sequences in the presumably active chromatin fraction.

d. Are Acetylated Histones Actually Associated with Active Genes? From the above discussion, it appears that acetylated histones are associated with DNA sequences in chromatin preferentially accessible to nucleases. Actively transcribed sequences display similar increased sensitivities to DNase, and it is tempting to conclude that they are associated with acetylated histones. However, other subsets of chromatin may also be included in those fractions hypersensitive to digestion. Newly replicated chromatin is an obvious example. It has already been documented that DNase I preferentially degrades newly synthesized DNA in the nucleus (*163*), and that micrococcal nuclease rapidly excises chromatin subunits containing replicating DNA (*164*). It has also been known for some time that newly synthesized histones can exhibit enhanced acetylation levels (*165*). In more general terms, it is possible to envision other processes besides transcription for which the less positively charged "arms" of acetylated histones might be functional: for example, to generate points of flexion in order to facilitate higher-order folding, or to mediate the binding of a specific H1 or nonhistone protein (NHP) to the nucleosome. Any such function might be unrelated to transcription, but the nucleosomes involved may be rendered DNase-hypersensitive by the presence of acetylated histones.

It is relevant to consider the point that there is a much higher proportion of histones capable of being acetylated and deacetylated than would appear to be necessary if this modification is involved only in the regulation of gene expression (*149*, *166*, *167*). Moreover, there are clearly two types of acetylation in all cells studied: one turning over very rapidly and the other metabolically more stable. Because of these considerations, the conclusion that highly acetylated histones are preferentially associated with "active" DNA sequences is not an irrefutable one. That is, the concurrence of highly acetylated histones and active DNA sequences in nuclease-released chromatin fractions may only be circumstantial.

e. Can the Acetylation of Histones Facilitate in Vitro RNA Transcription? In order to support the contention that acetylated histones are associated with "active" genes, it was of interest to determine whether this modification could facilitate RNA transcription *in vitro*. Studies using chemically acetylated chromatin have shown both a stimulation of elongation rate by (*142*), and an increase in number of initiation sites for (*145*), *E. coli* RNA polymerase. However, these findings are subject to question, as discussed in Section IV,A,2.

Because of the massive increase in acetylation levels, one might expect butyrate-treated cells to exhibit a modified spectrum of newly synthesized protein if acetylation does, indeed, affect RNA synthetic capacity. However, it appears that treated and control cells synthesize very similar proteins at similar relative rates (*117*, *149*, *167a*). In fact, any comparison of the actual endogenous or exogenous RNA polymerase activity in nuclei or chromatin from butyrate-treated versus control cells would be complicated when the pleiotropic effects of this inducer are considered (e.g., *168*).

Thus, a simpler system with which to approach this question was sought. *In vitro*-assembled chromatin consisting of SV40 DNA form DNA 1 and the core histones appeared to be an attractive alternative for two reasons: first, these complexes exhibit patterns of nuclease sensitivity characteristic of the nuclei from which the component histones are derived (*156*); second, transcription, by *E. coli* polymerase on reconstituted complexes has already been well characterized (*169*, *170*). Using this approach, it was possible to show that the highly acetylated core histones from butyrate-treated HeLa cells can inhibit transcription by *E. coli* polymerase and the calf thymus A and B polymerases as efficiently as the core histones from calf thymus or untreated HeLa cells (*156*, and Mathis, unpublished results). This observation was true for the *E. coli* polymerase even at elevated salt concentrations, when the electrostatic binding of histones to DNA would presumably be weakened.

Three points should be considered when interpreting the finding that histone acetylation can effect an increased accessibility to nucleases, but not to RNA polymerase. First, it is conceivable that only the most highly acetylated histones are involved in any mechanism to promote transcription. The proportion of these molecules is still very low in butyrate-treated cells [as it is known to be in all other cells studied (149, 166, 167)]. Stated in other terms, DNases might be able to distinguish a continuum of susceptibilities resulting in a detectable increase in the kinetics of digestion of nucleosomes containing a mixture of acetylation levels; however, RNA polymerase might respond only to the maximally acetylated state, and the percentage of these molecules would be too low to produce a detectable increase in the rate of transcription. Second, acetylation is known to be a dynamic process. Acetyl groups can turn over very rapidly, with a half-life as low as 2 to 3 minutes (166, 167). It is possible that any assay that deals with acetylation as a static phenomenon has no relevance to the *in vivo* situation. Third, acetylation may dictate DNase sensitivity, but other factors (e.g., absence of H1 or presence of nonhistone protein) may be involved in the facilitation of RNA synthesis. That is, acetylation may be necessary, but not sufficient to allow transcription. This point is reminiscent of that made above concerning DNase I sensitivity: DNase I may reflect the potential for transcription, but other factors are required before actual synthesis can proceed.

3. The High-Mobility-Group Proteins

The "high-mobility-group" (HMG) proteins are a now well-characterized subset of the nonhistone protein population (for review, see *171*). They may be extracted from chromatin either by 5% perchloric acid or 0.35 M NaCl. They are of relatively low molecular weight ($\sim 30,000$) and have an unusual amino-acid composition, containing approximately 25% basic and 30% acidic amino acids. Five such fractions from calf thymus chromatin have been studied extensively: HMGs 1, 2, 14, 17, and 20 (or ubiquitin). Proteins with somewhat similar properties have been identified in trout testis (*171*): "H6" (analogous to HMGs 14 and 17), "T" (analogous to HMGs 1 and 2), and "S" (ubiquitin). In addition, an erythrocyte-specific HMG (E) has been described (*172*). The discovery of HMG proteins in wheat and yeast (*173*) has established their wide distribution in eukaryotes; in fact, some similarity has been noted between HMGs and the prokaryotic "HU" proteins (*173*). The lack of substantial tissue and species specificity (*172, 174*) has suggested that the HMGs, like histones, play predominantly a structural role in chromatin organization. However, since they are present in amounts only 1–5% of the DNA content, it follows that they cannot be present on more than about 10% of the nucleosomes. Thus, it appears possible that they may play some structural role in the organization of a specific subpopulation of chromatin subunits, e.g., those associated with actively transcribed DNA sequences.

a. DNase I Digestion Studies. Several groups have sought to ascertain whether HMG proteins are preferentially solubilized along with "active" DNA sequences at early stages of DNase I digestion. As Table I indicates,

TABLE I
"High-Mobility-Group" (HMG) Proteins Released after DNase I Digestion

Tissue	Percent DNA acid-soluble	HMGs preferentially solubilized	Preparation of soluble fraction	References
1. Duck erythrocytes	Not stated	1, 2, E (14 and 17 not analyzed)	EDTA extraction; low speed centrifugation	118
2. Rat liver, rat thymus	10–20	None	±EDTA extraction; low speed centrifugation	175
3. Trout testis	10	H6 (analog of HMGs 14 and 17)	Low speed centrifugation	176
4. Fetal calf thymus	14	1 and 2; 14 and 17 much less selectively	Low speed centrifugation	177
5. Mouse brain	75	All four, but 14 and 17 more selectively	Low speed centrifugation	177
6. Chick erythrocytes	10	1, 2, E, 14, 17	EDTA extraction; low speed centrifugation	178

the results from these studies are somewhat contradictory. There is disagreement on whether and which HMGs are preferentially solubilized. The observation that only certain ones are released into the soluble fraction after digestion could merely reflect a lesser ability to aggregate, or signify that a particular one binds to digested chromatin subunits in a manner relatively easy to disrupt. Thus, at least some of the variation in the proteins released could be due to differences in the preparation of the soluble fractions. For example, it is known that EDTA treatment of nuclease-digested nuclei can release material not released without such treatment (176, 179). On the basis of these DNase I solubilization studies, it is not possible to conclude whether HMG proteins are associated with "active" chromatin.

Using a different approach, a role for certain HMG proteins in the maintenance of the DNase I hypersensitivity of transcribed DNA sequences in chromatin has been shown (178). Weintraub's group had previously shown (61) that the globin gene is preferentially digested by DNase I in chick erythrocyte but not brain chromatin. Weisbrod and Weintraub subsequently reported (178) that the globin genes are not preferentially sensitive in 0.35 M NaCl-washed chromatin from erythrocytes. However, when the 0.35 M eluate is added back to the washed chromatin, the sensitivity is restored.

The activity of the 0.35 M fraction could be attributed to two HMG proteins, probably the chicken equivalents of calf thymus HMGs 14 and 17.

The question immediately arose as to the mechanism of the HMG-mediated chromatin conformational change leading to the DNase I hypersensitivity of expressed genes: are the HMGs able to recognize specific gene sequences? or do they recognize some other component that delineates those chromatin regions to be converted to a more "open" conformation? The second alternative is supported by the finding that the 0.35 M erythrocyte eluate does not enhance the DNase I sensitivity of the globin gene in washed brain chromatin, but that the 0.35 M brain fraction promotes this gene's digestibility in washed erythrocyte chromatin. Thus, it was surmised that there exists another component (or components) responsible for conferring selectivity on the HMG-induced chromatin conformational change. This additional component apparently resides at the level of the individual chromatin subunit, because added HMGs restore the DNase I hypersensitivity of the globin gene in isolated 0.35 M salt-washed chromatin monomers (H. Weintraub, personal communication). When added back to washed chromatin, the HMGs could either recognize only those chromatin subunits having the additional component(s) or could distribute randomly among chromatin subunits, but induce an elevated DNase I sensitivity only in those containing the component(s). Since the hypersensitivity of the globin gene can be restored to its original value in 0.35 M salt-washed chromatin with approximately the same amount of HMGs as were removed (H. Weintraub, personal communication), it can be inferred that the added HMGs are actually capable of recognizing the additional component(s). The possibility that the HMG 14 + 17 fraction contains a minor constituent that influences the binding of HMGs 14 and 17 to chromatin has not yet been excluded.

Two very recent studies have established the generality of the finding that certain HMG proteins can promote DNase I sensitivity. First, the DNase I sensitivity of the endogenous viral RAV-O genome in chicken erythrocytes is also mediated by HMGs 14 and 17 (H. Weintraub, personal communication). RAV-O sequences are expressed as mRNA only at 1 to 5 copies per cell, and thus the HMGs seem also to be involved in regulating the chromatin conformation around genes rarely transcribed. Second, the DNA labeled by "nick-translation" of nuclear DNA is enriched in transcribed sequences and is highly susceptible to DNase I digestion. (see Section III,B and 75). When chromatin from "nick-translated" nuclei is washed with 0.35 M NaCl before digestion, this sensitivity is no longer apparent, but addition of the 0.35 M eluate restores about 80% of the original DNase I hypersensitivity (H. Cedar, personal communication).

Results that on the surface appear contradictory to the above have been published (*180*). In this study, the globin genes were preferentially digestible in chick erythrocyte nuclei washed with 0.6 M NaCl. At such a high ionic strength, HMG proteins would not be expected to bind to chromatin, but since the extractions were performed at pH 3.0 (instead of the usual neutral pH), this is not assured. Unfortunately, there was no assay for HMG, the concern being with the effect of H1 and H5 on DNase I sensitivity.

It seems quite clear, then, that HMGs 14 and 17 are somehow involved in the maintenance of the DNase I-hypersensitive conformation of "active" chromatin. It remains to elucidate the mechanism involved and to identify other necessary components.

b. Micrococcal Nuclease and DNase II Digestion Studies. The question whether HMG proteins are components of early-released, "active" nucleoprotein monomers (see Section III,C) has been approached most rigorously by Levy-Wilson and co-workers (*83, 119, 120, 137, 176, 177*), who have shown that the proteins solubilized after mild digestion of trout testis nuclei by micrococcal nuclease include H6 (the HMG 14 and 17 analog, see Section IV,A,3), HMG-T (analogous to HMGs 1 and 2) and S (ubiquitin) as well as core histones and small amounts of H1. Further fractionation yielded two types of nucleoprotein particles (MN-1 and MN-2), enriched in DNA sequences coding for cytoplasmic RNA-poly(A). The MN-1 particle, soluble in 0.1 M NaCl, contained a 140-base-pair DNA fragment, the core histones, and about one molecule of H6 per histone octamer. About 90% of the chromatin H6 was present in this particle. The MN-2 particle, insoluble in 0.1 M NaCl, contained DNA fragments ranging in size from 140 to 190 base-pairs, the core histones, and H1; 44% and 36% of MN-1 and MN-2 DNA, respectively, hybridized with a vast excess of cytoplasmic RNA-poly(A), indicating that both particles were highly enriched in transcribed DNA. Cross-hybridization experiments demonstrated that the DNA from MN-1 and MN-2 showed extensive sequence overlap, but that each particle also contained a subset of DNA sequences absent from the other. On the basis of these results, it was concluded that the HMG proteins can be major structural components of transcribing chromatin. H6 (the analog of HMGs 14 and 17) was apparently bound preferentially to transcriptionally competent chromatin subunit "cores," while HMG-T (analogous to HMGs 1 and 2) was associated with regions linking adjacent subunits (see also *180a*). However, it does not seem obligatory for "active" chromatin to be organized in this fashion, because the MN-2 nucleoprotein particles lacked H6; nevertheless, they were highly enriched in transcribed DNA sequences. A similar procedure for the isolation of HMG-enriched early-released nucleoprotein particles has been described recently (*180b*). Interestingly, in this case

HMG 14 and 17 (analogous to the trout H6) were not associated with early released "cores." However, there was no quantitation of the amount of co-isolating "active" DNA sequences in this study.

Studies (181) of the localization of HMG proteins within nucleoprotein particles generated by micrococcal nuclease digestion of rabbit thymus nuclei yielded results not entirely consistent with the above. Micrococcal nuclease did not preferentially excise rabbit thymus chromatin subunits containing HMGs 14 and 17 (in contrast with its rapid excision of H6-containing subunits in trout testis). HMGs 14 and 17 coelectrophoresed with chromatin core particles after digestion of nuclei to 14% DNA acid-solubility, but the core particles were devoid of these proteins when only 3% of the DNA had been degraded. Thus, HMGs 14 and 17 appeared to be associated with subunit "cores" (like H6), but they were components of subunits whose DNA resisted rapid trimming to a 145-base-pair fragment (unlike H6). Two populations of HMGs 1 and 2 were detected, one readily dislodged during micrococcal nuclease digestion, the other more tightly bound to chromatin subunits (181). It is not clear whether either type of HMG 1 and 2 was associated with regions linking adjacent chromatin subunits (as has been suggested for HMG-T).

The conclusion that certain HMG proteins are associated with chromatin subunit cores is at variance with the results of a study in which two-dimensional polyacrylamide gel electrophoresis was used to catalog those nonhistone proteins associated with nucleoprotein particles generated by limited micrococcal nuclease digestion of Ehrlich ascites tumor chromatin (182). In the first dimension, the various nucleoprotein particles were separated, including three monomer-sized particles. When the proteins from these particles were displayed in the second dimension, it was evident that no HMGs were associated with those particles containing a 140-base-pair DNA fragment. The extents of digestion (9% and 15% DNA rendered acid-soluble) were comparable with those in the other studies (5–10% in 83, 119, 120, 137, 176, 177, and 3 to 19% in 181). The source of this discrepancy over whether HMG proteins are associated with chromatin subunit cores remains unclear. Parenthetically, it should be noted that certain subnucleosomal particles that contained only DNA (25–30 base-pairs) and HMG proteins (not identified, but not HMGs 1 and 2) have been detected (182). The DNA from these subnucleosomal particles was enriched in sequences complementary to hnRNA.

DNase II has not been used extensively to probe the structure of HMG-containing chromatin, but the results obtained thus far are consistent with the view that the HMGs are structural components of transcribing chromatin. The isolation of an HMG-enriched nucleoprotein particle within the DNase II-released, Mg^{2+}-soluble fraction from trout testis nuclei has been reported (120). This particle contained all four core histones, H6 and HMG-T, and a DNA fragment of about 180 base-pairs. Unfortunately, the content of transcribed DNA sequences within this nucleoprotein particle was not quantitated.

In summary, micrococcal nuclease and DNase II digestion studies tend to support the conclusion that at least some HMG proteins are preferentially associated with transcribed chromatin regions. The exact localization of the different HMGs within "active" chromatin subunits is still unknown. In fact, the question as to whether HMGs are constituents of the core or linker may be inappropriate, since it has not been established that "active" chromatin subunits actually have cores and linkers (except for the case of the ovalbumin gene; see Section III,A).

c. *A Relationship between HMG Proteins, Transcribed DNA Sequences, and Acetylated Histones?* It has been convincingly demonstrated (178 and H. Weintraub, personal communication) that HMGs 14 and 17 play a role in the maintenance of the DNase I hypersensitivity of transcribed genes in chromatin. Given this conclusion, it would be expected that these proteins are preferentially solubilized after limited DNase I digestion of nuclei and that they are components of the chromatin fragments released after mild micrococcal nuclease or DNase II digestion. Further nuclease probing of the released chromatin fragments might also provide a clue to the organization of HMGs within "active" chromatin subunits. However, experiments using these approaches have not provided unequivocal confirmation of the association of HMGs with "active" sequences or a clear indication of exactly how HMG-containing chromatin subunits are organized. As already discussed, some of the discrepancies could be related to methods of fractionation after nuclease digestion. Certainly, some differences could also be due to the tissues used. Although HMGs appear to exhibit little tissue specificity (172, 174), reports of ribosylation of HMG by poly(ADP-ribose) (183–185), acetylation (186), and methylation (186a) suggest that HMG subpopulations do exist. These subpopulations could have a differential association with and different binding properties to "active" sequences, and they could conceivably vary in amount from tissue to tissue. It has already been suggested that in rabbit thymus nuclei there are two types of HMGs 1 and 2, differentiated on the basis of their response to micrococcal nuclease probing (181). One type is bound to chromatin subunits relatively tightly; the other is more loosely associated, being rapidly released from subunits at early times of micrococcal nuclease digestion. The latter fraction may correspond to an HMG 1 and 2 subpopulation extracted from chromatin with 0.14 M NaCl. One is led to question whether these HMGs ever participated in a nucleosome-like structure and whether they were even bound to chromatin *in vivo*.

In the presence of additional chromatin components, HMGs 14 and 17 confer preferential DNase I sensitivity on transcribed DNA sequences; histone acetylation seems also to be characteristic of accessible chromatin regions. It is possible that acetylated histones are the additional components, and that HMG proteins and acetylated histones somehow act in concert to modify chromatin structure. However, a DNase I hypersensitivity was characteristic of *in vitro*-assembled complexes of SV40 form I DNA and the core histones from butyrate-treated cells (*156*) and of 0.4 M NaCl-extracted, highly acetylated "minichromosomes" from wild-type polyoma (see Section IV,A,2,a). HMG proteins could not be involved in maintaining nuclease hypersensitivity in these two cases. However, it is conceivable that acetylated histones and HMG proteins generally do interact somehow to promote DNase I hypersensitivity, but that with closed-circular DNA molecules (as with the polyoma "minichromosomes" and SV40 reconstitutes) the requirement for HMG protein is lost. A second possibility is that there are (at least) two separate modes of promoting DNase I susceptibility.

4. Nonhistone Proteins

The chromosomal nonhistone proteins (NHPs) are a very heterogeneous group, consisting of macromolecules that vary widely in molecular weight, amino-acid composition, half-life, and DNA-binding properties (for reviews, see *115* and *116*). This group is known to contain proteins with enzymic and structural roles, and it has been hypothesized, partly on the basis of a well-documented tissue specificity, that it includes proteins that function in the regulation of gene expression. Several recent reports of changes in the levels or modifications of defined NHP fractions during cell differentiation have provided an impetus for studies on the role of nonhistone proteins in chromatin function (*187–189*). Unfortunately, since these studies merely describe an alteration in the protein complement of nuclei or bulk chromatin preparations, their significance with respect to gene expression remains unclear. In order to support convincingly the contention that a particular protein fraction plays a direct role in the regulation of chromatin transcription, it is necessary to demonstrate a selective association between this protein species and actively transcribed DNA sequences. Fractionation of chromatin by nuclease digestion may provide a means to establish this association.

DNase II probing has provided evidence that nonhistone proteins are major constituents of template-active chromatin. It was noted that the NHP/DNA ratio is highly elevated (compared with bulk chromatin or an "inactive" chromatin fraction) in the DNase II-released, Mg^{2+}-soluble, putatively "active" chromatin fraction from rat liver nuclei (*104, 122*). However, nonhistone proteins are not exclusively present in "active" chromatin; at least 90% of rat liver nucleosomes react with serum directed against total chromatin NHPs (*190*). Evidence that specific nonhistone proteins could be associated with actively transcribed DNA sequences comes from studies illustrating the preferential release of a subset of the nonhistone protein population after mild DNase I digestion (118, 191). Micrococcal nuclease probing has also provided an example of the association of specific NHPs with chromatin fractions likely to be enriched in transcribed DNA (192).

As discussed above, nuclease digestion can provide evidence for the association of a particular chromatin constituent with actively transcribed DNA, but cannot prove this association, because certain nontranscribed DNA sequences could also be nuclease-hypersensitive. Recently, two approaches have been described to demonstrate more directly an association between specific nonhistone proteins and specific DNA sequences. First, Elgin et al. prepared antibodies against defined molecular-weight fractions of the nonhistone protein population of Drosophila cells and subsequently visualized their binding to salivary gland polytene chromosomes by immunofluorescence (193). Particularly relevant is the demonstration of a chromosomal localization of antibodies directed against a 63,000-dalton protein fraction solubilized after limited DNase I digestion of Drosophila embryo nuclei (194). All salivary gland puffs appeared fluorescent, as did many nonpuffed loci known to puff at other times during third instar or prepupal stages of development. In addition, heat-shock-induced puffs were stained brightly, using this antiserum. These loci, not normally expressed during third instar or prepupal development, were not stained prior to induction.

The second approach involves the use of a filter-binding assay to select DNA sequences that specifically bind isolated NHPs (195). This technique has allowed the identification of an rDNA-binding nonhistone protein from Drosophila cells, and it will obviously be of interest similarly to select DNA sequences that bind the proteins released by mild nuclease digestion of chromatin. It should be kept in mind that this approach is applicable only to DNA-binding proteins, and will not be of value in characterizing NHPs that interact solely with the protein constituents of "active" chromatin subunits. However, the recent development of a histone–DNA–cellulose chromatography matrix (195a) may provide a means for the screening of cell extracts for chromatin-binding proteins.

Although perhaps associated with template-active chromatin, proteins released by mild nuclease digestion may not necessarily be involved in the generation or maintenance of the "open" conformation. For example, it is not sure that the protein fraction solubilized by mild DNase I digestion and associated with puffed loci (194) is, in fact, chromatin-bound. These proteins could just as well be RNA packaging proteins preferentially released during DNase I digestion by virtue of an association with nascent RNA, and expected to accumulate at puffed loci. In this respect, it should be noted that the antibodies directed against proteins isolated from nucleoplasmic ribonucleoprotein particles can decorate *Triturus* lampbrush loops, and that for some proteins this decoration is specific to certain loops (*196*). Thus, it is important to demonstrate that a given NHP actually binds to chromatin subunits. The electron microscopic observation of an increase in chromatin subunit diameter upon complexing NHP antibodies (*190*) may provide a method to visualize the binding of specific nonhistone proteins to "active" chromatin subunits. More convincing evidence that a given NHP is involved in the maintenance of an "active" chromatin conformation would come from the demonstration that its removal affects the nuclease susceptibility of transcribed genes in chromatin, as has already been shown for HMGs 14 and 17 (*178*).

In summary, there is highly suggestive evidence that template-active chromatin contains a specific population of nonhistone proteins. It has not yet been demonstrated that an individual NHP is a component of the chromatin subunit associated with a specific transcribed gene, but the technology to do this is available. It is also not clear that any NHP (except for the HMGs) is actually involved in specifying the altered chromatin conformation about "active" genes.

B. Protein Content of Isolated "Native" Chromatin

The most direct approach to study template-active chromatin is actually to isolate it. The ideal situation would be to obtain test-tube quantities of chromatin containing the entirety of only one transcription unit. The complexity of the eukaryotic genome has so far rendered this impossible. However, certain rather special genes or gene complexes do appear to be isolatable: the multiple, extrachromosomal ribosomal cistrons of amphibian oocytes and lower eukaryotes, and the SV40 transcription complex. Although still in the preliminary stages, studies on the protein composition of these isolated chromatins have provided some information on the histone complement, degree of postsynthetic histone modification, and nonhistone protein complement of transcribing chromatin.

1. Isolated rDNA-Chromatin

The ribosomal genes of certain eukaryotes are contained within multiple, extrachromosomal nucleoli. Well-known examples of this phenomenon are the amplified rDNA of amphibian oocytes (e.g., *Xenopus*) and the palindromic ribosomal cistrons of *Tetrahymena* and *Physarum*. Several groups have recognized the potential for isolating these genes in an intact, pure, chromatin form. The encasement of the ribosomal cistrons within nucleoli could be used as a basis of fractionation, as could their small size compared with unsheared bulk chromatin if isolated from unfractionated nuclei. Furthermore, rDNA-containing chromatin in these systems represents a higher proportion of the genome than most other genes: about 2% in *Tetrahymena* (*197*) and *Physarum* (*190*) and as much as 70% in *Xenopus* oocytes (*199*). An additional advantage in the case of *Tetrahymena* and *Physarum* is that there are methods to label rDNA specifically, providing a "tag" for following enrichment during fractionation (*55, 200*).

To date, the isolation and characterization of rDNA-containing chromatin has been the most successful in *Tetrahymena*. Jones (*201, 202*) has recently published a purification procedure that results in yields of 25 to 75% and a very high purity. When the proteins in this fraction were compared with those of bulk chromatin, Jones made the following observations:

1. All the major histone classes characteristic of bulk chromatin were also found in rDNA-containing chromatin. All five histones were present in the same relative ratios in both chromatin fractions; however, rDNA-containing chromatin had a histone/DNA ratio only 40% that of bulk chromatin.
2. The level of acetylation of histones H3 and H4 was very similar for bulk and rDNA-containing chromatin. H1 was less modified (probably owing to phosphorylation) in the latter fraction.
3. The rDNA-containing chromatin was highly enriched in nonhistone proteins.

Like any purification procedure, this one must be scrutinized for the possible selective loss of a subfraction of the ribosomal cistrons (note that the yield was 25 to 75%) or the generation of artifacts during the preparative procedure. Most disturbing is the low histone/DNA ratio. The polyamines spermine and spermidine were used to keep nuclei intact during the early steps of the isolation, and it was noticed that the sedimentation rate of the rDNA changed according to the levels of residual polyamine. As the author points out, the observation of a reduced histone/DNA ratio is surprising in light of the studies demonstrating that rDNA-containing and bulk chromatin in *Tetrahymena* exhibit very similar kinetics in micrococcal nuclease digestion (*53, 54*). One is led to question whether the polyamine could displace histones, perhaps a specifically modified subset, from chromatin subunits containing rDNA, and/or whether nonhistone proteins could form structures that protect rDNA from micrococcal nuclease digestion (for a discussion of possible problems with polyamines, see *100a*). Fortunately, Gocke et al. (*203*) have described a method to purify nucleoli from *Tetrahymena* that uses different extraction conditions. It will be of interest to see how the histones within these nucleoli compare with the histones from Jones's rDNA-containing chromatin (*201, 202*).

A procedure for purifying the amplified nucleoli from immature oocytes of *Xenopus* has been detailed (204), but the material is not as well characterized as the *Tetrahymena* rDNA-chromatin. What can be stated is that RNA polymerase I (A) activity, the "nicking-closing" enzyme, and the four core histones are all present.

A preliminary report of the isolation of rDNA-containing chromatin from *Physarum*, yielding ribosomal "minichromosomes" sedimenting at 100 S and containing RNA polymerase I (A) activity has appeared (205, 206). The histone content has not yet been characterized, but a 70,000-dalton phosphorylated nonhistone protein that can stimulate *in vitro* transcription of the rDNA "minichromosome" and that binds specifically to a defined region of rDNA has been identified (206a).

It is apparent, then, that rDNA-containing chromatin will prove to be a useful model for the study of actively transcribed chromatin. However, two disadvantages should be acknowledged. First, since the ribosomal genes are present in multiple copies, it cannot be stated with certainty that all, or perhaps any, of the genes isolated were being actively transcribed *in vivo*. Second, it may well be that the "active" ribosomal genes are not typical "active" genes. As discussed above, their appearance under the electron microscope is distinctive from that of "active" nonribosomal transcription units, and they may, therefore, exhibit a special structure.

2. Isolated Transcribing SV40 "Minichromosomes"

The purification of a chromatin fraction containing a specific unique-copy gene in its entirety appears beyond the scope of present technology. However, the isolation of the SV40 genome in a transcriptionally active state appears feasible and may offer the possibility of purifying a defined nonribosomal transcription unit. In lytically infected cells, SV40 DNA is found in association with cellular histones as a beaded "minichromosome" (e.g., 207–209). In late stages of infection, viral sequences are amplified and are actively transcribed. Thus, like the ribosomal genes, the SV40 genome offers the advantages for isolation of being a well-defined DNA sequence that is extrachromosomal and multiple-copy.

The isolation of the SV40 transcription complex has only progressed to the level of 5 to 20% purity (42). The most fruitful approach has been to extract minichromosomes from cells late in infection, allow them to extend *in vitro* nascent RNA chains initiated *in vivo*, and then use the change in sedimentation rate caused by the bound RNA to shift the transcription complexes away from the bulk "minichromosomes" in a sucrose gradient (42). This shift can be monitored by labeling the RNA chains extended *in vitro*. The purity of material from these experiments is not yet very high, but the ability to label transcription complexes specifically (by labeling their nascent

RNA) has provided some information on the structure of actively transcribing chromatin.

Cesium chloride buoyant density analysis of "minichromosomes" fixed after *in vitro* RNA synthesis has provided evidence that transcriptionally active complexes contain protein in amounts similar to those of inactive complexes (*42, 210*). Since elevated salt concentrations influenced the sedimentation of transcription complexes and bulk "minichromosomes" indistinguishably, it was surmised that the proteins associated with "active" complexes included histones (*42, 209*). That is, proteins on the transcription complexes had been "stripped" off at salt concentrations known to be characteristic of the histones. Furthermore, the proteins on "active" complexes exchanged onto competing naked DNA at the same ionic strength that histone cores exchanged from the bulk "minichromosomes" (*42*). That H1, in particular, may be associated with transcribing complexes was suggested by the fact that they underwent a striking conformational change with exposure to NaCl concentrations greater than 0.3 to 0.4 M (*209*). This structural transition has been observed for bulk "minichromosomes" and is considered to be the result of H1 removal (*209, 211*).

Further evidence for the presence of histones on the transcription complexes was provided by the electron microscopic visualization of nucleosome-like beads (*42*). The putative transcribing "minichromosomes" exhibited a compaction ratio and bead number similar to bulk "minichromosomes." As discussed by the authors, this last observation does not prove that nucleosomes are present on transcription complexes *in vivo*, since a structure originally extended *in vivo* could become compact through subsequent manipulation.

It is clear from the above discussion that the isolation and characterization of rDNA-containing chromatin and of the SV40 transcription complexes are in the infant stages. Nevertheless, certain points about the protein complement of template-active chromatin can be made. First, both these putatively "active" chromatin complexes seem to contain the core histones and H1. The histone/DNA ratio may be reduced in the case of *Tetrahymena* ribosomal genes, but is unknown for SV40 and for the *Physarum* and *Xenopus* ribosomal genes. Second, assuming that intact, template-active rDNA-containing chromatin was isolated, Jones's results indicate that high levels of histone acetylation need not be characteristic of "active" chromatin. Third, the enrichment for NHP in *Tetrahymena* rDNA-chromatin and the identification of a specific NHP that stimulates the *in vitro* transcription of *Physarum* rDNA-chromatin are consistent with the idea that nonhistone proteins are important factors in the regulation of chromatin transcription. It remains to be seen whether these points can be generalized to all "active" chromatin, and whether, in fact, some of the observations are attributable to isolation artifacts.

V. Conclusions, Conjectures, and Prospects

A. The Basic Structure of "Active" Chromatin: A Unique, Repeating, Altered Nucleosomal Structure?

Both micrococcal nuclease and DNase II digestion studies strongly suggest that "active" chromatin is organized in repeating subunits that share at least some properties with bulk nucleosomes. Although it is conceivable that proteins other than histones (e.g., HMGs) could generate a repeating subunit structure (60), most of the available data suggest that all four core histones are present within "active" chromatin. It is unknown at the present time whether they are present in the same stoichiometry as in bulk chromatin or whether some of them could be, at least in part, replaced by core histone variants or even nonhistone proteins. Replacement of histones by HMG proteins is, however, unlikely since "active" chromatin loses its DNase I hypersensitivity when HMG proteins are extracted. (see Section IV,A,3).

Whether the histones associated with "active" genes form a "core" around which the DNA is folded to yield a compact structure resembling the nucleosome of bulk chromatin is difficult to judge from the available biochemical and physicochemical studies. *In vitro* folding of structures extended *in vivo* could occur during nuclease digestions owing to change in the ionic environment or to release of chromatin components during the course of the digestions. A detailed analysis of the "subnucleosomal" particles generated by extensive digestion of various "active" chromatins with micrococcal nuclease will be helpful in revealing whether all "active" subunits are composed of well defined core particles and linker regions as are bulk nucleosomes.

Taking into account the possible artifacts generated during the necessary manipulations, the present electron microscopic studies of "active" nonribosomal chromatin (particularly those of Lamb and Daneholt; see Sections II,B and II,C) lead to the conclusion that nucleosome-like beads interspersed between the transcribing RNA polymerase molecules are present on moderately "active" transcription units, whereas they are absent from the maximally transcribed genes. It seems, therefore, that the compact nucleosomal structure may unfold locally during the transcriptional event and refold rapidly after the passage of the RNA polymerase molecule. In marked contrast, under similar experimental conditions, the transcribing ribosomal genes appear to be almost fully extended irrespective of their transcription rate. Whether the near absence of DNA compaction for nonmaximally transcribed rDNA transcription units is artifactual or not is unknown (see Section II), but in any case it reflects a different composition and organization for ribosomal and nonribosomal transcribing chromatin. The apparent unfolding of the nucleosome-like beads when transcription is taking place is in agreement with the *in vitro* studies showing that the eukaryotic (*212*) and prokaryotic (*169, 170*) RNA polymerases can transcribe through regions containing nucleosomes, but that histone core–DNA interactions have to be altered in order to allow efficient RNA chain initiation and elongation.

The finding of active genes within the micrococcal nuclease-generated bulk chromatin DNA repeat suggests, as previously discussed (see *1* for references), that the periodic structure of chromatin can be maintained about these genes in the absence of compact nucleosomes, presumably because the bound histones are still organized within "open" nucleosomes in such a way as to generate more accessible nuclease sites at original DNA repeat-length intervals. Whether half-nucleosomes (*213–215*) are involved in the formation of the "open," more "extended," structure of "active" chromatin is still conjectural, but it is interesting to note that the recent psoralen cross-linking experiments (*216*) suggest the presence of more accessible sites every 100 base-pairs in putatively "active" *Tetrahymena* ribosomal genes. The mechanisms responsible for the generation of an "open," more extended structure *in vivo* are unknown, but a role for electrostatic repulsion forces has recently been invoked (*217*). Possible mechanisms that could allow RNA polymerase read-through by nucleosome disassembly without dissociation of the histones from the DNA have recently been discussed (*1, 42, 169, 170, 217–219*). Of particular interest is the recent discovery (*195a, 219a*) that nucleosome-like structures can be assembled *in vitro* on single-stranded DNA, provoking speculations that, *in vivo*, half nucleosomes are preferentially associated with one DNA strand, and thus binding of histones to DNA would interfere minimally with opening of the DNA double-helix required for transcription. Although it appears that most of the active chromatin could be organized in repeating subunits, it should be kept in mind that there is no general agreement as to whether all active genes are similarly susceptible to micrococcal nuclease digestion (see Sections III,A and III,C). There are even reports claiming accelerated micrococcal nuclease degradation to nonhybridizable fragments for both ribosomal and nonribosomal genes, suggesting heterogeneity in the basic structure of "active" chromatin. Careful quantitative studies of the digestion products of a much greater number of ribosomal and nonribosomal genes transcribed at various rates are clearly required.

Further evidence that "active" chromatin may be characterized by more than one type of organization is provided by some unexpected or conflicting results from DNase I or DNase II digestion studies of presumably "active" genes (see Sections III,B and III,D). That more than one type of "active" chromatin organization may exist is also supported by the results of studies on genes that will be or have been transcribed (see Section III,B). For instance, the nonbeaded electron microscopic appearance of ribosomal genes before the onset of transcription is likely to reflect a subunit organization markedly different from that of transcriptionally "competent" (potentially transcribable) nonribosomal genes. The case of the heat shock versus developmental puffs in *Drosophila* provides further evidence for heterogeneity in "active" chromatin organization. Heat-shock loci are hypersensitive to DNase I in the induced state, but not the uninduced state either before or after the heat-shock period (see Section III,B). To the contrary, developmental loci, very active before heat shock, but dramatically reduced in transcriptional activity after the stimulus, are equally sensitive to DNase I in both states (78). Likewise, when exposed to a fluorescent antibody directed against a protein released early in DNase I digestion, the developmental loci stain according to their "competence" for transcription while heat-shock loci stain only when in the process of transcription (see Section IV,A,4).

Clearly, it is not possible to account for all the known properties of chromatin containing ribosomal and nonribosomal genes that will be, are, or have been transcribed by invoking a unique, altered nucleosomal organization for "active" chromatin. Isolation of pure chromatin fragments containing "active" ribosomal and nonribosomal genes and careful studies of their composition and organization are required before understanding the basis for the variability of "active" chromatin structure. The preparation of antibodies directed against proteins suspected of being "active" chromatin components, and their use to "fish out" "active" chromatin regions could be very useful in this respect.

B. Is Histone H1 Present in "Active" Chromatin?

Recent electron microscopic results make very unlikely the existence of higher-order structures in at least one type of transcribing chromatin (see Sections II,B and II,C; Lamb and Daneholt, personal communication). Since the existence of higher-order structures in chromatin can also be inferred from the release of nucleosome octamers and their multiples under appropriate micrococcal nuclease digestion conditions (220-221), it would be of great interest to probe the higher-order structure of "active" chromatin by such digestion studies. Because histone H1 is responsible at least in part for the existence of higher-order structures in bulk chromatin (for references, see 1), the problem of its presence in "active" chromatin is evidently posed. In this respect it should be noted that the increased susceptibility of "active" chromatin to micrococcal nuclease might be explained by the absence of H1, since the rate of release of chromatin subunits by micrococcal nuclease digestion of bulk chromatin is drastically increased after H1-depletion (*128*). However, it is worth remembering that removal of H1 and H5 histones from erythroid cell chromatin does not suppress the DNase I hypersensitivity of the globin genes (*180*).

From the conflicting results available at the present time (summarized in Section IV,A,1), it is obviously impossible to conclude whether or not histone H1 (and its variants) is present in "active" chromatin. Again, there is a need for careful well-controlled studies, and again, the use of antibodies directed against the various H1 variants should be very useful in elucidating whether they are associated with the isolated "active" subunits as they are with the isolated nucleosomes of bulk chromatin.

C. Factors Involved in the Nuclease Hypersensitivity of "Active" Chromatin

In spite of a great number of studies, the basis for the DNase I hypersensitivity of "active" chromatin remains unclear at the present time. HMG proteins 14 and 17 are required, but they appear to act in conjunction with other chromatin components (see Section IV,A,3, and *178*). What are these components? There is a vast body of circumstantial evidence that histone acetylation might be a feature of "active" chromatin (Section IV,A,2). However, most of the chromatin histones are available for acetylation to some degree, and only a fraction of the genome is preferentially digested by DNase I. Is a certain pattern of acetylation required in order for HMGs to bind and to promote DNase I hypersensitivity? Is there a heterogeneity in DNase I hypersensitivity? Does it always involve HMG proteins? For example, are there different types of DNase I hypersensitivity characteristic of "active" ribosomal and nonribosomal genes, as suggested by the absence of the 10-base ladder in digests of presumably "active" ribosomal genes and their presence in the DNase I digests of "active" nonribosomal genes isolated by the DNase II/Mg^{2+} technique (see Section III,B and III,D)? In this respect, it would be of great interest to see whether HMGs are required to maintain the DNase I hypersensitivity of chromatin from growing cells exposed to butyrate. Are HMG proteins necessary for maintaining the DNase I sensitivity of genes that will be transcribed or have been transcribed, or are other components involved? What is the role of HMG acetylation? All these studies are urgently required in order to decide whether there are alternative mechanisms to generate the DNase I hypersensitivity of "active" chromatin and whether, in fact, DNase I hypersensitivity is a property restricted to "active" chromatin.

DNase I hypersensitivity appears to be an intrinsic property of isolated "active" chromatin subunits (see Sections III,B and IV,A,3). However, similar studies should be carried out on a variety of "active" genes to resolve some of the present contradictions (see Section III,B), which might well be due to differences in isolation procedures resulting in some cases in losses of HMGs or other necessary NHP components. Obviously, the structural basis of the hypersensitivity to DNase I remains to be established. That it could be related to decreased interactions between the NH_2-terminal histone regions and the DNA is suggested by reports showing that core particles in which such interactions are decreased are digested more rapidly by DNase I (222, 223). How HMG proteins might act on core particles, possibly enriched in acetylated histones, to promote such decreased interactions is still an open question.

Acetylated histones and some HMGs are often (but not always) preferentially associated with the subunits released at early times of micrococcal nuclease or DNase II digestions, and these subunits are often enriched in transcribed sequences (Sections IV,A,2 and 3). There is no available evidence on whether there is also a causal relationship between the presence of HMG proteins and the preferential release of "active" subunits. Further studies are necessary to elucidate this point and, more generally, to determine whether the same chromatin components are responsible for DNase I, micrococcal nuclease, and DNase II hypersensitivity. Such studies should also shed some light on the possible location of the different HMG proteins within "active" subunits, an issue that is very confusing at the present time (see Section IV,A,3). Again, the question is raised whether the situation is not much more complicated than is currently believed and whether, for instance, additional components might not be required to confer increased accessibilities to micrococcal nuclease and DNase II on structures already DNase I hypersensitive. In this respect, it would be very interesting to know whether the chromatin subunits of DNase I-hypersensitive genes that will be or have been transcribed are also released preferentially by micrococcal nuclease.

D. Generation of the "Active" Chromatin Conformation. Modulation of Gene Activity via Changes in Chromatin Structure

The problem of the generation of an "active" chromatin conformation is central to the understanding of the mechanisms involved in the transcriptional regulation of gene expression in eukaryotic cells. From what is presently known about the structure of "active" chromatin, there are at least three levels at which gene transcription can be regulated. Since the "active" altered chromatin structure seems sometimes to precede transcription and to survive its disappearance, we can consider, for the sake of simplicity, that

STRUCTURE OF TRANSCRIBING CHROMATIN   47

three events should occur before a given gene can be transcribed (the first two events could occur either consecutively or simultaneously): (*a*) relaxation of the higher-order chromatin structure of a defined region; (*b*) conversion of the nucleosomes to an "active" form that is potentially transcribable ("competent"); (*c*) intervention of "factors" allowing RNA polymerase to bind and to initiate at a specific site.

How could a defined region of the higher-order chromatin structure be specifically recognized in a given cell type in order to enable its relaxation into an uncoiled "active" nucleosomal chain? The actual mechanisms involved in the triggering of these structural modifications and their propagation along the nucleosomal chain are totally unknown and might require a technical breakthrough to be unraveled if three-dimensional structures are involved. However, some interesting speculations have been put forward (*224*), and some points deserve to be stressed. The interphase nucleus, which should be viewed as a three-dimensional problem, appears by many criteria to contain highly ordered microdomains (*225*). It is possible that such domains reflect reproducible arrangements of specific chromosomes in the interphase nucleus and that such arrangements might be cell type-specific, clonally inherited, and involved in cell determination. Evidence for such cell type-specific three-dimensional arrangements is still lacking, but close pairing between homologous chromosomes is known to occur in *Drosophila* interphase nuclei (*226*) and may be important for gene expression, as suggested by the existence of "transvection effects" (see *226* for references). A specific three-dimensional architecture could be important in allowing "uncoiling factors" to reach easily the genome regions that should be expressed and, to the contrary, in preventing them from reaching the genome regions that should remain silent. It is worth recalling the observations (*149, 167*) of differential accessibility of histones to acetylase and deacetylase enzymes, suggesting the existence of special nucleosome environments within the cell. The well known position effects in *Drosophila* chromosomes (*227, 228*) might also be due to such specific spatial arrangements.

The specificity of the three-dimensional architecture might itself reside in a specific composition of the various chromosome domains, perhaps related to nucleosome heterogeneity (i.e., nonrandom distribution of histone variants and nonhistone proteins, arrays of nucleosomes with a given DNA repeat length). Mechanisms have been proposed as to how such specific chromatin structures could be directly inherited (*113, 215, 219a, 224, 229, 229b*) in order to transfer the "epigenetic information" that might be generated by nucleosome heterogeneity. Are DNA modifications, e.g., DNA methylation (*230–230b*), involved in such mechanisms? (Incidentally, the possibility that DNA modifications could play a role in chromatin structure and the regulation of gene expression may limit the questions that can be asked by using cloned DNAs.) If there is such "epigenetic information" in chromatin structure, other mechanisms should also operate, since the chromatin structure about cloned ribosomal genes injected as naked DNA into *Xenopus* germinal vesicles has the same "active" configuration as "native" active ribosomal transcription units, whereas the plasmid vector DNA is compacted in nucleosomes (Section II) (45). Furthermore, the spacer ribosomal DNA is in an extended form, although apparently it is not transcribed. These results suggest very strongly either that the transcription itself is generating the altered "active" configuration (the spacer DNA being transcribed at a very reduced rate) or that both spacer and rRNA-coding DNA sequences contain sequences that can interact with "factors" triggering the "active" chromatin conformation. Along the same lines, it would be very interesting to know whether the mouse erythroleukemia cell-fusion-mediated activation of the α-globin gene in human nonerythroid cells (231) involves rearrangements of the three-dimensional structure of human chromatin.

The possible relationship, if any, between the microdomains observed in interphase nuclei and the chromatin DNA domains revealed by biochemical (232–235) and electron microscopic (236) studies is unknown. It is also unknown whether these chromatin DNA domains are differentially organized in different cell types, but the question whether their length is differentially organized in different cell types is now open to direct experimentation (237). Using *Drosophila* cells (237), it should also be possible to determine whether, during gene activation, the uncoiling of a chromatin DNA domain precedes the conversion of nucleosomes into the "active" DNase I-hypersensitive state. More generally, it will be possible to establish how the three events—loss of higher-order structure, alteration of nucleosomal structure, and transcription—are related when a gene is turned on or turned off. The interrelationships are likely to be quite complicated; for example, one can envisage that histone acetylation is either the cause or the effect of each of the first two steps (126, 158, 217) and the cause of the third. We have assumed in the above discussion that the uncoiling of higher-order chromatin structures and the generation of the altered nucleosomal structure are the causes, not the consequences, of transcription itself. Such an assumption is certainly justified by the bulk of the data available, but the other possibility is not excluded (see above).

Is the initiation of transcription guided in any way by the location of nucleosomes on the DNA? Although there is no compelling evidence for nucleosome phasing in general (see 1 for references), it has not been excluded that some genome regions involved in the regulation of transcription (notably initiation) could be more readily accessible to regulatory factors owing to the absence of nucleosomes. Such a possibility is indeed suggested by recent studies showing that in bulk nontranscribed SV40 "minichromosomes," the region corresponding to the origin of replication, and presumably to the initiation sites for RNA transcription, is more accessible to restriction endonuclease (238, 239) and DNase (239a, 240, 240a). Similarly, there are specific preferential micrococcal nuclease cleavage sites in chromatin DNA of *Drosophila* tissue culture cells (58), and a preferred DNase I cleavage site in chicken conalbumin chromatin (230b), which could be interpreted as indicating a nonrandom location of some nucleosomes with respect to DNA sequence. In this respect, it is interesting that nucleosomes assembled *in vitro* might be nonrandomly located on the SV40 genome (241, 241a) as well as on other DNA fragments of defined sequence (242, 243). However, it does not appear that the association of histones with DNA to form nucleosomes is imperative for the RNA polymerase B molecules to initiate RNA chains specifically, since the adenovirus genome is not associated with histones late in infection (244).

Although an impressive amount of information is now available on the structure of transcriptionally active chromatin, one is left with the feeling that few solid conclusions can be drawn. It is clear that this field of chromatin research has avenues still to be explored, but what is equally evident is that there is a need to step back and solidify the basis for many popular, but not fully substantiated, beliefs.

Acknowledgments

We are greatly indebted to Mrs. Colette Kutschis and Edith Badzinski for their help in preparing the manuscript. We thank all those who sent us preprints of their work. D. Mathis was a fellow of the Damon Runyon Walter Winchell Cancer Fund. Our laboratory is supported by grants from the CNRS, the INSERM, the DGRST, and the Fondation pour la Recherche Médicale Française.

References

1. P. Chambon, *CSHSQB* **XLII**, 1209 (1977).
2. R. S. Gilmour, in "Receptors and Hormone Action" (O'Malley, ed.), Vol. 1, p. 331. Academic Press, New York, 1977.
3. T. Pederson, *Int. Rev. Cytol.* **55**, 1 (1978).
4. *CSHSQB* **XXXVIII** (1973).
5. R. D. Kornberg, *ARB* **46**, 931 (1977).
6. J. O. Thomas, in "International Review of Biochemistry" (B. F. Clark, ed.), Vol. 17, p. 181 Univ. Park Press, Baltimore, Md., 1978.
6a. R. A. Garrett, in "International Review of Biochemistry" (R. E. Offord, ed.), Vol. 25, p. 179. Univ. Park Press, Baltimore, 1979.
7. G. Felsenfeld, *Nature* **271**, 115 (1978).
7a. D. M. Lilley and J. F. Pardon, *AR Genet.* **13**, 197 (1979).
8. R. D. Kornberg, *Science* **184**, 868 (1974).
9. J. B. Rattner and B. A. Hamkalo, *J. Cell Biol.* **81**, 453 (1979).
10. A. L. Olins and D. E. Olins, *J. Cell Biol.* **81**, 260 (1979).

11. U. Scheer and H. Zentgraf, *Chromosoma* **69**, 243 (1978).
12. O. L. Miller and B. R. Beatty, *Science* **164**, 955 (1969).
13. O. L. Miller and A. H. Bakken, *Acta Endocrinol. Suppl.* **168**, 155 (1972).
14. V. E. Foe, L. E. Wilkinson, and C. D. Laird, *Cell* **9**, 131 (1976).
15. V. E. Foe, *CSHSQB* **XLII**, 723 (1977).
16. S. L. McKnight, M. Bustin, and O. L. Miller, Jr., *CSHSQB* **XLII**, 741 (1977).
17. R. H. Reeder, S. L. McKnight, and O. Miller, *CSHSQB* **XLII**, 1174 (1977).
18. U. Scheer, M. F. Trendelenburg, and W. W. Franke, *Exp. Cell Res.* **80**, 175 (1973).
19. U. Scheer, M. F. Trendelenburg, G. Krohne, and W. W. Franke, *Chromosoma* **60**, 147 (1977).
20. N. Angelier and J. C. Lacroix, *Chromosoma* **51**, 323 (1975).
21. H. Spring, G. Krohne, W. W. Franke, U. Scheer, and M. F. Trendelenburg, *J. Microsc. Biol. Cell.* **25**, 107 (1976).
22. W. W. Franke, U. Scheer, M. F. Trendelenburg, H. Spring, and H. Zentgraf, *Cytobiology* **13**, 401 (1976).
23. W. W. Franke, U. Scheer, H. Spring, M. F. Trendelenburg, and G. Krohne, *Exp. Cell Res.* **100**, 233 (1976).
24. W. W. Franke, U. Scheer, M. Trendelenburg, H. Zentgraf, and H. Spring, *CSHSQB* **XLII**, 755 (1977).
25. C. D. Laird, L. E. Wilkinson, V. E. Foe, and W. Y. Chooi, *Chromosoma* **58**, 169 (1976).
26. F. Puvion-Dutilleul, J. P. Bachellerie, J. P. Zalta, and W. Bernard, *J. Microscop. Biol. Cell.* **30**, 183 (1977).
27. C. L. F. Woodcock, L. L. Y. Frado, C. L. Hatch, and L. Ricciardiello, *Chromosoma* **58**, 33 (1976).
28. R. M. Grainger and R. C. Ogle, *Chromosoma* **65**, 115 (1978).
29. M. F. Trendelenburg, U. Scheer, H. Zentgraf, and W. W. Franke, *JMB* **108**, 453 (1976).
30. M. F. Trendelenburg, H. Zentgraf, W. W. Franke, and J. B. Gurdon, *PNAS* **75**, 3791 (1978).
31. U. Scheer, *Cell* **13**, 535 (1978).
32. D. Rungger, M. Crippa, M. F. Trendelenburg, U. Scheer, and W. W. Franke, *Exp. Cell Res.* **116**, 481 (1978).
33. W. W. Franke and U. Scheer, *Phil. Trans. R. Soc. London B* **283**, 333 (1978).
34. S. L. McKnight and O. L. Miller, *Cell* **8**, 305 (1976).
35. S. L. McKnight, N. L. Sullivan, and O. L. Miller, *This Series* **19**, 313 (1976).
36. F. Puvion-Dutilleul, E. Puvion, and W. Bernhard, *J. Ultrastruct. Res.* **63**, 118 (1978).
37. D. Villard and S. Fakan, *C.R. Acad. Sci. D* **286**, 777 (1978).
38. M. Matsuguchi, F. Puvion-Dutilleul, and G. Moyne, *J. Gen. Virol.* **42**, 443 (1979).
39. S. Busby and A. Bakken, *Chromosoma* **71**, 249 (1979).
40. M. M. Lamb and B. Daneholt, *Cell* **17**, 835 (1979).
41. C. D. Laird and W. Y. Chooi, *Chromosoma* **58**, 193 (1976).
42. P. Gariglio, R. Llopis, P. Oudet, and P. Chambon, *JMB* **131**, 75 (1979).
43. S. C. R. Elgin, L. A. Serunian, and L. M. Silver, *CSHSQB* **XLII**, 839 (1977).
44. M. Jamrich, A. L. Greenleaf, F. A. Bautz, and E. K. F. Bautz, *CSHSQB* **XLII**, 389 (1977).
45. M. F. Trendelenburg and J. B. Gurdon, *Nature* **276**, 292 (1978).
45a. T. Moss and M. L. Birnstiel, *NARes* **6**, 3733 (1979).
45b. D. Rungger, H. Achermann, and M. Crippa, *PNAS* **76**, 3957 (1979).
46. E. Lacy and R. Axel, *PNAS* **72**, 3978 (1975).
47. R. Axel, *This Series* **19**, 355–371 (1976).
48. A. Garel and R. Axel, *PNAS* **73**, 3966 (1976).
49. A. Garel and R. Axel, *CSHSQB* **XLII**, 701 (1977).

50. A. Garel, R. Weinstock, R. Sweet, H. Cedar, and R. Axel, in "The Cell Nucleus" (H. Busch, ed.), Vol. VI, pp. 149–180. Academic Press, New York, 1978.
51. M. T. Kuo, C. G. Sahasrabuddhe, and G. F. Saunders, PNAS 73, 1572 (1976).
52. R. Reeves and A. Jones, Nature 260, 495 (1976).
53. D. J. Mathis and M. A. Gorovsky, Bchem 15, 750 (1976).
54. P. W. Piper, H. Celis, K. Kaltoft, J. C. Leer, O. F. Nielsen, and O. Westergaard, NA Res 3, 493 (1976).
55. M. J. Butler, K. E. Davies, and I. O. Walker, NARes 5, 667 (1978).
56. J. M. Gottesfeld and D. A. Melton, Nature 273, 317 (1978).
57. M. Bellard, F. Gannon, and P. Chambon, CSHSQB XLII, 779 (1977).
58. C. Wu, Y. C. Wong, and S. C. R. Elgin, Cell 16, 807 (1979).
59. J. Stalder, T. Seebeck, and R. Braun, BBA 561, 452 (1979).
60. L. F. Liu and J. C. Wang, Cell 15, 979 (1978).
61. H. Weintraub and M. Groudine, Science 193, 848 (1976).
62. R. D. Camerini-Otero, B. Sollner-Webb, R. H. Simon, P. Williamson, M. Zasloff, and G. Felsenfeld, CSHSQB XLII, 57 (1977).
63. H. Weintraub, in "Cell Cycle and Cell Differentiation" (J. Reinert and H. Holtzer, eds.), Vol. 7, p. 27. Springer-Verlag, Berlin and New York, 1975.
64. D. J. Mathis and M. A. Gorovsky, CSHSQB XLII, 773 (1977).
65. J. Stadler, T. Seebeck, and R. Braun, EJB 90, 391 (1978).
66. S. J. Flint and H. M. Weintraub, Cell 12, 783 (1977).
67. E. I. Frolova and E. S. Zalmanzon, Virology 89, 347 (1978).
68. E. I. Frolova, E. S. Zalmanzon, E. M. Lukanidin, and G. P. Georgiev, NARes 5, 1 (1978).
69. C. B. Chae, T. K. Wong, and R. A. Gadski, BBRC 83, 1518 (1978).
70. M. Groudine, S. Das, P. Neiman, and H. Weintraub, Cell 14, 865 (1978).
71. A. Panet and H. Cedar, Cell 11, 933 (1977).
72. M. Breindl and R. Jaenisch, Nature 277, 320 (1979).
73. J. Paul, E. J. Zollner, R. S. Gilmour, and G. D. Birnie, CSHSQB XLII, 597 (1977).
74. B. Levy-W. and G. H. Dixon, NARes 4, 883 (1977).
75. A. Levitt, R. Axel, and H. Cedar, Dev. Biol. 69, 496 (1979).
76. P. Chambon, C. Benoist, R. Breathnach, M. Cochet, F. Gannon, P. Gerlinger, A. Krust, M. LeMeur, J. P. Lepennec, J. L. Mandel, K. O'Hare, and F. Perrin, in "From Gene to Protein: Transfer in Normal and Abnormal Cells." Academic Press, 1979.
77. A. Garel, M. Zolan, and R. Axel, PNAS 74, 4867 (1977).
78. H. Biessmann, S. Wadsworth, B. Levy-W., and B. J. McCarthy, CSHSQB XLII, 829 (1977).
79. D. M. Miller, P. Turner, A. W. Nienhuis, D. E. Axelrod, and T. V. Gopalakrishnan, Cell 14, 511 (1978).
80. R. D. Palmiter, E. R. Mulvihill, G. S. McKnight, and A. W. Senear, CSHSQB XLII, 639 (1977).
81. N. S. Young, E. J. Benz, Jr., J. A. Kantor, P. Kretschmer, and A. Nienhuis, PNAS 75, 5884 (1978).
81a. D. Lohr and L. Hereford, PNAS 76, 4285 (1979).
82. K. S. Bloom and J. N. Anderson, Cell 15, 141 (1978).
83. B. Levy-Wilson and G. H. Dixon, PNAS 76, 1682 (1979).
84. E. M. Johnson, V. G. Allfrey, E. M. Bradbury, and H. R. Matthews, PNAS 75, 1116 (1978).
85. E. M. Johnson, H. R. Matthews, V. C. Littau, L. Lothstein, E. M. Bradbury, and V. G. Allfrey, ABB 191, 537 (1978).
86. R. Reeves, Science 194, 529 (1976).

87. R. Reeves, *EJB* **75**, 545 (1977).
88. R. Reeves, *Bchem* **17**, 4908 (1978).
89. K. J. Vavra and M. A. Gorovsky, submitted (1979).
90. J. M. Gottesfeld, *Methods Cell Biol.* **XVI**, 421 (1977).
91. R. J. Billing and J. Bonner, *BBA* **281**, 453 (1972).
92. J. M. Gottesfeld, W. T. Garrard, G. Bagi, R. F. Wilson, and J. Bonner, *PNAS* **71**, 2193 (1974).
93. J. M. Gottesfeld, G. Bagi, B. Berg, and J. Bonner, *Bchem* **15**, 2472 (1976).
94. J. M. Gottesfeld and G. A. Partingon, *Cell* **12**, 953 (1977).
95. K. Marushige and J. Bonner, *PNAS* **68**, 2941 (1971).
96. J. Bonner, W. T. Garrard, J. M. Gottesfeld, D. S. Holmes, J. S. Sevall, and M. Wilkes, *CSHSQB* **XXXIII**, 303 (1973).
97. C. B. Kimmel, S. K. Sessions, and M. C. MacLeod, *JMB* **102**, 177 (1976).
98. T. Pederson and J. S. Bhorjee, *Bchem* **14**, 3238 (1975).
99. A. F. Lau, R. W. Ruddon, M. S. Collett, and A. J. Faras, *Exp. Cell Res.* **111**, 269 (1978).
100. R. B. Wallace, S. K. Dube, and J. Bonner, *Science* **198**, 1166 (1977).
100a. M. A. Bellett and T. J. Hall, *NARes* **6**, 2929 (1979).
101. R. S. Gilmour, P. R. Harrison, J. D. Windass, N. A. Affara, and J. Paul, *Cell Diff.* **3**, 9 (1974).
102. K. J. Hardy, J. F. Chu, K. Sakuma, and L. S. Hnilica, *Int. J. Biochem.* **9**, 357 (1978).
103. D. Hendrick, P. Tolstoshev, and D. Randlett, *Gene* **2**, 147 (1977).
104. J. M. Gottesfeld and P. J. G. Butler, *NARes* **4**, 3155 (1977).
105. J. Bonner, J. M. Sala-Trepat, W. R. Pearson, and J. R. Wu, in "The Cell Nucleus" (H. Busch, ed.), Vol. IV, pp. 369–407. Academic Press, New York, 1978.
106. R. B. Wallace, in "The Cell Nucleus" (H. Busch, ed.), Vol. VI, pp. 35–73. Academic Press, New York, 1978.
107. M. Savage and J. Bonner, *Methods Cell Biol.* **XVIII** 1 (1978).
108. A. J. Mac Gillivray and D. Rickwood, in "The Cell Nucleus" (H. Busch, ed.), Vol. VI, pp. 263–304. Academic Press, New York, 1978.
109. G. D. Birnie, *Methods Cell Biol.* **XVIII**, 23 (1978).
110. J. R. Tata and B. Baker, *JMB* **118**, 249 (1978).
111. A. J. Varshavsky, Y. V. Ilyin, and G. P. Georgiov, *Nature* **250**, 602 (1974).
112. C. B. Chae, *Bchem* **13**, 1110 (1974).
113. Newrock, K. M., C. R. Alfageme, R. V. Nardi, and L. H. Cohen, *CSHSQB* **XLII**, 421 (1977).
114. E. M. Johnson and V. G. Allfrey, in "Biochemical Actions of Hormones" (G. Litwack, ed.), Vol. V, pp. 1–51. Academic Press, New York, 1978.
115. S. C. R. Elgin and H. Weintraub, *ARB* **44**, 725 (1975).
116. G. S. Stein, J. L. Stein, and J. A. Thomson, *Cancer Res.* **38**, 1181 (1978).
117. L. Sealy and R. Chalkley, *NARes* **5**, 1863 (1978).
118. G. Vidali, L. C. Boffa, and V. G. Allfrey, *Cell* **12**, 409 (1977).
119. B. Levy-W. and G. H. Dixon, *NARes* **5**, 4155 (1978).
120. B. Levy-W, W. Connor, and G. H. Dixon, *JBC* **254**, 609 (1979).
121. J. R. Davie and E. P. M. Candido, *PNAS* **75**, 3574 (1978).
122. I. L. Goldknopf, M. F. French, Y. Daskal, and H. Busch, *BBRC* **84**, 786 (1978).
123. M. T. Chong, W. T. Garrard, and J. Bonner, *Bchem* **13**, 5128 (1974).
124. H. H. Ohlenbusch, B. M. Olivera, D. Tuan, and N. Davidson, *JMB* **25**, 299 (1967).
125. Y. V. Ilyin, A. Y. Varshavsky, U. N. Mikelsaar, and G. P. Georgiev, *EJB* **22**, 235 (1971).
126. D. Nelson, M. E. Perry, and R. Chalkley, *NARes* **6**, 561 (1979).
127. M. Renz, P. Nehls, and J. Hozier, *PNAS* **74**, 1879 (1977).
128. M. Noll and R. Kornberg, *JMB* **109**, 393 (1977).

128a. G. M. Lawson and R. D. Cole, *Bchem* **18**, 2160 (1979).
129. S. C. Albright, P. P. Nelson, and W. T. Garrard, *JBC* **254**, 1065 (1979).
130. A. Zweidler, *Methods Cell Biol.* **XVII**, 223 (1978).
131. C. Von Holt, W. N. Strickland, W. F. Brandt, and M. S. Strickland, *FEBS Lett.* **100**, 201 (1979).
132. N. L. Bafus, S. C. Albright, R. D. Todd, and W. T. Garrard, *JBC* **253**, 2568 (1978).
133. I. L. Goldknopf and H. Busch, in "The Cell Nucleus" (E. Busch, ed.), Vol. VI, pp. 149–180. Academic Press, New York, 1978.
134. I. L. Goldknopf and H. Busch, *PNAS* **74**, 864 (1977).
135. I. L. Goldknopf, M. F. French, R. Musso, and H. Busch, *PNAS* **74**, 5492 (1977).
136. H. G. Martinson, R. True, J. B. E. Burch, and G. Kunkel, *PNAS* **76**, 1030 (1979).
137. D. C. Watson, B. Levy-W., and G. H. Dixon, *Nature* **276**, 196 (1978).
138. V. G. Allfrey, in "Chromatin and Chromosome Structure" (H. J. Li and R. A. Eckhardt, eds.), p. 167. Academic Press, New York, 1977.
139. G. H. Dixon, in "Organization and Expression of Chromosomes" (V. G. Allfrey, E. K. F. Bautz, B. J. McCarthy, R. T. Schimke and A. Tissières, eds.), p. 197. Dahlem Konferenzen, Berlin, 1976.
140. V. G. Allfrey, R. Faulkner, and A. E. Mirsky, *PNAS* **51**, 786 (1964).
141. K. R. Yamamoto and B. M. Alberts, *ARB* **45**, 721 (1976).
142. C. Marushige, *PNAS* **73**, 3937 (1976).
143. J. Bonner, W. T. Garrard, J. Gottesfeld, D. S. Holmes, J. S. Sevall, and M. Wilkes, *CSHSQB* **XXXVIII**, 303 (1973).
144. C. K. Shewmaker, B. N. Cohen, and T. E. Wagner, *BBRC* **84**, 342 (1978).
145. H. Oberhauser, A. Csordas, B. Puschendorf, and H. Grunicke, *BBRC* **84**, 110 (1978).
146. T. K. Wong and K. Marushige, *Bchem* **15**, 2041 (1976).
146a. L. O. Tack and R. T. Simpson, *Bchem* **18**, 3110 (1979).
147. M. G. Riggs, R. G. Whittaker, J. R. Neumann, and V. M. Ingram, *CSHSQB* **XLII**, 815 (1977).
148. E. P. M. Candido, R. Reeves, and J. R. Davie, *Cell* **14**, 105 (1978).
149. L. S. Cousens, D. Gallwitz, and B. M. Alberts, *JBC* **254**, 1716 (1979).
150. H. K. Hagopian, M. G. Riggs, L. A. Swartz, and V. M. Ingram, *Cell* **12**, 855 (1977).
151. L. Sealy and R. Chalkley, *Cell* **14**, 115 (1978).
152. M. Perry, D. Nelson, M. Moore, and R. Chalkley, *BBA* **561**, 517 (1979).
153. L. C. Boffa, G. Vidali, R. S. Mann, and V. G. Allfrey, *JBC* **253**, 3364 (1978).
154. G. Vidali, L. C. Boffa, E. M. Bradbury, and V. G. Allfrey, *PNAS* **75**, 2239 (1978).
155. R. Reeves and E. P. M. Candido, *FEBS Lett.* **91**, 117 (1978).
156. D. J. Mathis, P. Oudet, B. Wasylyk, and P. Chambon, *NARes* **5**, 3523 (1978).
157. D. A. Nelson, M. Perry, L. Sealy, and R. Chalkley, *BBRC* **82**, 1346 (1978).
158. R. T. Simpson, *Cell* **13**, 691 (1978).
159. B. S. Schaffhausen and T. L. Benjamin, *PNAS* **73**, 1092 (1976).
160. D. Nelson, W. M. Perry, and R. Chalkley, *BBRC* **82**, 356 (1978).
161. B. Levy-Wilson, D. C. Watson, and G. H. Dixon, *NARes* **6**, 259 (1979).
162. J. R. Davie and E. P. M. Candido, *JBC* **252**, 5962 (1977).
163. E. H. Shelton, J. Kang, P. M. Wassarman, and M. L. DePamphilis, *NARes* **5**, 349 (1978).
164. D. B. Jump, T. R. Butt, and M. Smulson, *Bchem* **18**, 983 (1979).
165. A. Ruiz-Carillo, L. J. Wangh, and V. G. Allfrey, *Science* **190**, 117 (1975).
166. V. Jackson, A. Shires, R. Chalkley, and D. K. Granner, *JBC* **250**, 4856 (1975).
167. M. Moore, V. Jackson, L. Sealy, and R. Chalkley, *BBA* **561**, 248 (1979).
167a. P. Rubestein, L. Sealy, S. Marshall, and R. Chalkley, *Nature* **280**, 692 (1979).
168. M. S. Kanungo and M. K. Thakur, *BBRC* **87**, 266 (1979).
169. B. Wasylyk, G. Thevenin, P. Oudet, and P. Chambon, *JMb* **128**, 411 (1979).

170. P. Williamson and G. Felsenfeld, *Bchem* **17**, 5695 (1978).
171. G. H. Goodwin, J. M. Walker, and E. W. Johns, in "The Cell Nucleus" (H. Busch, ed.), Vol. VI, p. 181. Academic Press, New York, 1979.
172. R. Sterner, L. C. Boffa, and G. Vidali, *JBC* **253**, 3830 (1978).
173. S. Spiker, J. K. W. Mardian, and I. Isenberg, *BBRC* **82**, 129 (1978).
174. A. Rabbani, G. H. Goodwin, and E. W. Johns, *BJ* **173**, 497 (1978).
175. G. H. Goodwin and E. W. Johns, *BBA* **519**, 279 (1978).
176. B. Levy-W., N. C. W. Wong, and G. H. Dixon, *PNAS* **74**, 2810 (1977).
177. B. Levy-W. and G. H. Dixon, *Can. J. Biochem.* **56**, 480 (1978).
178. S. Weisbrod and H. Weintraub, *PNAS* **76**, 630 (1979).
179. M. M. Sanders, *J. Cell Biol.* **79**, 97 (1978).
180. B. Villeponteaux, L. Lasky, and I. Harary, *Bchem* **17**, 5532 (1978).
180a. E. H. Peters, B. Levy-Wilson, and G. H. Dixon, *JBC* **254**, 3358 (1979).
180b. J. B. Jackson, J. M. Pollock, Jr., and R. L. Rill, *Bchem* **18**, 3739 (1979).
181. C. G. P. Mathew, G. H. Goodwin, and E. W. Johns, *NARes* **6**, 167 (1979).
182. V. V. Bakayev, T. G. Bakayeva, V. V. Schmatchenko, and G. P. Georgiev *EJB* **91**, 291 (1978).
183. A. Caplan, M. G. Ord, and L. A. Stocken, *BJ* **174**, 475 (1978).
184. N. C. W. Wong, G. G. Poirier, and G. H. Dixon, *EJB* **77**, 11 (1977).
185. C. P. Giri, M. H. P. West, and Mark Smulson, *Bchem* **17**, 3495 (1978).
186. R. Sterner, G. Vidali, R. L. Heinrikson, and V. G. Allfrey, *JBC* **253**, 7601 (1978).
186a. L. C. Boffa, R. Sterner, G. Vidali, and V. G. Allfrey, *BBRC* **89**, 1322 (1979).
187. J. Lough and V. M. Ingram. *Exp. Cell Res.* **114**, 349 (1978).
188. M. Lunadei, P. Matteucci, L. Ullu, R. Gambari, G. B. Rossi, and A. Fantoni, *Exp. Cell Res.* **114**, 468 (1978).
189. F. Keppel, B. Allet, and H. Eisen, *PNAS* **4**, 653 (1977).
190. M. Bustin, D. Goldblatt, and R. Sperling, *Cell* **7**, 297 (1976).
191. N. Defer, M. Crepin, C. Terrioux, J. Kruh, and F. Gros, *NARes* **6**, 953 (1979).
192. J. Neumann, R. Whittaker, B. Blanchard, and V. Ingram, *NARes* **5**, 1675 (1978).
193. L. M. Silver and S. C. R. Elgin, in "The Cell Nucleus" (H. Busch, ed.), Vol. V, pp. 215–262. Academic Press, New York, 1979.
194. J. E. Mayfield, L. A. Serunian, L. M. Silver, and S. C. R. Elgin, *Cell* **14**, 539 (1978).
195. H. Weideli, P. Schedl, S. Artavanis-Tsakonas, R. Steward, R. Yuan, and W. J. Gehring *CSHSQB* **XLII**, 693 (1977).
195a. K. B. Palter and B. M. Alberts, *JBC*, in press (1980).
196. J. Sommerville, C. Crichton, and D. Malcolm, *Chromosoma* **66**, 99 (1978).
197. M. C. Yao and M. A. Gorovsky, *Chromosoma* **48**, 1 (1974).
198. A. Zellweger, U. Ryser, and R. Braun, *JMB* **64**, 681 (1972).
199. R. Reeder, in "Ribosomes" (M. Nomura, A. Tissières and P. Lengyel, eds.), p. 489. Cold Spring Harbor Lab., New York, 1974.
200. J. Engberg, J. R. Nilsson, R. E. Pearlman, and V. Leick, *PNAS* **71**, 894 (1974).
201. R. W. Jones, *BJ* **173**, 145 (1978).
202. R. W. Jones, *BJ* **173**, 155 (1978).
203. E. Gocke, J. C. Leer, O. F. Nielsen, and O. Westergaard, *NARes* **5**, 3993 (1978).
204. T. Higashinakagawa, H. Wahn, and R. H. Reeder, *Dev. Biol.* **55**, 375 (1977).
205. T. Seebeck, G. D. Kuehn, and R. Braun, *J. Cell Biol.* **79**, 107a (1978).
206. T. Seebeck, J. Stalder, and R. Braun, *Bchem* **18**, 484 (1979).
206a. G. D. Kuehn, H. U. Affolter, V. J. Atmar, T. Seebeck, U. Gubler, and R. Braun, *PNAS* **76**, 2541 (1979).
207. J. D. Griffith, *Science* **187**, 1202 (1975).
208. M. Bellard, P. Oudet, J. E. Germond, and P. Chambon, *EJB* **70**, 543 (1976).

209. U. Müller, H. Zentgraf, I. Eicken, and W. Keller, Science **201**, 406 (1978).
210. M. H. Green and T. L. Brooks, NARes **4**, 4279 (1977).
211. A. J. Varshavsky, S. A. Nedospasov, V. V. Schmatchenko, V. V. Bakayev, P. M. Chumackov, and G. P. Georgiev, NARes **4**, 3303 (1977).
212. B. Wasylyk, P. Oudet, and P. Chambon, EJB, in press (1980).
213. H. Weintraub, A. Worcel, and B. Alberts, Cell **9**, 409 (1976).
214. P. Oudet, C. Spadafora, and P. Chambon, CSHSQB **XLII**, 301 (1977).
215. R. Tsanev, in "The Cell Nucleus" (H. Busch, ed.), Vol. IV, pp. 107-133. Academic Press, New York, 1978.
216. T. R. Cech and K. M. Karrer, submitted (1979).
217. A. Mirzabekov and A. Rich, PNAS **76**, 1118 (1979).
218. D. M. J. Lilley, Cell Biol. Int. Rep. **2**, 1 (1978).
219. A. Stein, JMB **130**, 103 (1979).
219a. K. B. Palter, V. E. Foe, and B. M. Alberts, Cell **18**, 451 (1979).
220. T. R. Butt, D. B. Jump, and M. B. Smulson, PNAS **76**, 1628 (1979).
221. W. H. Strätling, V. Muller, and H. Zentgraf, Exp. Cell Res. **117**, 301 (1978).
222. J. P. Whitlock, Jr. and R. T. Simpson, JBC **252**, 6516 (1977).
223. J. P. Whitlock, Jr. and A. Stein, JBC **253**, 3857 (1978).
224. B. Alberts, A. Worcel, and H. Weintraub, in "The Organization and Expression of the Eukaryotic Genome" (E. M. Bradbury and K. Javaherian, eds.), pp. 165-191. Academic Press, New York, 1977.
225. J. Sedat and L. Manuelidis, CSHSQB **XLII**, 331 (1977).
226. A. Garcia-Bellido and F. Wandosell, Mol. Gen. Genet. **161**, 317 (1978).
227. W. K. Baker, Adv. Genet. **14**, 133 (1968).
228. E. B. Lewis, Adv. Genet. **3**, 73 (1950).
229. H. Weintraub, S. J. Flint, I. M. Leffak, M. Groudine, and R. M. Grainger, CSHSQB **XLII**, 401 (1977).
229a. H. Weintraub, NARes **7**, 781 (1979).
229b. M. Weidman and A. Levine, Cell **18**, 439 (1979).
230. R. Holliday and J. E. Pugh, Science **187**, 226 (1975).
230a. J. L. Mandel and P. Chambon, NARes, in press (1980).
230b. M. T. Kuo, J. L. Mandel, and P. Chambon, NARes, in press (1980).
231. A. Deisseroth and D. Hendrick, PNAS **76**, 2185 (1979).
232. P. R. Cook and I. A. Brazell, J. Cell Sci. **19**, 261 (1975).
233. T. Ide, M. Nakane, K. Anzai, and T. Andoh, Nature **258**, 445 (1975).
234. T. Igo-Kemenes and H. G. Zachau, CSHSQB **XLII**, 109 (1977).
235. C. Benyajati and A. Worcel, Cell **9**, 393 (1976).
236. J. R. Paulson and V. K. Laemmli, Cell **12**, 817 (1977).
237. C. Wu, P. M. Bingham, K. J. Livak, R. Holmgren, and S. C. R. Elgin, Cell **16**, 797 (1979).
238. A. J. Varshavsky, O. Sundin, and M. Bohn, NARes **5**, 3469 (1978).
239. A. J. Varshavsky, O. Sundin, and M. Bohn, Cell **16**, 453 (1979).
239a. O. Sundin and A. Varshavsky, JMB **132**, 535 (1979).
240. W. A. Scott and D. J. Wigmore, Cell **15**, 1511 (1978).
240a. W. Waldeck, B. Föhring, K. Chowdhury, P. Gruss, and G. Sauer, PNAS **75**, 5964 (1979).
241. B. A. J. Ponder and L. V. Crawford, Cell **11**, 35 (1977).
241a. B. Wasylyk, P. Oudet, and P. Chambon, NARes **7**, 705 (1979).
242. M. V. Chao, J. Gralla, and H. G. Martinson, Bchem **18**, 1068 (1979).
243. R. T. Simpson and P. Künzler, NARes **6**, 1387 (1979).
244. C. Kédinger, O. Brison, F. Perrin, and J. Wilhelm, J. Virol. **26**, 364 (1978).

Ligand-Induced Conformational Changes in Ribonucleic Acids

HANS GÜNTER GASSEN

*Fachgebiet Biochemie der Technischen
Hochschule, Darmstadt
Darmstadt, Germany*

I. Introduction 57
II. The Role of the Ribosomal Protein S1 in Adjusting the Conformation of either 16 S RNA or mRNA 58
   A. Function of Protein S1 in the Ribosomal Cycle 58
   B. Complex Formation of Protein S1 with Oligo- and Polynucleotides ... 60
   C. Function of Protein S1 in the Formation of the 30 S Initiation Complex and in the Elongation Cycle 64
   D. The Biochemical Function of Protein S1 as a Ribosomal Protein Remains Unclear 70
III. Induced Conformational Changes in tRNA 71
   A. Introduction 71
   B. Conformational Flexibility in the Anticodon Loop 73
   C. Induced Allosteric Conformational Changes in tRNA 76
   D. Conformational Changes in tRNA May Be Triggered by Small Molecules but Always Require a Protein 80
   E. Kinetic Data Are Necessary to Understand the Selection of the Cognate tRNA by the Programmed Ribosome 81
References 82

I. Introduction

Proteins, functioning as enzymes or carrier molecules, may change their tertiary structure on binding a ligand of low molecular weight. Alteration of the protein's conformation is accompanied most often by a change in defined biochemical properties—for example, an increase in catalytic activity, or a higher affinity for the substrate. If the inducing ligand is bound far from the active center, the structural transition is called an allosteric conformational change. Two well-known examples are the molecular carrier hemoglobin and the enzyme aspartate transcarbamoylase (1, 2). One may ask whether induced conformational changes are unique to proteins, or whether nucleic acids may behave in a similar way.

Instead of listing the numerous examples in which a ligand is thought to induce a conformational change in a nucleic acid—numerous, especially if ligands include mono- and divalent cations—I concentrate in the following on two examples, both of which describe partial reactions in ribosome-dependent protein synthesis. The first is protein S1, which alters the conformation of nucleic acids, and, besides, functions as a host-donated subunit in Qβ replicase in the ribosomal cycle (3). The second is the structural changes in tRNA that occur especially in codon–anticodon complex formation (4).

II. The Role of the Ribosomal Protein S1 in Adjusting the Conformation of either 16 S RNA or mRNA

A. Function of Protein S1 in the Ribosomal Cycle

1. COMMON ASPECTS

One of the intriguing problems in protein biosynthesis is the recognition of the correct initiation sequence of the mRNA by the 30 S ribosomal subunit (5). The formation of the four-component initiation complex (30 S) · (IF-2 · GTP)· (fMet-tRNA)·mRNA defines the triplet reading frame, differentiates between initiator AUG or GUG and internal methionine-decoding AUG, and discriminates in polygenic mRNAs among the cistrons to be translated. Furthermore, translational control (e.g., after perturbation of the cell cycle by viral infection) occurs mainly at the level of initiation complex formation (6).

Besides the small ribosomal subunit, the initiator tRNA (tRNA^{fMet}), and the three initiation factors (IF-1, -2, -3), the ribosomal protein S1, which may be called a pseudo-initiation factor, plays a pivotal role in the handling of the mRNA by the ribosome (7).

A second fascinating problem in protein synthesis is the high precision by which an aminoacyl-tRNA is selected by a codon when bound to the aminoacyl site of the ribosome. In both reactions, the ribosomal protein S1 acts possibly as the ligand that alters the conformation of the nucleic acid so that double-strand formation with a second nucleic acid is optimized.

2. OCCURRENCE OF PROTEIN S1

The ribosomal protein S1, by far the largest protein of the 30 S ribosomal subunit, was identified by different groups in the course of the isolation and characterization of the ribosomal proteins (8, 9). Isolated ribosomes contain 0.4 copy of S1 per 30 S, and 0.3 copy per 70 S ribosome (10, 11). However, polysomes active in protein synthesis contain stoichiometric amounts of S1 (12). Also, S1 is one of the host-donated subunits of Qβ replicase, being required for the synthesis of the minus strand of Qβ mRNA (13).

During work on bacterial initiation factors, a protein—the "interference factor" i—that inhibits the translation of phage RNA was isolated (14, 15). The physical and functional identity of this interference factor with the ribosomal protein S1 was documented later (*16*). S1 can be removed from either 30 S ribosomal subunits or 70 S ribosomes by 1 M KCl or by dialysis against 1 mM MgCl$_2$ solution (*17, 18*). From DEAE-cellulose, it elutes as free S1 and in part as an S1 RNA complex (*19*). Tritium-labeled S1 exchanges freely with S1 bound to the 30 S subunit, and so-called S1-depleted 30 S ribosomes (30 S-1) can be functionally reconstituted with free S1 at an appropriate Mg^{2+} concentration (*20*).

3. LOCATION OF PROTEIN S1 WITHIN THE RIBOSOME, AND ITS FUNCTION IN TRANSLATION

If protein S1, when bound to the 30 S subunit, maintains its extended form, it should stretch as a narrow band over the whole 30 S surface (*21*). Cross-linking experiments locate parts of the S1 in the so-called "16 S-RNA 3'-end cluster," which consists of the 3' end of 16 S RNA, the proteins S1, S2, S4, S7, S11, S13, S18, S19, S21, and IF-2 and IF-3 (*22–25*). Removal of the 50-nucleotide 3'-terminal fragments of 16 S RNA from the 30 S by cleavage *in situ* with colicin E3 leads to the loss of S1 by the 30S (*26*). This indicates that this fragment contains a binding site for S1, or holds the ribosome in a conformation optimized for S1 binding. Immunoelectron microscopy studies locate this cluster, which is involved functionally in decoding the mRNA, below the head of the 30 S ribosome (Fig. 1) (*27, 28*).

Since S1 can be cross-linked to poly(s^{4}U), which can substitute for poly(U) in poly(Phe) synthesis, physical contact between the mRNA and S1 is highly plausible (*29*).

The presence of S1 in 30 S subunits is indispensable for the formation of 30 S initiation complexes with natural mRNA and phage RNA, and furthermore is required for the translation of the mRNA into a defined protein (*7, 30*). Optimum poly(U)-dependent poly(Phe) synthesis also requires S1, whereas the effect on poly(A)-dependent poly(Lys) synthesis is less pro-

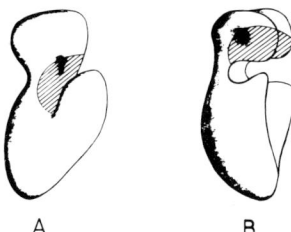

A   B

FIG. 1. Location of the mRNA decoding center on the 30 S ribosomal subunit. The hatched area represents the part of the ribosome termed "16 S RNA end cluster." The accessible parts of proteins involved in the decoding seem to be located in this part of the 30 S ribosome (A) in the Lake model and (B) in the Stöffler model (*27, 28*).

nounced (7, 31). Since S1 does not markedly influence the AUG-dependent binding of fMet-tRNA to 30 S ribosomes (32, 33), it was postulated that protein S1 as a helper stabilizes double-strand formation between the 3' end of 16 S RNA and the precistronic sequence of the mRNA—10 to 12 bases upstream from the AUG initiation codon (34, 35). I will come to a detailed description of the implication of S1 in the "Shine and Dalgarno complex" after dealing with the binding characteristics of S1 to isolated nucleic acids and oligo- or polynucleotides (36).

B. Complex Formation of Protein S1 with Oligo- and Polynucleotides

1. Physical Characteristics of Protein S1

The amino-acid composition of the protein as determined by different investigators is given in Table I (37–40). It contains a noticeably high amount of acidic amino acids (24%) and hydrophobic amino acids (44%) and tryptophan (not in the table).

TABLE I
AMINO-ACID COMPOSITION OF PROTEIN S1 AS PREPARED AND DETERMINED IN DIFFERENT LABORATORIES[a]

	Composition as given in reference			
Amino acid	37	38	39	40
Aspartic acid	11.2	11.6	11.8	11.6
Threonine	4.6	4.3	4.6	5.5
Serine	4.7	4.4	4.7	4.5
Glutamic acid	12.9	14.1	14.2	14.3
Proline	2.0	1.9	2.0	2.3
Glycine	8.5	8.9	9.7	9.7
Alanine	9.1	9.3	8.7	9.7
Valine	11.3	11.3	12.0	10.2
Methionine	2.0	1.0	0.6	1.5
Isoleucine	5.7	5.3	5.1	5.6
Leucine	8.6	8.2	8.3	8.2
Tyrosine	1.7	1.1	0.6	1.0
Phenylalanine	2.7	3.1	3.1	2.8
Histidine	1.9	1.5	1.7	1.1
Lysine	7.8	8.2	7.5	7.6
Arginine	5.4	5.5	5.5	4.3

[a] Amino-acid composition is expressed in mole-percent. The variation in the amino-acid composition and the range in the molecular weights (68,000–75,000) demonstrate best that S1 is a protein not easy to prepare and to handle.

The N-terminal 24-member sequence is Met-Thr-Glu-Ser-Phe-Ala-Gln-Leu-Phe-Glu-Glu-Asp-Leu-Lys-(Glu)-$^{Leu}_{Thr}$-Glu-Asn-(Arg)-Pro-$^{Gly}_{Asp}$-Gly *(41)*. Laughrea and Moore calculated from circular dichroism (CD) measurements that the secondary structure of S1 consists of 35–44% β-sheet structure and 10–15% α-helix *(42)*. The molecular weight as determined by dodecyl-sulfate gel electrophoresis and sedimentation equilibrium is in the range of 65,000 to 73,000, and the sedimentation coefficient is 3.2 S *(43–45)*. Low-angle X-ray scattering gave an axial ratio of 10:1. This allows one to calculate an average length of 20 nm, which equals the widest diameter of the 30 S ribosome *(42)*. Isoelectric focusing of S1 revealed a value of 4.8 for the isoelectric point. Recently, two different physical techniques indicated that protein S1 has two nonidentical structural domains with a hinge region between *(46, 47)*.

2. GENERAL FEATURES OF RNA·S1 COMPLEXES

The complex formed between S1 and poly(U) adsorbs to nitrocellulose filters and is the basis for an assay in the isolation of S1 *(48)*. In addition to single-stranded homopolynucleotides, S1 binds to 16 S RNA, 23 S RNA, tRNA, and φX174 DNA *(49, 50)*. Complex formation to polymers of pyrimidine nucleotides, especially if their secondary structure is of low thermal stability, like poly(U), is preferred over the binding to polymers of purine nucleotides, such as poly(A) *(31, 51)*. The association constant, as measured with oligouridylates and oligoadenylates, is chain-length dependent and approaches its optimum value at a chain length of $n = 12$–15 ($K_{Ass} = 3 \times 10^7$ M^{-1}). Short oligonucleotides exceed a stoichiometry of one, whereas more than one S1 is bound to polyuridylates of average chain length above 100 *(52)* (Table II).

The binding site on the S1 thus accommodates at least 12–15 nucleotides. Draper and von Hippel, observing the quenching of the tryptophan fluorescence in the presence of oligo- or polynucleotides, interpret their data as indicating two nucleic acid binding sites for S1. Site I binds either single-stranded DNA or RNA, whereas site II binds only single-stranded RNA, yet with a much higher affinity *(53, 54)*. The binding constants for the oligonucleotide·S1 complex at 25°C determined by this method differ from the values obtained by other techniques *(52)*. Complex formation is found furthermore with poly(rA)·poly(dT) and 2 poly(U)·poly(A), showing that the binding capacity of S1 is not restricted to single-stranded nucleic acids *(55)*.

A wide variety of homopolynucleotides containing different bases react with S1; thus the nature of heterocycle plays no major role in binary complex formation *(52)*. However, the dodecanucleotide G-A-U-A-C-C-U-C-C-U-U-A at the 3' end of 16 S RNA loses its capacity to bind to S1 after formaldehyde treatment, which points to a special role of the cytidines and adeno-

TABLE II

ASSOCIATION CONSTANTS FOR THE COMPLEXES OF OLIGONUCLEOTIDE AND PROTEIN S1[a]

Chain length	Association constant × 10^4 M^{-1}		Ligand bound per S1	
	Oligo (U)	Oligo (A)	For Oligo (U)	For Oligo (A)
5	5	1	2.5	—
7	23	6.3	1.3	—
9	110	35	0.9	1.0
12	2200	150	1.0	1.1
14	2700	260	0.9	1.1
16	3300	—	0.9	—
18	3700	—	0.9	—
~30	4000	—	0.8	—

[a] Binding constants were determined by equilibrium dialysis up to a chain-length of 14, then by adsorption of the complexes to nitrocellulose filters (31,52). Data were obtained from a Scatchard plot by linear regression.

sines in the recognition process (34). A dominant role of the 2′-hydroxyl for the recognition can be excluded, because $(U)_{10}$ and $(dT)_{10}$ show the same binding characteristics (52). This leaves the negatively charged polyphosphate lattice of the polynucleotide as the major element for electrostatic interaction with the protein.

On the protein side, there is evidence that the lysyl, histidinyl, and tryptophanyl side chains are involved in the recognition of the nucleic acid (56). If more than three lysines out of 30–40 per S1 are modified by reductive methylation of their ϵ-NH_2 groups to ϵ-NMe_2 groups, S1 becomes inactive in poly(U) binding (57). This points toward electrostatic interactions between the ϵ-NH_2's of the lysyl residues of the protein and the polyphosphate lattice of the nucleic acid. Alkylation of one of the two cysteinyl sulfhydryl groups does not influence the binding properties of S1, but appears to interfere with the so-called "RNA unwinding" function of S1 (58).

3. Unfolding of Polynucleotides by S1

Mainly two techniques have been used to observe changes in the polynucleotide structure as a result of S1 binding: the increase in ultraviolet (UV) absorbance due to the unstacking of bases, and the determination of differences in the CD spectra, which reflect changes in the interaction of the bases in polynucleotides (55). Lately, in addition, electron microscopy and nuclear magnetic resonance (NMR) have been used to elucidate the unfolding properties of S1 (59, 60). With these techniques, it was observed that helical poly(U), poly(C,U), and neutral or acidic poly(C) are converted to their thermally denatured forms by the binding of stoichiometric amounts of the protein. S1 cannot unfold double-helical poly(C) at 5.5°C, but does so at 22°C. It was estimated that one S1 molecule unstacks 20–40 nucleotides in a polynucleotide (55, 59).

The CD spectra of a number of triple and double-helical polynucleotides, with the exception of poly(A)·poly(dT), are not altered by S1 (55). On the contrary, S1 prevents the formation of a poly(A)·poly(U) double-helix by binding to the single-stranded chain (59). In the presence of S1, poly(A)·poly(dT) undergoes a transition to a new structure that has a CD spectrum unlike that of the thermally denatured form. Intercalated ethidium bromide is released from poly(A)·poly(dT) by S1, confirming the occurrence of a conformational rearrangement of the double strand (55). The translational inhibitor aurintricarboxylic acid completely inhibits the action of S1 on polymers of pyrimidine nucleotides, but has no effect on the conformational pertubation of poly(A)·poly(dT) (55).

Comparable effects are not exclusive to S1, but are found with the ribosomal proteins S3 and S21 as well (J. van Duin, personal communication).

If the reactive sulfhydryl group of one cysteine (the second one is buried in the tertiary structure) is modified by N-ethylmaleimide, S1 still binds to RNA, but no longer has the ability to unfold or melt the secondary structure of a nucleic acid (30).

Complex formation between S1 and homopolynucleotides depends on Mg^{2+} ions in buffers of low NaCl concentration (< 40 mM) and becomes independent of Mg^{2+} at or above 80 mM NaCl. Increasing the concentration from 0.1 M to 1.0 M NaCl gives a linear decrease in the association constant with the logarithm of the ionic strength (62).

From the linear relationship of log K_{Ass} to log $[Na^+]$, one can calculate that a U_{12}·S1 complex is stabilized by three charge–charge interactions, whereas for A_{12}·S1 only two electrostatic interactions exist (63).

The formation of the U_{13}·S1 complex is pH-dependent with a point of inflection at pH 10.5. This pH-dependence again points toward a participitation of the lysyl residues in complex formation (31).

A number of oligonucleotides obtained from ribonuclease T1 hydrolyzates of either Qβ RNA, 16 S RNA, or 23 S RNA bind tightly to S1. The two fragments from Qβ RNA could represent S1 binding sites, when S1 is a subunit of the Qβ replicase (64, 65). The function of the tightly binding oligonucleotide from the 23 S RNA is unclear at the present (66). Complex formation may have occurred by chance during isolation, since this oligonucleotide shows a stretch of six pyrimidine nucleotides. The dodecanucleotide

TABLE III

RNA Fragments from Qβ RNA or rRNA That Binds to Protein S1[a]

Original RNA molecule	S1 binding fragment	Reference
3' End of 16 S RNA	A-U-C-A-C-C-U-C-C-U-U-A$_{OH}$	34
Near the 3' end of Qβ RNA	A-A-U-A-A-A-U-U-A-U-C-A-C-A-A-U-U-A-C-U-C-U-U-A-C-G	64
Region preceding coat cistron of Qβ RNA	U-A-U-C-U-U-U-U-A-U-U-A-A-C-C-C-A-A-C-G	65
23 S RNA, L1 binding site	A-A-A-A-A-C-C-C-U-U-U-A-C-A-A-U-G	66

[a] Since the fragments were obtained by ribonuclease-T1 treatment of either Qβ RNA or ribosomes, all four sequences are rich in pyrimidines as expected. This is an argument against specificity, since S1 binds preferentially to poly(Y) (Y = a pyrimidine nucleoside), especially poly(U). There are, however, other pyrimidine-rich stretches in Qβ RNA that do not bind S1 (76).

representing the 3' end of the 16 S RNA and the precistronic sequence from the coat protein gene of Qβ replicase may represent true S1 binding fragments (Table III).

Since natural RNAs are highly ordered three-dimensional structures, not too much information may be obtained from the mere one-dimensional sequences of S1-binding oligonucleotides.

C. Function of Protein S1 in the Formation of the 30 S Initiation Complex and in the Elongation Cycle

1. S1 as a "Helper" Protein to Stabilize the Double Strand in the Precistronic Region

Shine and Dalgarno, in sequencing the 3'-terminal end of 16 S RNA from *Escherichia coli*, pointed out that this part of the 16 S RNA could form up to eight base-pairs with the precistronic sequence of phage RNA (36). This yields a plausible mechanism to explain the recognition of the initiation codon AUG by the initiation complex (Fig. 2). Close to 80 sequences of ribosomal binding sites for mRNA are now known; they contain, on the average, 4.5 matching bases roughly 10 nucleotides upstream from the initiator AUG (67). However, no correlation of number of complementary bases with efficiency in initiation complex formation, especially with the

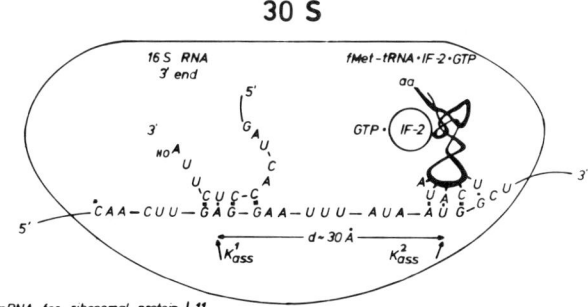

FIG. 2. Selection of the initiation region of an mRNA by the 30 S ribosomal subunit and the fMet-tRNA·(IF-2·GTP) complex. Specific recognition of the initiator AUG, and thus the setting of the reading frame, may be determined by the existence of two binding sites with a defined distance in between.

polygenic phage RNAs, exists so far. Since, furthermore, at least six to seven base-pairs appear to be needed to obtain an RNA·RNA hybrid stable at 25°C, additional factors are considered necessary to stabilize that type of double strand (68). The ideal candidate for this function is still the protein S1.

The nonanucleotide A-G-A-G-G-A-G-G-U, a fragment of Qβ RNA that is complementary to the 3′-terminal sequence of 16 S RNA, binds tightly to the ribosome and is released as a complex with the 49-mer of 16 S RNA after digestion with cloacin DF-13 (67, 69). A mixture of this nonanucleotide and G-G-G-A-G-A-A-G inhibits the binding of Qβ mRNA to 70 S ribosomes up to 78%, whereas the binding of pAUG is stimulated by 42%. The same oligonucleotide inhibits the binding of fMet-tRNA to 70 S ribosomes when directed by Qβ RNA but not by the AUG trinucleotide (70).

From the results described above, it can be concluded that indeed initiation complex formation is stabilized by the formation of two short double-stranded regions at a defined distance to each other: the 3′ end of the 16 S RNA and the precistronic sequence on the one side, and the initiation codon and anticodon of the initiator tRNA on the other side (Fig. 2).

However, in the elongation step of protein synthesis, it is necessary to translocate the mRNA with respect to the ribosome, which means that the two types of association in the 70 S ribosome are to be released. The double strand between the AUG and tRNA^{fMet} is broken by the elongation factor (EF-2 · GTP)-dependent release of the initiator-tRNA from the elongation complex (5). The double strand in the precistronic region can easily be weakened by a conformational change of the 3′ end of the 16 S RNA. The protein S1 is very probably involved in this conformational change.

S1 binds to the terminal 16 S RNA dodecanucleotide, and it is proposed that S1 unwinds, or stabilizes, the last 12 nucleotides in the single-stranded form as depicted in Fig. 3 (34). Within the isolated 49-nucleotide 3'-terminal fragment of the 16 S RNA, at least in solution, the dodecanucleotide exists even in the absence of S1 as a single-stranded area, as shown by NMR measurements (60, 71) (Fig. 4).

Steitz et al. investigated the binding of the initiation sequences of the three cistrons from R17 RNA to the 30 S ribosome in the presence and in the absence of S1 and initiation factors. The initiation sequence of the cistron for the maturation protein contains seven bases complementary to 16 S RNA, a number that is reduced to four in the respective sites for the coat protein and the replicase. The presence of S1 had two effects: it depressed the binding of ribosomes to spurious sites, and it increased the binding of coat and replicase cistrons. This led again to the conclusion that S1 may align the bases of the 16

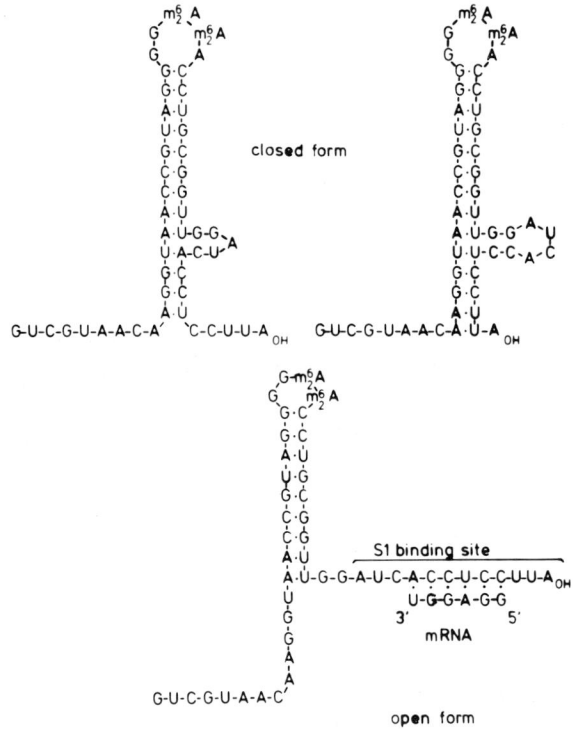

FIG. 3. RNA·RNA interaction between mRNA and 16 S RNA and the role of protein S1 in stabilizing the double-strand formed. According to Dahlberg and Dahlberg, the 3' end of 16 S RNA may exist in an open or a closed form (34). The presence of the mRNA (precistronic region) induces the transition to the open form. Protein S1 either holds the 16 S RNA in a conformation optimal for mRNA recognition or, by binding to the mRNA·16 S RNA hybrid, stabilizes the double strand.

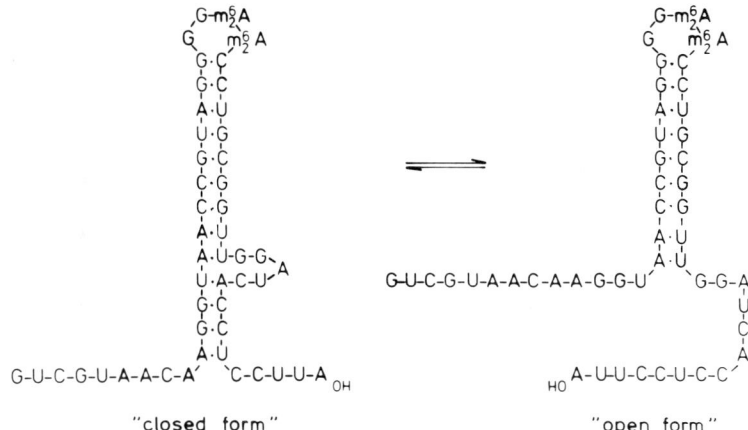

FIG. 4. Secondary structure model of a 49-nucleotide fragment from the 3' end of 16 S RNA. Nuclear magnetic resonance data obtained by Baan et al. point toward an open structure of this 16 S RNA fragment in solution (60). Solution data, however, may be far from the situation on the ribosome where, at least in part, the RNA exists in a protein environment (60).

S RNA in a specific configuration that facilitates stable interaction with the mRNA (72, 73).

Further evidence for the requirement of S1 comes from the melting curves of a R17 RNA gene with a seven base-pair complementary region, and the lambda P_R promoter transcript, which forms nine base-pairs with the 16 S RNA, giving melting points of only 32°C and 37°C, respectively. Since R17 RNA can be translated at 70°C in *Bacillus stearothermophilus*, it is suggested that even with a complementary run of nine base-pairs, double-strand formation without helper proteins is not stable enough to result in efficient initiation complex formation (74).

2. PROTEIN S1 UNFOLDS THE INITIATION REGION OF AN mRNA FOR INITIATION COMPLEX FORMATION

The translation of phage RNA is completely dependent on the presence of S1, whereas with either polyuridylate or polyadenylate the requirement for S1 can be released by raising the Mg^{2+} concentration (7, 15, 75, 76). MS2 RNA treated with aqueous formaldehyde can be translated by S1-lacking ribosomes. In addition, the inhibitory effect of excess protein S1 disappears (77). The chemistry of formaldehyde treatment has been reviewed, and it is suggested that only 10 nucleotides per RNA react with the formaldehyde (78). From the melting curves of both modified and unmodified RNAs in the presence of Mg^{2+}, it can be concluded that the formaldehyde-treated RNA loses the property of cooperative melting. Thus the major conclusion from these experiments is that S1 binds to the mRNA and alters its tertiary structure so that the initiation site of the mRNA becomes available for binding to the ribosome.

This explanation of the function of S1 is further supported by the finding that S1 is required for effective translation of the coat cistron of F2 RNA by *B. stearothermophilus* ribosomes, which themselves contain no S1 (79). Thus the biological role of S1 may be recognition and unfolding of the mRNA initiation site. An attractive feature of this hypothesis is that the dual functions of S1 in initiation and phage RNA replication may be explained by a single mechanism: opening of a hitherto buried site to allow for the specific binding of either Qβ replicase or 30 S ribosomes to an RNA initiation sequence (76).

The requirement of S1 in translation is not restricted to phage RNA, but it is evident that it is also required for the synthesis of T4-induced enzymes and bacterial anthranilate synthetase (80).

There is further evidence, using anti-S1 to inhibit initiation complex formation, that at 37°C recognition and specific binding of the coat protein cistron of MS2 RNA is solely a function of native 30 S ribosomes. Anti-S1 does not inhibit translation once the initiation complex is formed (77).

In summary, S1 recognizes the initiator region of an mRNA and transforms its tertiary structure to the ribosome binding type of conformation. However, this requires that the preinitiation structures of all bacterial mRNAs have a common recognition site for S1.

3. PROTEIN S1 ADJUSTS THE CODON FOR PROPER MATCHING WITH THE ANTICODON OF THE tRNA

The formation of the binary complex poly(U)·30 S subunit within a wide Mg^{2+} range (2–30 mM) does not depend on protein S1, at least not when examined by the nitrocellulose filter assay. On the other hand, Phe-tRNA binding to the 30S·poly(U) complex at 12–16 mM Mg^{2+} requires the presence of S1 (31).

Subunits from which protein S1 is removed (called "30S−1" ribosomes) can be restored in their capacity to bind Phe-tRNA by the addition of a stoichiometric amount of S1. In the binding reaction, excess S1 is not inhibitory; this is in contrast to poly(Phe) synthesis, where a twofold excess of S1 reduces the amount of poly(Phe) by 50%. If polyuridylate is replaced by short oligouridylates (U_3 to U_8), only a partial functional reconstitution of the S1-depleted 30 S ribosomes by the addition of S1 is found. This seems to indicate that a polynucleotide·S1 complex is formed first, which then binds to the depleted 30 S ribosomes (31). On agarose/acrylamide gels, 30 S subunits can be separated from 30 S ribosomes lacking S1. The 30 S particle migrates more slowly on gels and in sedimentation in sucrose gradients (81).

This could indicate that the overall structure of 30 S is changed when protein S1 is removed. Even the AUG-mediated binding of fMet-tRNA to the 30 S ribosomal subunit is promoted by this protein (82), in contrast to earlier reports (32). The S1 effect becomes even more pronounced when AUG is replaced by AUGA, which is more effective then AUG in fMet-tRNA binding to the 30 S subunit (82). However, the binding of Lys-tRNA to oligo(A)- or poly(A)-programmed ribosomes does not require S1 (7, 31). Since oligo- and polyadenylate, under the experimental conditions of the binding test, are in a single-stranded, stacked conformation, one is tempted to conclude that S1 is required only when an oligonucleotide with an unstable secondary structure is used as mRNA analog.

Thus S1 could adjust the conformation of a codon located in the aminoacyl site of the ribosome so that it forms the corresponding helix and thus binds to the anticodon helix of the AA-tRNA. This conformational adjustment could be brought about by electrostatic interactions with the RNA backbone, since S1 complex formation with nucleic acids is stabilized by two to three electrostatic interactions (54) (Fig. 5).

One has to keep in mind that, in AA-tRNA binding to programmed ribosomes, S1 can be replaced by an increase in Mg^{2+} concentration. This

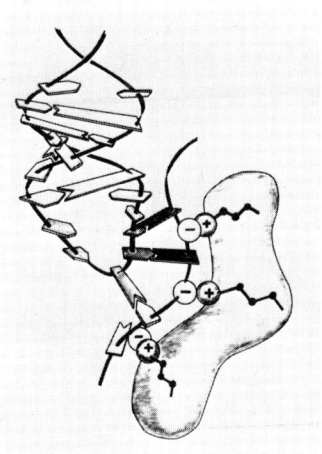

FIG. 5. Possible interaction between the ribosomal protein S1 and the codon when part of the 30 S ribosomal decoding site. Two to three electrostatic interactions between the charged groups of both molecules could hold the codon in a conformation optimized for RNA·RNA double-strand formation. As obvious from the three-dimensional structure of the tRNA, conformational rigidity in the anticodon comes from the anticodon loop conformation, whereas the codon in its transient complex with the 30 S ribosome must be stabilized by interaction with the 16 S RNA or with ribosomal proteins.

may even be so in the polymerization reaction. However, since mRNA-dependent protein synthesis is inhibited by excess Mg^{2+} (> 16 mM), the interplay of S1 and Mg^{2+} cannot be tested in the polymerization reaction (31).

D. The Biochemical Function of Protein S1 as a Ribosomal Protein Remains Unclear

S1, an acidic protein, can accommodate a segment of a polynucleotide up to 12–15 nucleotides long. By preferential binding to single-stranded nucleic acids of low thermal stability, it alters the conformation of the oligonucleotide so that it resembles the random-coil structure. The binding to double-stranded or triple-stranded polynucleotides is of a different nature and not well understood at the moment. Complex formation at low ionic strength is Mg^{2+}-dependent, and two to three positively charged amino-acid side chains, probably of lysines, interact electrostatically with the polyphosphate backbone. The binding of S1 to polynucleotides is stoichiometric and alters the conformation of up to 30 nucleotides (76).

Within the translation process, protein S1 seems to be required for optimal function of all steps in initiation and elongation. One may group the proposed S1 functions as follows: (a) S1 holds the 30 S subunit in a defined and active conformation; (b) S1, by binding mainly to the phosphate lattice of the 16 S RNA, allows and stabilizes double-strand formation with the mRNA in either the precistronic region or the aminoacyl site or both; (c) S1 binds to the initiation region of an mRNA and unfolds (unwinds) the structure of that region so that binding to the 30 S subunit can occur.

The first proposed function may be very appropriate and correct, as much experimental evidence substantiates it. However, from a mechanistic point of view it is not rewarding, as it does not lead us to a better understanding of protein–nucleic acid recognition.

Concerning the second concept, solid experimental support is scarce, yet there are many speculations that take into account the "sense and logic" of a biological system. From a chemist's view, this is the most satisfying explanation: the positive and stable charge-lattice of the protein forces, through multifold electrostatic interactions, a hitherto flexible part of the 16 S RNA into a defined and stable conformation optimized for mRNA recognition. The neutralization of the charges within this region of the nucleic acid will, in addition, reduce the charge–charge repulsion and thus contribute to the stabilization of the RNA's conformation. This explanation of S1 function would fit the model as outlined in Fig. 3.

The third proposal is also supported by experimental observations. However, since they are obtained in a complicated system—with a 100,000 g supernatant and not with purified factors—the observations may be explained in different ways. The only general conclusion to be drawn from this interpretation is that proteins may alter the nucleic acid conformation on binding.

III. Induced Conformational Changes in tRNA

A. Introduction

Transfer RNAs (tRNA) play an important role in protein synthesis, where they function as adaptor molecules between the triplet codon and the cognate amino acid to be inserted into a growing peptide chain. Besides protein biosynthesis, tRNA is involved in many other biological processes, especially in cell membrane synthesis and gene regulation. The existence of an RNA as an adaptor between mRNA and protein was postulated by Crick five years before even the proper name was created and three years before "soluble" RNA was identified as transfer RNA (83-85). Transfer RNA was the first RNA to be sequenced (86),[1] the first RNA to be crystallized, and the first functioning RNA of known three-dimensional structure (87, 88).[2]

Like hemoglobin for protein chemists, tRNA has become the favorite child for nucleic acid chemists in their experiments with various chemical and physical procedures. tRNA has been over- and undermethylated, nucleosides or bases have been excised or added, and it has been subjected to a tremendous variety of chemical modifications.[3] Defined fragments of native tRNA have been made—here even a new name, the "dissected molecule," was created—and reassociated (89).

Lately the attack on the anatomy of tRNA by chemical treatment has abated somewhat, yielding to a battery of different techniques, such as tritium exchange, sedimentation velocity, differential melting, NMR, electron spin resonance, Raman spectroscopy, fluorescence, CD, laser light-scattering, low-angle X-ray scattering, and X-ray analysis.

The results of these different experimental approaches, using biochemical, chemical, or physical methods, have appeared in numerous publications, which are again summarized in about 40 review articles. The two I want to mention especially are "The Structures and Functions of Transfer RNA" by J. P. Goddard (90), which contains a complete file of references up to 1978,[1] and the article by P. E. Cole and D. Crothers entitled "Conformational Changes in tRNA" (91). The latter review gives an excellent introduction into the proper use of physical measurements to follow con-

---

[1] See Singhal and Fallis in Vol. 23 of this series, and Munns and Liszewski in this volume [Ed.].
[2] See articles by Kim (Vol. 17), Kearns (Vol. 18), and Clark (Vol. 20) in this series [Ed.].
[3] See Sprinzl and Cramer in Vol. 22 of this series [Ed.].

formational transitions and evaluates critically the results available up to about 1977 on structural changes in tRNA.

Since the two crystal structures of tRNA^{Phe}, independently determined by two groups from even different crystal forms, are surprisingly similar, many dreams were ruined and more than one ingenious model (90a) of the three-dimensional structure of tRNA as derived from chemical modification experiments or nuclease digestion data was quietly removed from the stage. Now the classical debate, well known from protein chemistry, started with tRNA: "Are there differences between the crystal structure and the solution conformation in tRNA?" Measurements by NMR confirmed the identity of the tRNA structure in a crystal and in solution, and so far there has appeared no compelling evidence, either from chemical procedures or physical measurement, hinting at a widely different tRNA three-dimensional (3-D) structure in solution (92). However, there is increasing evidence that tRNA functioning in biochemical processes is not a static structure, but may exist in different conformations optimized toward its specific function (93).

Before I can deal with induced conformational changes in tRNA, I summarize the biochemical role of tRNA in protein biosynthesis. Transfer RNAs are recognized by two different classes of proteins: those that bind only to their cognate tRNAs (like aminoacyl synthetases), and those, like the elongation factor EF-Tu, that form complexes with all aminoacylated tRNAs. Transfer RNA may exist as either aminoacyl- or peptidyl-tRNA, and several isoacceptor tRNAs may exist for only one amino acid (94, 95). Thus tRNAs must possess grossly similar 3-D structures in order to be recognized by the elongation factor EF-Tu and to fit into the AA-tRNA site of the ribosome. On the other hand, differences in 3-D structure must exist, so that AA-tRNA synthetase can differentiate among the various amino-acid-specific tRNAs. The large number of modified nucleosides (up to 20 per tRNA)[1] and the "tertiary" interactions among nucleotides may be necessary to enable a nucleic acid to fulfill such a variety of functions (96).

Concerning the "density" of the modified nucleosides, two extremes exist for the two main functional centers of tRNA: the anticodon region, guaranteeing selectivity, shows a large variety of modified nucleosides, while the acceptor end, always engaged in the same type of ester bond formation (side-chain selectivity is left to the synthetase) is always simply C-C-A. Besides a complicated tertiary structure and many modified nucleosides, there is a third way to ease the correlation of structure with the many tRNA functions. Like proteins, tRNA may undergo induced structural transitions and there may be one defined tRNA conformation for every biochemical process (97).

In the last five years, conformational changes in tRNA have become an active field of discussion and research. In the following sections, I review first the newer results on induced structural changes in tRNA, focusing on codon-induced changes, and then emphasize a fron ter aspect, namely evidence for allosteric conformational changes in nucleic acids.

B. Conformational Flexibility in the Anticodon Loop

Experimental evidence for conformational changes in tRNA, especially the mechanism of thermal unfolding, and the effects of divalent cations, ionic strength, and pH, was recently extensively reviewed (91). Here I restrict myself to structural changes that occur when complementary oligonucleotides are bound to a single-stranded region of the tRNA, e.g., the anticodon loop (98–100).

Different anticodon loop conformations of tRNA have been attributed to the two principal functional states of tRNA in protein synthesis, namely as aminoacyl-tRNA or as peptidyl-tRNA (101). According to a model for the anticodon loop conformation, the anticodon triplet exists in two different conformations, the "3'-stacked" conformation, and the "5'-stacked" form. In both crystal structures, the anticodon exists as the 3'-stacked conformer (87, 88).

The temperature dependence of the fluorescence of the unusual base wye (formerly "Y-base") in yeast tRNA^{Phe} shows slow structural transitions characterized by a monomolecular "all-or-none effect," which can be explained by a transition from the 3'-stacked (the more stable one) to the 5'-stacked anticodon structure (102). With the conformational flexibility in the anticodon region of the tRNA, the question arises whether conformational transitions in one domain of the molecule can lead to conformational changes in other parts.[4]

If two tRNAs are bound during protein synthesis adjacent to each other and to the mRNA forming a continuous hexanucleotide helix (Fig. 7), the conformation of the peptidyl-tRNA must differ from that of the aminoacyl-tRNA (103). The width of the anticodon loop in the crystal structure is 20 Å, whereas the length of a codon, a trinucleoside diphosphate, in mRNA is only 10 Å. Therefore, only a geometric arrangement of an mRNA·(tRNA)$_2$ complex as shown in Fig. 6 is possible. The switch from one anticodon loop conformation to another has been postulated to be a dynamic process related to mRNA translocation (103).

If we take a closer look at the anticodon loop structure to understand the exposure of the three nucleotides for double-strand formation, the following becomes evident. The fourth base in the 3' direction is blocked in most tRNAs by chemical modification, whereas the constant U$_{33}$ in the 5' direction is prevented from stacking on the anticodon by a hydrogen bond be-

---

[4] See Wells et al. in this volume [Ed.].

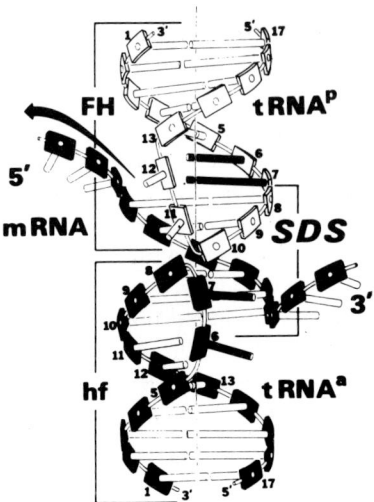

FIG. 6. The basic translation complex, according to Woese (*103*). The complex consists of an mRNA (SDS = sextuplet duplex structure) and two tRNA molecules. The "tRNA^p" in the peptidyl site (FH) adopts the 3' stacked conformation, whereas the "tRNA^a" in the aminoacyl site (hf) is seen in the 5' stacked conformation.

tween the N-3 of U_{33} and the phosphate backbone (P36) (*87, 88*). Chemical modification of groups capable of hydrogen bond formation prevents base-pairing with the mRNA in an irreversible way, whereas the unfavorable positioning of the uracil of U_{33} may be reversible depending on the functional state of the tRNA. The constant U_{33}, held in its steric position by only one hydrogen bond, is an obvious candidate for a conformational change in the anticodon loop structure. Experimental evidence that U_{33} is available for base-pairing in contrast to the crystal structure comes from studies with oligonucleotides complementary to the anticodon (*104, 105*). Tetranucleotides terminating in adenosine show a 10-fold higher association constant with their cognate tRNA compared to the anticodon complementary trinucleoside diphosphates. Control tetranucleotides terminating in cytidine show no increased affinity for the anticodon region of a tRNA (*106*).

The higher affinity of these tetranucleotides for the anticodon of a tRNA can be explained in three ways: (*a*) additional base-pairing between the fourth nucleoside and the constant uridine; (*b*) base stacking between the uridine and the fourth base; and (*c*) stabilization of the tetranucleotide conformation by the fourth nucleoside (the "dangling end" effect) (*107*). The most likely explanation is a stacking type of interaction between U_{33} and the fourth codon nucleoside, since purine nucleoside shows the same properties as adenosine (*82*).

The high efficiency of UUCA or A in binary complex formation with their cognate tRNAs should be reflected in a better stimulation of AA-tRNA binding to the ribosome, either in the amount of tRNA bound or in the quantity of oligonucleotide needed to obtain saturation in AA-tRNA binding. When the coding properties of these oligonucleotides were examined, a surprising effect emerged. In spite of its increased association constant with tRNA^{Phe}, UUCA is inactive in stimulating the binding of Phe-tRNA to 30 S ribosomes. A similar effect is found with A_4/Lys-tRNA and GUAA/Val-tRNA (Fig. 7). Thus tetranucleotides terminating in adenosine stabilize a tRNA conformation that cannot be bound to the 30 S ribosome. With 70 S ribosomes, the effects are less pronounced, but UUC is still a more effective codon than UUCA (108). With the ternary complex AA-tRNA·(EF-Tu·GTP), the binding of the triplet is strengthened, whereas the binding of the tetranucleotide is weaker. The correct AA-tRNA conformation thus is stabilized by EF-Tu·GTP (93).

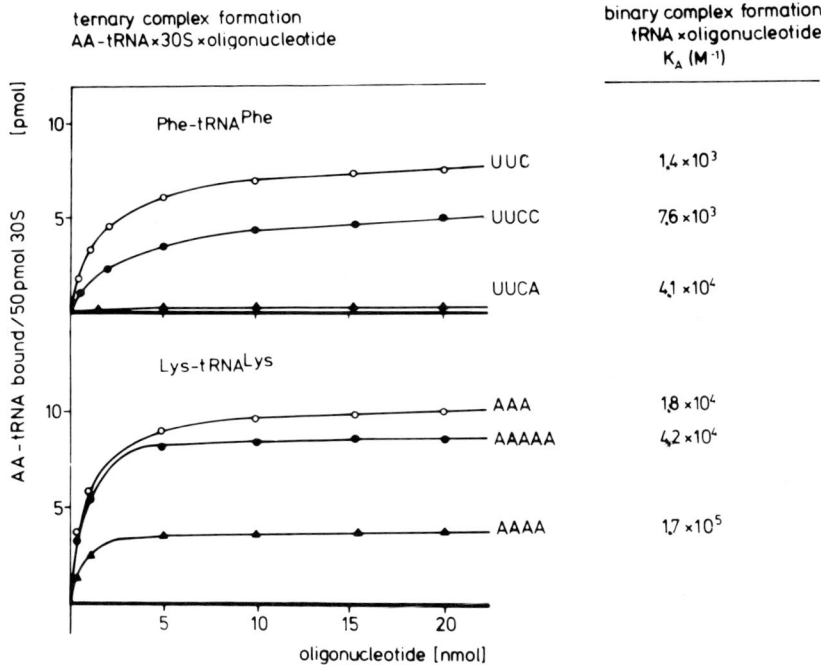

FIG. 7. Efficiency of tetranucleotides with 3'-terminal adenosine in binary complex formation and in their coding capacity. Oligonucleotides terminating in adenosine show a 10-fold higher affinity to their cognate tRNA as compared to tetranucleotides with a pyrimidine in the 3'-terminal position. However, they do not stimulate the binding of the cognate tRNA to the ribosome. Thus, an oligonucleotide that binds effectively to the anticodon region of a tRNA is not always a codon (93).

The initiator tRNA (tRNA^{fMet}) also shows a higher affinity to AUGA as compared to AUG (105). In contrast to AA-tRNA, AUGA stimulates the binding of fMet-tRNA to 30 S ribosomes more effectively than AUG, especially in the presence of the initiation factor IF-2·GTP (109). With 70 S ribosomes the situation is reversed. AUG is more effective than AUGA. Thus tRNAs should exist in two basic conformations with respect to the anticodon loop. With a tRNA functioning in the elongation process, an exposed U_{33} means an inactive tRNA conformation with the initiator tRNA; however, the conformation with the U_{33} exposed represents the active conformation, e.g., the one bound to the 30 S ribosome. Furthermore, a tetranucleotide interaction of the initiator tRNA with the 30 S mRNA complex blocks the binding of an elongator tRNA to the 30 S initiation complex. Addition of the 50 S ribosomal subunit removes (IF-2)·GDP from the ribosome. Without the protein's stabilization, the fMet-tRNA switches back to triplet recognition of the mRNA (AUG solely). Thus the first elongator codon becomes available, and the AA-tRNA can be bound to the 70 S aminoacyl site (82).

C. Induced Allosteric Conformational Changes in tRNA

1. Other Effectors besides the Codon

Since the Mg^{2+} concentration is one of the most important parameters in protein synthesis, the binding of Mg^{2+} to tRNA has been studied in great detail. In the crystal structure, four strong binding sites for Mg^{2+} exist (110–114). The number of binding sites for the solution structure is still under dispute, ranging from 1 to 7 strong binding sites and 20 to 25 weaker binding sites (91). Binding of Mg^{2+} to the nonnative tRNA is cooperative, and Mg^{2+} acts as an allosteric effector to convert tRNA from one conformation to another (for a detailed treatment of this issue, see 91, 115). Furthermore, the aminoacyl group—no ligand, but a covalent attachment to the 3' end of the tRNA—has long been suspected of exerting an allosteric influence on the tRNA conformation (91, 116).[3]

A conformational difference between the free tRNA and aminoacyl-tRNA could ease the discrimination between the two species by the elongation factor (EF-Tu·GTP). Positive and negative evidence exists for an aminoacylation-dependent conformational change, with negative evidence from CD studies (117–122), partial nuclease digestion (123), low-angle X-ray scattering (124), NMR and Raman spectroscopy studies (117, 125–127). Positive evidence comes from other CD studies (128), different binding of oligonucleotides to AA-tRNA (129, 130), dye and steroid binding (129, 131), and recently from NMR measurements and from laser light scattering (*116, 132*).

Pongs *et al.* find no effect of aminoacylation on the tRNA conformation as monitored by oligonucleotide binding to the TΨCG and dihydrouridine (D) loops, using an aminoacylated tRNA with a stable amide instead of an ester bond (*133*).[5] Dvorak *et al.*, on the other hand, report that complementary oligonucleotides bind to the TΨCG and dihydrouridine loop in Phe-tRNA^{Phe} but not in tRNA^{Phe} (*131*). With laser light scattering, it was found that the diffusion coefficient varies linearly with the extent of aminoacylation (*132*). Deacylation at 10 mM Mg^{2+} leads to a 14% calculated increase in the translational diffusion coefficient. A substantial change in shape is required to produce such a large change in diffusion coefficient. In contrast to earlier results, there is now positive evidence from NMR studies that aminoacylation changes the type of interaction of guanosine-53 with ribothymidine-54 in tRNA^{Phe} (*116*).

If these observations are confirmed, it still is difficult to rationalize from the crystal structure how the addition of an amino acid to the free and flexible C-C-A terminus will alter the gross tRNA conformation. However, there are notions that, in native tRNA, the C-C-A end comes much closer to the anticodon loop, as suggested from the crystal structure of tRNA^{Phe} (*134*).

2. CODON-INDUCED STRUCTURAL TRANSITIONS IN tRNA

The base complementarity between the constant oligonucleotide TΨCG (or TΨGA) in tRNA and the ^{43}CGAA^{46} in the 5 S RNA of the 50 S subunit led to the suggestion that tRNA is bound via three or four base-pairs to the 5 S RNA of the 50 S subunit (*135, 136*). This idea was supported by inactivation of the 50 S subunit by N-oxidation of two adenosines within the 5 S RNA with perphthalic acid, and by the inhibition of AA-tRNA binding to 70 S ribosomes by TΨCG (*137–139*). However, there are two facts that contradict this simple explanation. In native tRNA, the TΨCG sequence is not available for complementary oligonucleotide binding nor to chemical modification (*140, 141*), but, as evident from the crystal structure, it is involved in tertiary interactions (*87, 88*). Furthermore, other tRNAs do not compete for the binding of the cognate tRNA to 70 S ribosomes, at least not under conditions of a nitrocellulose filter assay (*142*). One solution to this discrepancy is to define a codon as an allosteric effector that alters the conformation of the tRNA from the nonbinding type to the (ribosome) binding type (*91, 93*).

In a preliminary way, we showed by equilibrium dialysis that a labeled

---

[5]See Krayevsky and Kukhanova in Vol. 23 [Ed.].

oligonucleotide CGAA can be bound only in the presence of the appropriate codon (here U_8) to tRNA^{Phe} (98). We concluded by analogy that CGAA binds to the TΨCG loop in a tRNA–codon complex. With low oligonucleotide concentration, the unfolding of the tRNA requires the 30 S ribosomal subunit, [Mg^{2+}] > 10 mM, and the elongation factor (EF-Tu·GTP). However, with oligonucleotide concentrations above the apparent association constant of the U_8·tRNA^{Phe} complex (that is, in the millimolar range), the change in the tRNA structure is achieved solely by the codon and [Mg^{2+}] > 12 mM. The stoichiometry of the reaction is such that even in excess of codon, only 5–10% of the tRNA binds CGAA. At Mg^{2+} below 5 mM, the rearrangement of the tRNA conformation is dependent on elongation factor (EF-Tu·GTP), whereas GTP can be replaced by guanylyl imidodiphosphate. In the absence of (EF-Tu·GTP), a sigmoidal CGAA-binding curve with respect to [Mg^{2+}] is obtained with half-saturation at 6 mM [Mg^{2+}]. Since the binding of Phe-tRNA to 70 S ribosomes shows the identical type of Mg^{2+} dependence [e.g., sigmoidal without (EF-Tu·GTP) and hyperbolic in the presence of the factor], it can be concluded that the new tRNA conformation contains an additional binding site for the 50 S ribosomal subunit (98, 99).

Equilibrium dialysis requires incubation times of 6 hours or longer, which always leads to a partial denaturation of proteins or to hydrolysis of oligonucleotides by ribosome-associated nucleases. Therefore, we looked for a different experimental approach to follow codon-induced conformational changes in tRNA.

Escherichia coli tRNA^{Lys} contains, in the "wobble" position of the anticodon, the modified uridine nucleotide 2-thio-5-methylaminomethyluridine (s^{2}mnm^{5}U), which causes a residual UV absorbance at 300 nm (143, 144). This absorption is decreased if the s^{2}mnm^{5}U forms a double strand, e.g., when a codon anticodon complex is formed. The binding constants for A_3·tRNA^{Lys} (K_A = 0.24 × 10^5 M^{-1}) and A_4·tRNA^{Lys} (K_A = 2.5 × 10^5 M^{-1}) as determined by this method are very close to those obtained by equilibrium dialysis. In the presence of CGA (CGAA was replaced by the triplet, since CGAA binds to the anticodon of lysine tRNA), the apparent binding constant of A_3 to tRNA^{Lys} is raised 10-fold, to 2.5 × 10^5 M^{-1}. From the scheme depicted in Fig. 8, it may be calculated that the binding constant of CGA to the binary complex A_3·tRNA^{Lys} is about 0.2 × 10^5 M^{-1} whereas the binding of CGA to free tRNA is less than 0.01 × 10^5 M^{-1}. Under appropriate conditions, A_3 binding to lysine tRNA can be induced directly by the addition of CGA (100).

These observations confirm our earlier results, obtained by equilibrium dialysis, showing that double-strand formation between codon and anticodon triggers an allosteric transition in the tRNA conformation, which allows the binding of CGA, presumably to the TΨCG loop of the tRNA. If the location

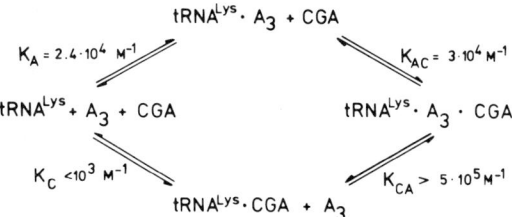

FIG. 8. Scheme for the simultaneous binding of CGA and AAA (A_3) to lysine tRNA from *Escherichia coli*. The scheme lists the apparent binding constants for the individual complexes. It is important to note that CGA can be bound only to the $A_3 \cdot tRNA^{Lys}$, not to free tRNA, and that the presence of CGA increases the codon–anticodon complex formation 10-fold (*100*).

of the binding site is correct, double-strand formation in the anticodon region alters the conformation of the tRNA some 60–70 Å away.

Additional evidence for codon-dependent conformational changes of tRNA structure comes from chemical modification experiments (*145, 146*). The kinetics of guanosine modification by kethoxal differ between $tRNA^{Lys}$ and its binary complex $A_4 \cdot tRNA^{Lys}$. In the binary complex, m^7G and, to a lesser extent, G_{57} in the TΨCG loop, react faster than they do in free tRNA. From such observations, it is concluded that tRNA may exist in different conformations, one of which is stabilized by the presence of the cognate codon. The structural changes derived from these experiments occur at or near the "extra" loop (*146*).

Fluorescence polarization measurements with a tRNA containing an ethidium bromide at either the anticodon loop or the dihydrouridine loop indicate that both loops are more rigid when the tRNA is bound to the poly(U)·70 S ribosome complex as compared to free ribosomes. The different dihydrouridine conformation in the Phe-tRNA·70 S·poly(U) complex or the Phe-tRNA·70 S complex is either the consequence of poly(U)·$tRNA^{Phe}$ complex formation, or results from the different type of binding of Phe-tRNA to a polyuridylate-coded ribosome (*147, 148*).

Nuclear magnetic resonance seems presently to be the method of choice for following ligand-dependent structural transitions in tRNA; however, data from NMR measurements are rather controversial on that issue at present (*116, 126*). Geerdes *et al.* investigated the effect of binding of oligonucleotides complementary to the anticodon of $tRNA^{Phe}$ (*126*). Using the methyl resonances of the wye at the 3' side of the anticodon as a probe, they proved that the binding constants determined with NMR are within the experimental error equal to values found with equilibrium dialysis (*126*). Evidence is given by ^{31}P NMR that the conformation of the anticodon loop is influenced by binding of complementary oligonucleotides. The conformational changes in the anticodon loop are propagated into the anticodon stem of the tRNA.

However, in the 360 MHz NMR spectra of the methyl protons of tRNA^{Phe} before and after addition of UUCA, no changes in the methyl resonances of thymidine 54 are found, so that a disrupture of the A_{58}-T_{54} hydrogen bonds is not likely. Furthermore, no ring-current shifts of hydrogen-bonded imino protons of the base pairs $G_{19} \cdot C_{56}$, $T_{54} \cdot A_{58}$, and $G_{18} \cdot \Psi_{55}$ were detected, comparing tRNA^{Phe} and UUCA·tRNA^{Phe}. Replacing UUCA by UUC, the correct codon (if a codon is defined as a trinucleoside diphosphate stimulating the binding of a cognate tRNA to the ribosome), codon·anticodon complex formation affects resonances of the tRNA far from the anticodon site and induces a general broadening of other resonances. These experimental results are explained by dimerization of tRNA molecules. But even this interpretation indicates that "long range" interactions exist between the anticodon sites and sites in the tRNA more than 20 Å away (126, 127).

Sprinzl et al. doing the same type of experiment come to somewhat different conclusion (116). Observing the high-field proton-NMR spectral region of the methylene and methyl groups of h_2U, m_2^2A, m^2A, and m^7G at different temperatures, the effect of aminoacylation, removal of the wye, and codon–anticodon interaction on the tertiary structure of yeast tRNA^{Phe} was investigated. The results of these studies are as follows. Upon aminoacylation, the stacking between G_{53} and T_{54} in tRNA^{Phe} changes. Removal of wye from the anticodon loop of yeast tRNA^{Phe} weakens the thermal stability of the tertiary interactions. The interaction of two complementary anticodons in the absence of proteins and of ribosomes stabilizes the tertiary structure. No codon-dependent rearrangement of the tertiary structure of yeast tRNA^{Phe} was observed (116).

Physical experiments with the tryptophan-suppressor tRNA also show that conformational interactions exist between the anticodon loop and remote parts of the tRNA, a G_{24} to A_{24} mutation in the dihydrouridine loop leading to a recognition of the "stop" codon UGA instead of UGG by the unaltered anticodon of tRNA$^{Trp}_{Su}$ (149–151).

Summarizing the different results on codon-induced structural transitions, it is evident that codon–anticodon double-strand formation alters the conformation of the tRNA. It forces the hitherto flexible anticodon loop into one defined conformation and alters parts of the tRNA structure remote from the anticodon loop. The exact confc mation of the tRNA in the codon·tRNA complex is still undetermined.

D. Conformational Changes in tRNA May Be Triggered by Small Molecules but Always Require a Protein

To our knowledge, the nucleic acids in cells never exist free but are always in complexes with proteins. For example, tRNA is bound either to aminoacyl synthetases, to the elongation factor EF-Tu, or to the ribosome.

Thus the discussion of induced conformational changes in nucleic acids should be understood as taking place in protein·nucleic acid complexes.

If the nucleic acid may exist in a number of conformers, it may be the task of the protein to select one defined conformer. Very likely this is the case in the complex (EF-Tu·GTP)·AA-tRNA in which (EF-Tu·GTP) binds only to AA-tRNA, but not to tRNA (152, 153). However, only EF-Tu·GTP, but not EF-Tu·GDP, can bind to the AA-tRNA, so a conformational difference should exist between the two protein complexes (154). Although a large conformational difference between the two forms of EF-Tu is not detected by competitive labeling and other studies, considerable physicochemical evidence for a more localized conformational difference between EF-Tu·GDP and EF-Tu·GTP has been reported (155). From the rate of cleavage of an accessible peptide bond by trypsin, and the rate of exchange between bound and free guanine nucleotides, it appears that EF-Tu is an allosteric protein whose conformation is modulated by GDP and GTP (155). Thus there may be a whole cascade of conformational transitions in the course of the binding of an AA-tRNA to the ribosome.

The binding of GTP induces a conformational change in the elongation factor leading to specific binding of AA-tRNA by EF-Tu·GTP. The mRNA codon bound to the ribosomal aminoacyl-tRNA site acts as an allosteric effector and changes the AA-tRNA structure within the ternary complex. This mechanism avoids competition of the noncognate AA-tRNA existing in the cytoplasm as AA-tRNA·(EF-Tu·GTP) with the cognate tRNA for the ribosomal binding site. Thus one allosteric effector, GTP, alters the conformation of the EF-Tu to enable the protein to select aminoacylated tRNA from tRNA, and the second ligand, the codon, adapts the conformation of the ternary complex to fit the aminoacyl site.

E. Kinetic Data Are Necessary to Understand the Selection of the Cognate tRNA by the Programmed Ribosome

If we define a codon as an allosteric effector that alters the AA-tRNA structure within the ternary complex AA-tRNA·(EF-Tu·GTP) from the nonbinding and thus noncompeting type to the ribosome-binding type structure, one may propose the following mechanism of tRNA recognition by the mRNA·70 S ribosome complex.

The major difference between cognate and noncognate AA-tRNA binding to the mRNA programmed ribosome must be the rate dissociation constant, which should be up to 10^5 times that of the noncognate tRNA, whereas the association rate constant is similar for cognate and noncognate tRNAs. These kinetic constants were measured using tRNA pairs with cognate anticodons (156, 157). Only the cognate tRNA stays sufficiently long in the aminoacyl site of the ribosome to allow for a structural change in the tRNA that results in a tight binding of the tRNA, especially to the 50 S ribosomal subunit. Thus the rate constant for the conformational change of the tRNA represents the critical parameter for the selective and efficient binding of a cognate tRNA to the programmed ribosome. If too slow, even the precious cognate tRNA—precious since selected out of 100—would diffuse away; if too fast, a noncognate tRNA becomes bound to the ribosomes (e.g., miscoding occurs). The many modified nucleosides and the complicated tertiary structure may act as a "fine-tuning" to adjust this rate constant for the conformational change between the rate dissociation constants of cognate and noncognate tRNAs.

Besides the anticodon, the C-C-A end is a second single-stranded region within the tRNA that may be converted into a double strand when the tRNA becomes bound to the ribosome. These two single-strand/double-strand transitions may stabilize tRNA attachment to the ribosome. Yet, hydrogen bonding between complementary bases is not the "ultima ratio" for nucleic acid complexes. Nucleic acids are bifunctional molecules forming hydrogen bonds on one side but possessing a regular lattice of permanent negative charges from their phosphodiester groups at the other side. Whatever alters the geometry of this lattice will alter the recognition pattern of a nucleic acid for a protein or a second nucleic acid.

Acknowledgments

The author is grateful to Mrs. E. Rönnfeldt for help in preparing the manuscript. The author's own experimental work reported was supported by grants from the Deutsche Forschungsgemeinschaft and the Fonds der Chemischen Industrie.

References

1. M. F. Perutz, *ARB* **48**, 327 (1979).
2. J. A. Cohlberg, V. P. Pigiet, and H. K. Schachman, *Bchem* **11**, 3396 (1972).
3. T. Blumenthal and G. C. Carmichael, *ARB* **48**, 525 (1979).
4. U. Schwarz, H. M. Menzel, and H. G. Gassen, *Bchem* **15**, 2484 (1976).
5. M. Grunberg-Manago and F. Gros, *This Series* **20**, 209 (1977).
6. P. H. J. Herrlich, S. H. Rahmsdorf, S. H. Pai, and M. Schweiger, *PNAS* **71**, 1088 (1974).
7. G. van Dieijen, C. J. van der Laken, P. H. van Knippenberg, and J. van Duin, *JMB* **93**, 351 (1975).
8. P. Lengyel, in "Ribosomes" (M. Nomura, A. Tissieres and P. Lengyel, eds.), p. 13. Cold Spring Harbor Lab., New York, 1974.
9. E. Kaltschmidt and H. G. Wittmann, *PNAS* **67**, 1276 (1970).
10. M. Noll, B. Hapke, M. Schreier, and H. Noll, *JMB* **75**, 281 (1973).
11. S. J. S. Hardy, *Mol. Gen. Genet.* **140**, 253 (1975).
12. P. H. van Knippenberg, P. J. J. Hooykaas, and J. van Duin, *FEBS Lett.* **41**, 323 (1974).
13. R. Kamen, M. Kondo, W. Römer, and Ch. Weissmann, *EJB* **31**, 44 (1972).
14. M. Revel, in "Molecular Mechanism of Protein Biosynthesis" (H. Weissbach and S. Pestka, eds.), p. 245. Academic Press, New York, 1977.
15. G. Jay and R. Kaempfer, *JBC* **250**, 5749 (1975).
16. Y. Groner, Y. Pollack, H. Berissi, and M. Revel, *Nature NB* **239**, 16 (1972).

17. A. J. Wahba, M. J. Miller, A. Niveleau, T. A. Landers, G. C. Carmichael, K. Weber, D. A. Hawley, and L. I. Slobin, *JBC* **249**, 3314 (1974).
18. M. Tal, M. Aviram, A. Kanarek, and A. Weiss, *BBA* **281**, 381 (1972).
19. R. Linde, N. Q. Khanh, and H. G. Gassen, in "Methods in Enzymology" (K. Moldave and L. Grossman, eds.), Vol. 60, part H, p. 417. Academic Press, New York, 1979.
20. M. Laughrea and P. B. Moore, *JMB* **112**, 399 (1977).
21. L. Giri and A. R. Subramanian, *FEBS Lett.* **81**, 199 (1977).
22. R. A. Kenner, *BBRC* **51**, 932 (1973).
23. A. P. Czernilowsky, C. G. Kurland, and G. Stöffler, *FEBS Lett.* **58**, 281 (1975).
24. A. Bollen, R. L. Heimark, A. Cozzone, R. R. Traut, J. W. B. Hershey, and L. Kahan, *JBC* **250**, 4310 (1975).
25. R. L. Heimark, L. Kahan, K. Johnston, J. W. B. Hershey, and R. R. Traut, *JMB* **105**, 219 (1976).
26. J. Argetsinger-Steitz, in "Biological Regulation and Development" (R. F. Goldberger, ed.), Vol. 1, p. 349. Plenum, New York, 1979.
27. G. Stöffler and H. G. Wittmann, in "Molecular Mechanism of Protein Synthesis" (H. Weissbach and S. Pestka, eds.), p. 117. Academic Press, New York, 1977.
28. J. A. Lake, M. Pendergast, L. Kahan, and M. Nomura, *J. Cell Biol.* **67**, 231a (1975).
29. I. Fiser, K. H. Scheit, G. Stöffler, and E. Küchler, *BBRC* **60**, 1112 (1974).
30. A. Kolb, J. M. Hermoso, J. O. Thomas, and W. Szer, *PNAS* **74**, 2379 (1977).
31. R. Linde, N. Q. Khanh, R. Lipecky, and H. G. Gassen, *EJB* **93**, 565 (1979).
32. W. Szer, J. M. Hermoso, and S. Leffler, *PNAS* **72**, 2325 (1971).
33. G. van Dieijen, P. H. van Knippenberg, and J. van Duin, *EJB* **64**, 511 (1976).
34. A. E. Dahlberg and J. E. Dahlberg, *PNAS* **72**, 2940 (1975).
35. J. Argetsinger-Steitz and K. Jakes, *PNAS* **72**, 4734 (1975).
36. J. Shine and L. Dalgarno, *PNAS* **71**, 1342 (1974).
37. E. Kaltschmidt and H. G. Wittmann, *PNAS* **67**, 1276 (1970).
38. G. R. Craven, P. Voynow, S. J. S. Hardy, and C. G. Kurland, *Bchem* **8**, 2906 (1969).
39. P. B. Moore, R. R. Traut, H. Noller, P. Pearson, and H. Delius, *JMB* **31**, 441 (1968).
40. A. R. Subramanian, C. Haase, and M. Giessen, *EJB* **67**, 591 (1976).
41. A. R. Subramanian, B. Wittmann-Liebold, A. W. Geissler, G. Stöffler, and M. Green, *FEBS Lett.* **99**, 357 (1979).
42. M. Laughrea and P. B. Moore, *JMB* **112**, 399 (1977).
43. H. Inouye, Y. Pollack, and J. Petre, *EJB* **45**, 109 (1974).
44. H. Labischinski and A. R. Subramanian, *EJB* **95**, 359 (1979).
45. L. P. Visentin, S. Hasnain, W. Gallin, K. G. Johnson, D. W. Griffith, and A. J. Wahba, *FEBS Lett.* **79**, 258 (1977).
46. Y. G. Chu and C. R. Cantor, *NARes* **6**, 2363 (1979).
47. P. B. Moore and M. Laughrea, *NARes* **6**, 2355 (1979).
48. M. Smolarsky and M. Tal, *BBA* **213**, 401 (1970).
49. J. M. Hermoso and W. Szer, *PNAS* **71**, 4708 (1974).
50. P. T. Li, T. Shea, S. Ellis, and T. H. Conway, *EJB* **98**, 155 (1979).
51. M. J. Miller, A. Niveleau, and A. J. Wahba, *JBC* **249**, 3803 (1974).
52. R. Lipecky, J. Kohlschein, and H. G. Gassen, *NARes* **4**, 3627 (1977).
53. D. E. Draper and P. H. van Hippel, *JMB* **122**, 321 (1978).
54. D. E. Draper and P. H. van Hippel, *JMB* **122**, 339 (1978).
55. D. G. Bear, R. Ng, D. van der Veer, N. P. Johnson, G. Thomas, T. Schleich, and H. Noller, *PNAS* **73**, 1824 (1976).
56. H. F. Noller, C. Chang, G. Thomas, and J. Aldridge, *JMB* **61**, 669 (1971).
57. N. Q. Khanh, R. Lipecky, and H. G. Gassen, *BBA* **521**, 476 (1978).

58. J. O. Thomas, A. Kolb, and W. Szer, *JMB* **123**, 163 (1978).
59. W. Szer, J. M. Hermoso, and M. Boublik, *BBRC* **70**, 957 (1976).
60. R. A. Baan, C. W. Hilbers, R. van Charldorp, E. van Leerdam, P. H. van Knippenberg, and L. Bosch, *PNAS* **74**, 1028 (1977).
62. M. Record, J. Charles, P. Anderson, and I. M. Lohmann, *Q. Rev. Biophys.* **11**, 103 (1978).
63. R. Lipecky and H. G. Gassen, in preparation.
64. A. W. Senear and J. Argetsinger-Steitz, *JBC* **251**, 1902 (1976).
65. S. Goelz and J. Argetsinger-Steitz, *JBC* **252**, 5177 (1977).
66. A. Krol, C. Branlant, J. P. Ebel, and L. P. Visentin, *FEBS Lett.* **80**, 225 (1977).
67. J. Argetsinger-Steitz, *PNAS* **70**, 2605 (1973).
68. J. Gralla and D. M. Crothers, *JMB* **73**, 497 (1973).
69. T. Tanaguchi and C. Weissmann, *JMB* **128**, 481 (1979).
70. T. Tanaguchi and C. Weissmann, *Nature* **275**, 770 (1978).
71. R. A. Baan, R. van Charldorp, E. van Leerdam, P. H. van Knippenberg, L. Bosch, J. F. M. de Rooij, and J. H. van Boom, *FEBS Lett.* **71**, 351 (1976).
72. J. Argetsinger-Steitz and D. A. Steege, *JMB* **114**, 545 (1977).
73. J. Argetsinger-Steitz, A. J. Wahba, M. Laughrea, and P. B. Moore, *NARes* **4**, 1 (1977).
74. J. Argetsinger-Steitz, *in* "The Ribosomes" (J. Davis, ed.), in press, 1980.
75. J. van Duin and C. G. Kurland, *Mol. Gen. Genet.* **109**, 169 (1970).
76. G. van Dieijen, Ph.D. Dissertation (proefschrift), Univ. of Leiden, 1977.
77. G. van Dieijen, P. H. van Knippenberg, and J. van Duin, *EJB* **64**, 511 (1976).
78. M. Y. Feldman, *This Series* **13**, 1 (1973).
79. S. Isono and K. Isono, *EJB* **56**, 15 (1975).
80. G. van Dieijen, P. H. van Knippenberg, J. van Duin, B. Koekman, and P. H. Pouwels, *Mol. Gen. Genet.* **153**, 75 (1977).
81. A. E. Dahlberg, *JBC* **249**, 7673 (1974).
82. M. Schmitt, U. Manderschied, H.-E. Wollny, and H. G. Gassen, *EJB*, submitted (1980).
83. F. H. C. Crick, quoted by M. B. Hoagland, *in* "The Nucleic Acids" (E. Chargaff and J. N. Davidson, eds.), Vol. 3, p. 401. Academic Press, New York, 1960.
84. E. H. Allen and R. S. Schweet, *BBA* **39**, 185 (1960).
85. M. B. Hoagland, M. L. Stephenson, J. F. Scott, J. F. Hecht, and I. Zamecnik, *JBC* **231**, 241 (1958).
86. R. W. Holley, J. Apgar, G. A. Everett, J. T. Madison, M. Marquisse, S. H. Morell, J. R. Penswick, and A. Zamir, *Science* **147**, 1462 (1965).
87. G. J. Quigley, A. J. H. Wang, N. C. Seeman, F. C. Suddath, A. Rich, J. L. Sussman, and S. H. Kim, *PNAS* **72**, 4866 (1975).
88. A. Jack, J. E. Ladner, and A. Klug, *JMB* **108**, 619 (1976).
89. A. A. Bagev, I. Fodor, A. D. Mirzabekov, A. I. Krutilina, I. Li, and V. D. Axelrod, *Biologiya* **1**, 754 (1967).
90. J. P. Goddard, *Prog. Biophys. Mol. Biol.* **32**, 233 (1977).
90a. D. M. Abraham, *J. Theoret. Biol.* **30**, 83 (1971).
91. D. M. Crothers and P. E. Cole, *in* "Transfer RNA" (S. Altman, ed.), p. 196. MIT Press, Cambridge, Ma., 1978.
92. D. R. Kearns, *This Series* **18**, 91 (1976).
93. A. Möller, U. Manderschied, R. Lipecky, S. Bertram, M. Schmitt, and H. G. Gassen, *in* "tRNA Monograph" (J. Abelson, P. Schimmel, and D. Söll, eds.). Cold Spring Harbor Lab., New York, 1979.
94. D. Riesner and R. Römer, *in* "Physico-chemical Properties of Nucleic Acids" (F. Duchesne, ed.), Vol. II, p. 237. Academic Press, New York, 1973.

95. D. H. Gauss, F. Grüter, and M. Sprinzl, *NARes* **6**, r1 (1979).
96. A. Rich and V. L. RajBhandary, *ARB* **45**, 805 (1976).
97. C. G. Kurland, R. H. Rigler, M. Ehrenberg, and C. Blomberg, *PNAS* **72**, 4248 (1975).
98. U. Schwarz, R. Lührmann, and H. G. Gassen, *BBRC* **56**, 817 (1974).
99. U. Schwarz, H. M. Menzel, and H. G. Gassen, *Bchem* **15**, 2484 (1976).
100. A. Möller, U. Wild, D. Riesner, and H. G. Gassen, *PNAS* **76**, 3266 (1979).
101. W. Fuller and A. Hodgson, *Nature* **215**, 817 (1967).
102. C. Urbanke and G. Maass, *NARes* **5**, 1551 (1978).
103. C. Woese, *Nature* **226**, 817 (1970).
104. O. C. Uhlenbeck, J. Baller, and P. Doty, *Nature* **225**, 508 (1970).
105. G. Högenauer, F. Turnowsky, and F. M. Unger, *BBRC* **46**, 2100 (1972).
106. F. H. Martin, O. C. Uhlenbeck, and P. Doty, *JMB* **57**, 201 (1971).
107. N. R. Kallenbach and H. M. Berman, *Q. Rev. Biophys.* **10**, 138 (1977).
108. A. Möller, U. Schwarz, R. Lipecky, and H. G. Gassen, *FEBS Lett.* **89**, 263 (1978).
109. U. Manderschied, S. Bertram, and H. G. Gassen, *FEBS Lett.* **90**, 162 (1978).
110. R. Römer and R. Hach, *EJB* **55**, 271 (1975).
111. A. Stein and D. C. Crothers, *Bchem* **15**, 160 (1976).
112. A. A. Schreier and P. R. Schimmel, *JMB* **86**, 601 (1974).
113. M. Cohn, A. Danchin, and M. Grunberg-Manago, *JMB* **39**, 199 (1969).
114. J. M. Wolfson and D. A. Kearns, *JACS* **96**, 3653 (1974).
115. M. Bina-Stein and A. Stein, *Bchem* **15**, 3912 (1976).
116. P. Davanloo, M. Sprinzl, and F. Cramer, *Bchem*, in press (1980).
117. G. J. Thomas, M. C. Chen, R. Lord, P. S. Kotsiopoulos, T. R. Tritton, and S. C. Mohr, *BBRC* **54**, 570 (1973).
118. L. Beres and J. Lucas-Leonard, *Bchem* **12**, 3998 (1973).
119. H. Hashizume and K. Imahori, *J. Biochem.* **61**, 738 (1967).
120. A. Bernardi and G. L. Cantoni, *JBC* **244**, 1468 (1969).
121. A. I. Adler and G. D. Fasman, *BBA* **204**, 183 (1970).
122. E. Wickstrom, *BBRC* **43**, 976 (1971).
123. U. J. Hänggi and H. G. Zachau, *EJB* **18**, 496 (1971).
124. J. Ninio, V. Luzzati, and M. Yaniv, *JMB* **71**, 217 (1972).
125. Y. R. Wong, B. R. Reid, and D. R. Kearns, *PNAS* **70**, 2193 (1973).
126. H. A. M. Geerdes, Ph.D. thesis (proefschrift), Univ. Nijmegen, 1979.
127. H. A. M. Geerdes, J. H. van Boom, and C. W. Hilbers, *FEBS Lett.* **88**, 27 (1978).
128. K. Watanabe and K. Imahori, *BBRC* **45**, 488 (1971).
129. R. C. Chin and C. Kidson, *PNAS* **68**, 2448 (1971).
130. A. Danchin and M. Grunberg-Manago, *FEBS Lett.* **9**, 327 (1970).
131. D. J. Dvorak, C. Kidson, and R. C. Chin, *JBC* **251**, 6730 (1976).
132. R. Potts, M. J. Fournier, and N. C. Ford, Jr., *Nature* **268**, 563 (1977).
133. O. Pongs, P. Wrede, V. A. Erdman, and M. Sprinzl, *BBRC* **71**, 1025 (1976).
134. J. Fresco, in "tRNA Monograph" (J. Abelson, P. Schimmel, and D. Söll, eds.). Cold Spring Harbor Lab., New York, 1979.
135. G. G. Brownlee, F. Sanger, and B. G. Barrell, *Nature* **215**, 735 (1967).
136. B. R. Jordan, *JMB* **55**, 423 (1971).
137. V. A. Erdmann, S. Fahnestock, K. Hijo, and M. Nomura, *PNAS* **68**, 2932 (1971).
138. V. A. Erdmann, M. Sprinzl, and O. Pongs, *BBRC* **54**, 942 (1973).
139. J. Ofengand and C. Henes, *JBC* **244**, 6241 (1969).
140. D. H. Gauss, F. von der Haar, A. Maelicke, and F. Cramer, *ARB* **40**, 1045 (1971).
141. O. Pongs, R. Bald, and E. Reinwald, *EJB* **32**, 117 (1973).
142. J. G. Levin, *JBC* **245**, 3195 (1970).

143. J. A. McCloskey and S. Nishimura, *Acc. Chem. Res.* **10**, 403 (1977).
144. W. Hillen, E. Egert, H. J. Lindner, and H. G. Gassen, *FEBS Lett.* **94**, 361 (1978).
145. R. Wagner and R. Garrett, *FEBS Lett.* **85**, 291 (1978).
146. R. Wagner and R. Garrett, *EJB* **97**, 615 (1979).
147. J. M. Robertson, M. Kahan, W. Wintermeyer, and H. G. Zachau, *EJB* **72**, 117 (1976).
148. R. H. Fairclough, C. Cantor, W. Wintermeyer, and H. G. Zachau, *JMB* **132**, 557 (1979).
149. D. Hirsh, *JMB* **58**, 439 (1971).
150. R. H. Buckingham, *NARes* **3**, 965 (1976).
151. R. H. Buckingham and C. G. Kurland, *PNAS* **74**, 5496 (1977).
152. D. L. Miller and H. Weissbach, *in* "Nucleic Acid Protein Recognition" (H. J. Vogel, ed.), p. 409. Academic Press, New York, 1977.
153. P. Y. Kaziro, *BBA* **505**, 95 (1978).
154. H. Wolf, G. Chinali, and A. Parmeggiani, *EJB* **75**, 67 (1977).
155. J. Douglass and T. Blumenthal, *JBC* **254**, 5383 (1979).
156. H. J. Grosjean, D. Söll, and D. M. Crothers, *JMB* **103**, 499 (1976).
157. H. J. Grosjean, S. de Henau, and D. M. Crothers, *PNAS* **75**, 610 (1978).

Replicative DNA Polymerases and Mechanisms at a Replication Fork

Robert K. Fujimura and
Shishir K. Das[1]

Biology Division, Oak Ridge National Laboratory[2]
and The University of Tennessee—Oak Ridge Graduate School of Biomedical Sciences
Oak Ridge, Tennessee

I. General Properties of Replicative DNA Polymerases	87
II. Overview of Chain Elongation at a Replication Fork	90
III. Mechanisms of DNA Polymerase Action at a Replication Fork	92
A. Unwinding Action	92
B. Processive Action	94
C. Base Selection and Editing	96
IV. Concluding Remarks on DNA Polymerase and the Replication Complex	102
References	104

In what follows, we discuss mechanisms of DNA synthesis at a replication fork, with emphasis on the replicative DNA polymerases. We compare the properties of some of these enzymes, particularly in relation to processes of strand unwinding, translocation along templates, and accurate incorporation of nucleotides, as well as the effects of other protein factors involved in these processes. Our objective is to arouse more interest in systematic studies on these DNA polymerases with respect to their function at a replication fork. Enzymes involved in DNA repair, other than those involved in correction of replicative errors, are not considered.

I. General Properties of Replicative DNA Polymerases

A definitive trait of replicative DNA polymerase is the presence of a conditional lethal mutant in the structural gene of the enzyme that leads to a marked reduction in DNA synthesis. Such mutants were first found in bacteriophages T4 and T5 (1). It was observed that DNA polymerases extracted from cells infected with certain temperature-sensitive mutants of T4 and T5

---

[1] Supported by a research assistantship granted by The University of Tennessee.
[2] Operated by Union Carbide Corporation under contract W-7405-eng-26 with the U. S. Department of Energy.

are more unstable at higher temperature than are the polymerases of the wild type. T4 amber mutants of the same genetic locus are also unable to induce detectable amounts of T4 DNA polymerase. These observations suggest that it is possible to identify DNA polymerases specifically involved in replication by isolation of their conditional lethal mutants. Probably the most famous result of the application of this approach was the identification of the replicative DNA polymerase of *Escherichia coli*. First, DeLucia and Cairns (2) isolated a viable mutant of *E. coli* with a very low level of DNA polymerase I, which raised the possibility that another DNA polymerase is the replicative enzyme. Subsequently, two other DNA polymerases were found in *E. coli*; one of them, designated *E. coli* DNA polymerase III, was temperature-sensitive when prepared from strains having the thermosensitive mutation in the dna-E locus (3). Thus DNA polymerase III is the polymerase essential for DNA replication.

Conditional lethal mutants for the structural genes for eukaryotic DNA polymerases have not yet been isolated. DNA polymerase α is generally regarded as the replicative DNA polymerase, because its level of activity is highest in the most prolific stage of cell growth or in the DNA synthesis phase of synchronized cells. However, these characteristics may arise from other causes (4), and thus there is the possibility that some other DNA polymerase is the replicative enzyme. Some animal viruses induce a DNA polymerase that is essential for DNA replication (4).

Table I lists the properties of most of the well-characterized replicative DNA polymerases. All the purified ones require a "primer-template" for replication, and they elongate the primer in the $5' \rightarrow 3'$ direction (5). As can be seen from Table I, other properties differ considerably among the polymerases. Most of the prokaryotic DNA polymerases have a $3' \rightarrow 5'$ exonuclease activity associated with them, which is thought to be the same activity involved in primer-template-dependent hydrolysis ("turnover") of dNTP to dNMP (29). Most preparations of purified DNA polymerase α are free of exonuclease activity. The DNA polymerase induced by Herpes simplex virus I has a $3' \rightarrow 5'$ exonuclease activity associated with it (30).

Most purified DNA polymerases are not capable of strand displacement. T5 DNA polymerase (9), *E. coli* DNA polymerase I (31), and the fragment of *E. coli* DNA polymerase I without $5' \rightarrow 3'$ exonuclease activity (32) are capable of strand displacement. Some preparations of DNA polymerase α synthesize from DNase I-treated DNA, which implies strand displacement, but recent work with highly purified α with better defined primer-template suggests that it cannot displace strands (24a,b).

All the replicative DNA polymerases tested so far are capable of translocation along the template concomitantly with nucleotide incorporation. However, as shown in Table I, only T5 DNA polymerase is capable of translocation for an extended distance; this is called "processiveness" (25).

TABLE I
PROPERTIES OF REPLICATIVE DNA POLYMERASES[a]

	T5		E. coli pol III		T4		T7		α		Bacillus subtilis pol III	
Molecular properties												
Molecular weight	96,000	(6)	140,000	(12)	114,000	(17)	87,000	(19b)	149,000?	(22)	166,000	(26)
Number of subunits	1	(9)	≥6	(14,15)	1	(17)	2	(19,19b,19c)	≥ 2	(21–23b)	1	(26)
Homogeneous	Yes	(9)	Yes?	(12,14)	Yes	(17)	Yes	(19b)	Yes?	(21–23b)	No	(26)
Repl⁻ mutant?	Yes	(1,8)	Yes	(3)	Yes	(1)	Yes	(18)	No	—	Yes	(27,28)
Template primer												
Primed single strand	+	(7)	±	(12)	+	(17)	+	(18)	+	(21,24c)	—	(27)
Nicked circular duplex	+	(9)	—	(12)	—	(17)	—	(20)	—	(24,24b)	?	—
Capped duplex	+	(7)	+	(12)	+	(17)	+	(18)	+	(21,24b)	+	(27)
De novo synthesis	—	(7)	—	(12)	—	(17)	—	(18)	—	(21,24b)	—	(26)
Functions												
Polymerization: 5'→3'	+	(7)	+	(12)	+	(17)	+	(18)	+	(21)	+	(26)
Exonuclease: 3'→5'	+	(11)	+	(13)	+	(17)	+	(18)	—?	(21,23b)	+	(26)
dNTP→dNMP turnover	+	(10,11)	+	(13)	+	(17)	+	(18)	—	(21)	?	—
Exonuclease: 5'→3'	—	(11)	+	(13)	—	(17)	—	(18)	—	(21)	—	(26)
Strand displacement	+	(9)	—	(12)	—	(17)	—	(18)	—	(24)	—	(26)
Processiveness:nucleotide added/E., 37°C	160	(25)	?	—	12	(25)	?	—	8,11	(25,24)	7?	(27)

[a] References are given in parentheses.

These comparisons show that the basic mechanisms for nucleotide incorporation by DNA polymerases are the same, but DNA polymerases have varied capabilities in their interactions with primer-template DNA. As shown in Section II, the current model for the replication of duplex DNA is applicable to all DNA polymerases, and the mechanisms of replication are too complex to permit any of these polymerases to accomplish the processes unassisted. This means that other protein factors take part in these processes either directly, in a complex with a polymerase or as part of a replication complex at a growing fork of a DNA, or indirectly by more subtle means.

II. Overview of Chain Elongation at a Replication Fork

A replication fork is a Y-like structure that is created at the growing point of a replicating double-helical DNA. According to a currently popular model of replication (5, 33, 34) (Fig. 1), DNA replication initiates at a unique site called an "origin of replication." Each unit of replication (replicon) has, by definition, one unique origin of replication. Almost as soon as replication is initiated, a replication fork is created, which moves away from the origin either unidirectionally or bidirectionally. Figure 1 shows the unidirectional mode of replication. For the bidirectional mode, another replication fork moves symmetrically in the opposite direction.

DNA synthesis at a replication fork may be divided into two phases: (*a*) replication in the same direction as a replication fork (leading strand); (*b*) replication away from a replication fork (lagging strand). Both syntheses occur in a 5'-to-3' direction by successive addition of nucleotides to the

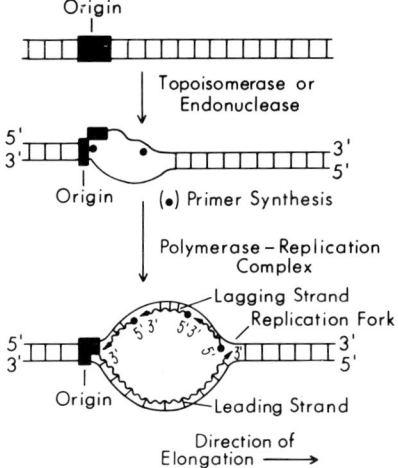

FIG. 1. Replication near the origin for unidirectional elongation.

3′-OH end of the most recently incorporated nucleotide. The leading strand synthesis may be continuous from a unique origin of replication. As this elongation process continues, the double-helix ahead of the fork is unwound, creating a single-stranded region in the opposite lagging strand (34b). This region is replicated discontinuously, and later the short pieces (Okazaki fragments) thus produced are joined to make one continuous strand. Okazaki fragments are initiated at multiple and less specific sites as single-stranded regions are exposed.

The initiation process at the origin of replication is not yet clearly understood (shown as a black box in Fig. 1). The mechanisms involved have been reviewed by Tomizawa and Selzer (35); here we summarize them briefly. The mechanisms by which proteins essential for initiation recognize the origin of replication are not known, even though the nucleotide sequences of DNA at the origins of replication for some systems are known. The initiation process appears to involve some kind of "topoisomerase," an enzyme involved in formation and relaxation of superhelical DNA structures, or a specific endonuclease. Such an enzyme "nicks" a DNA molecule at its origin, causing local unwinding of DNA. The free end, if it has a 3′-OH terminus, may act as the primer for replicative DNA polymerase, or a primer may be synthesized at an unwound region by an RNA polymerase, or by a "primase" specifically involved in primer synthesis. The primer could be synthesized *in situ* or at some other locus on DNA and transferred to the origin. As soon as a primer is situated at the origin, the replication complex containing DNA polymerase elongates the primer in 5′-to-3′ direction.

The initiation of DNA synthesis on the lagging strand (Okazaki fragment) proceeds from a primer synthesized *de novo*. The enzyme for primer synthesis may be an RNA polymerase, as in the case of phage M13 DNA synthesis (36), or it may be a specific primase, such as the dna G protein of *E. coli* or the gene 4 protein of T7 (37, 37b, 37c). In the case of dna G protein, ribonucleoside triphosphates are not obligatory precursors *in vitro* (38, 39). The precursors can be deoxyribonucleoside triphosphates. Thus primers may not necessarily be oligoribonucleotides *in vivo*. The process by which primer synthesis switches to DNA synthesis is unknown. Perhaps only the first nucleotide of the primer must be a ribonucleotide. Subsequently, chance incorporation of deoxyribonucleotide to a primer may cause switching to DNA synthesis by a replication complex. Primers characterized from several systems are short and of different lengths (two to eight nucleotides) (35), which makes this proposal more probable.

Once a DNA polymerase takes over the chain elongation process, the most efficient mechanism is probably processive, that is, by translocation of the same polymerase in the replication complex to the end of the template. For the translocation to occur on the leading strand, the strand ahead of the growing chain must be displaced efficiently. There also have to be efficient methods for selecting correct precursor nucleotides ("correct" in the Watson–Crick sense). For these three processes, DNA polymerase is intrinsically involved to a greater or lesser degree as will become clear from subsequent discussions.

III. Mechanisms of DNA Polymerase Action at a Replication Fork

A. Unwinding Action

T5 DNA polymerase is an example of the type of replicative DNA polymerase that is capable of strand displacement as a primer is being elongated (9). With "nicked" circular duplex DNA the displacement reaction is quite efficient, but with linear duplex DNA the strands are displaced only a short distance. Displaced strands become double-stranded, and this appears to inhibit further displacement and synthesis (40). This displacement reaction does not require ATP as an energy source. Apparently, nucleotide incorporation causes enough motive force or nucleotide hydrolysis releases enough energy for the polymerase to translocate and break hydrogen bonds between base-pairs. For more efficient unwinding of linear duplex DNA, additional factors are apparently required.

T4 DNA polymerase is a polymerase that binds to a nick in a duplex DNA, but is incapable of strand displacement (31). However, in the presence of "helix-destabilizing" protein, it does displace strands to some extent without ATP as an energy source (41). The DNA may be in dynamic equilibrium between a partially melted structure and a Watson–Crick helical structure (42). When the nicked region of DNA is in a partially melted structure, the destabilizing protein binds preferentially to that region, causing it to remain unwound. This enables a polymerase to synthesize and to translocate in that region. T4 DNA polymerase in the presence of the destabilizing protein will displace a strand and synthesize preferentially at an (A+T)-rich region (41). This is consistent with the above interpretation. Three additional proteins (from genes 45, 44, and 62, the last two in a complex) are required for more efficient strand displacements (43). These additional proteins (in the presence of ATP) facilitate the movement of the polymerase along the template. A combination of these proteins with DNA polymerase causes the polymerase to translocate into double-helical regions, forcing displacement of the strand. It is not clear whether ATP is required for this unwinding process. No displacement reaction was observed in the absence of DNA synthesis.

T7 DNA polymerase requires gene 4 protein for efficient strand displacement (44, 45). The reaction occurs concomitantly with synthesis on duplex DNA and is accompanied by hydrolysis of NTPs to NDPs. All dNTPs and rNTPs are hydrolyzed except rGTP. The moles of dNTPs and rNTPs hydrolyzed per mole of dNMP incorporated is 4.2. At least part of the energy released by hydrolysis appears to be used for the unwinding reaction (44). Gene 4 protein differs from the helix-destabilizing protein in that it does not have a high affinity for single-stranded DNA, and only a catalytic amount is required (20).

Most of the studies of DNA replication with *E. coli* DNA polymerase III *in vitro* have used circular viral DNAs. Among these, the ϕX174 DNA system requires the largest number of gene products from the *E. coli* host, and thus is considered to be closest to the host DNA replication system (36). For the unwinding process in this system, "rep" protein is required (46). Rep protein is capable of causing unwinding independently of the DNA synthesis reaction (47). To unwind partially single-stranded DNA, rep protein requires only the helix-destabilizing protein and ATP (48). For ϕX174 RF-I, a duplex circular DNA, ϕX gene A protein is required as well (47). The gene A protein cleaves the (+) strand at the origin of replication and remains bound at the 5' end of the break. Apparently the rep protein forms a complex with the gene A protein and is thereby guided to the nicked position of the DNA for unwinding (49). Yarranton and Gefter (48) showed that a single-stranded region is required for the initial binding of rep protein in the absence of the gene A protein. They proposed that, in the presence of bound ATP, the leading region of the rep protein undergoes a conformational change that increases the affinity of that region for the duplex, causing the protein to move to the duplex region. As ATP is hydrolyzed, the conformation of the leader region changes back to the initial form, forcing disruption of a base-pair and thereby causing unwinding.

In addition to rep protein, *E. coli* has at least three other unwinding enzymes, and T4 infection induces another one (50-53). These bind preferentially to single-stranded region of DNA and require ATP for their action. They have not yet been shown to be essential for DNA replication.

Another protein that may take part in unwinding of DNA during replication is DNA "gyrase," which produces negative superhelical turns in a closed circular DNA molecule in the presence of ATP (54). It is required for some circular duplex (55) and linear duplex DNA synthesis (56). Champoux (57) speculated that the superhelical turns it forms may provide the motive force for helix unwinding.

These observations and conjectures suggest that whether or not the polymerase can displace strands depends on how readily it can move along the template during the replication process. Helix-destabilizing proteins are required to keep single-stranded region from rewinding. Other protein factors also facilitate movement of DNA polymerase along the chain.

B. Processive Action

During a polymerization reaction, DNA polymerase is in one of two places: (a) at the 3′ end of the still-growing chain (i.e., functioning); or (b) at the end of the template or dissociated from it (nonfunctioning). In one process of elongation, the enzyme at the active primer end binds dNTPs, catalyzes condensation of a nucleotide, and then translocates along the template. Alternatively, after condensation of nucleotides, the enzyme does not move but dissociates from the primer-template. If the enzyme translocates in stepwise fashion almost to the end of template, it is called "processive." If it does not, and dissociates after each nucleotide addition, it is called "nonprocessive" or "distributive." If the process is somewhere between these two extremes, we call it "quasi-processive" (25).

Recently the degrees of processiveness of DNA polymerases were compared with $(dA)_{300} \cdot (dT)_{10}$ as the primer-template (25). The products of the reaction were washed thoroughly after adsorption on oligo(dT)-cellulose to remove nucleotides. Chain length of the products was determined by sequential digestion of the products by micrococcal nuclease and spleen phosphodiesterase. The chain length is given by the ratio of (nucleoside + nucleotide)/nucleoside. The reactions were carried out under conditions where the probability of a primer molecule being utilized by the enzyme more than once is negligible. The results are shown in Table II. As can be seen, only T5 polymerase was processive under these *in vitro* polymerase assay conditions. The effective length of the template, assuming primers to be positioned randomly over it, was about 150, indicating that T5 polymerase translocated to about the end of template. This is in agreement with earlier work in which processiveness was determined from kinetic data (58, 59). All the other polymerases tested were also able to translocate, but added only 12 nucleotides or less. Thus they were quasi-processive. Fisher *et al.* (24) found, by the technique described below (60), that DNA polymerase α from KB cells has a processiveness of 11 at 37°C with activated KB DNA.

Another quantitative method for measuring processiveness was developed by Bambara *et al.* (60). An equation was derived based on an idea that the ratios of the rates of DNA synthesis in the presence of all four dNTPs to the rates in the absence of one, two, or three of the four dNTPs give a quantitative relative measure of processiveness. A large excess of primer termini over enzyme molecules was used so that there was no detectable free enzyme in the reaction mixture. The cycling time of the enzyme from one active primer terminus (3′-OH end) to another primer terminus was controlled by addition of the same DNA with 3′-phosphate termini, which compete with active primer termini for the enzyme. The results indicate that the processiveness of *E. coli* DNA polymerase I is affected by DNA struc-

TABLE II
COMPARISON OF PROCESSIVENESS OF DNA POLYMERASES[a]

Enzyme	Chain length[b]	
	25°C	37°C
T5 DNA polymerase	146 ± 16	161 ± 8
T4 DNA polymerase	3 ± 1.0	12 ± 0.8
Escherichia coli DNA polymerase I	12 ± 1.0	12.3 ± 0.8
Calf thymus α-polymerase	11 ± 0.9	8 ± 1.0
Calf thymus β-polymerase	9 ± 1.0	10 ± 1.2

[a] DNA synthesis with each polymerase was determined under usual in vitro conditions for each polymerase (25). $(dA)_{300} \cdot (dT)_{10}$ was used as the primer-template with [^{3}H]dTTP as the labeled precursor. The primer molecules were in large excess over the enzyme molecules, and the average number of additions per available primer during the reactions was kept at < 0.3. After the reaction, the products were adsorbed onto oligo(dT)-cellulose, washed thoroughly to remove nucleotides, eluted, and sequentially digested by micrococcal nuclease and spleen phosphodiesterase. The amounts of nucleotides and nucleosides in the digests were determined by adsorption to DE-81 paper disks under conditions where adsorption of nucleosides is negligible. The adsorbed fraction was eluted and shown to be 3'-dTMP by paper chromatography.

[b] The chain length was obtained by the ratio of (nucleoside + nucleotide)/nucleoside. Maximal deviations are indicated.

ture, ionic strength, and temperature. With "gapped" DNA, the polymerase was slightly more processive than with "nicked" DNA. It incorporated 188 nucleotides per primer with poly(dA-dT) as primer-template. Processiveness decreased with ionic strength and temperature.

Previous to the work discussed above, processiveness was determined from the kinetics in the presence of "challenger" polymers, sometimes followed by chromatography or electrophoresis (61–63). These approaches gave only very approximate values. Uyemura et al. (64) used the ends of λ DNA to study the extent of processiveness, but the shortness of the template limited the measurement.

The degree of processiveness may affect the maximum rate of elongation. Most of the enzymes studied are quasi-processive (25). For these to become fully processive, additional factors are probably required, as shown for T4 DNA polymerase (33). The elongation process by this enzyme is processive in the presence of five other proteins essential for elongation, and the rate of elongation becomes that of the in vivo rate, about 1000 nucleotides per second at 37°C (33). Purified E. coli DNA polymerase III is very inefficient in chain elongation through single-stranded regions (12). The longer the single-stranded region, the slower the initial rate of polymerization, while the extent of synthesis remains constant. Apparently other proteins are required to make it translocate more efficiently. The properties of some of these proteins have been reviewed (65).

C. Base Selection and Editing

1. Three Processes Involved

During DNA-replication, the accuracy of the copies may be maintained by three mechanisms operating sequentially: (a) proper selection of nucleotides at the 3'-OH end of primers; (b) "editing" of nucleotides as they are incorporated; (c) elimination (post-editing) of "wrong" nucleotides after replication.

In studies of the mechanisms ensuring accurate replication of DNA, T4 DNA polymerase has been the most systematically employed. Mutants of gene 43, the structural gene for T4 DNA polymerase, with mutator or antimutator phenotypes, were shown to have changes in either (a) the nucleotide selection step, or (b) in the editing step. Mutants L88 and CB120 are considered as mutator and antimutator, respectively, owing to change in stage (a) (66, 67). Mutants L56 and L98 are mutators, and L141 and L42 are antimutators owing to changes in stage (b) (68, 69).

Stage (b) has been the most intensely studied mechanism with respect to accurate copying of templates (70-72). The accuracy of replication is so high that most have considered it unlikely that it can be attained in a single step. Thus, when Brutlag and Kornberg (73) showed preferential removal of improperly base-paired nucleotides by the $3' \rightarrow 5'$ exonuclease associated with E. coli polymerase I, two-step selection and editing of bases were immediately proposed and widely accepted. It was found that a mutator phenotype correlates with a low ratio of $3' \rightarrow 5'$ exonuclease to polymerase, and an antimutator correlates with a high ratio of these two activities. Yet these correlations were observed without ambiguity only with some of T4 DNA polymerase mutants (66-68). The best examples of these are strain L56 for a mutator and strain L141 for antimutator (68, 69).

These findings with mutants of T4 DNA polymerase may be extended to other prokaryotic polymerases with associated $3' \rightarrow 5'$ exonucleases. Some mutants of E. coli DNA polymerase III have a mutator-antimutator phenotype (74); no extensive biochemical studies seem to have been carried out with them. For DNA polymerase α, which does not have an associated $3' \rightarrow 5'$ exonuclease (Table I), a multistep model of base selection and editing may not be applicable. However, a $3' \rightarrow 5'$ exonuclease has been found in crude preparations of α that may serve in editing of incorporated nucleotides. Recently, a mammalian DNA polymerase, called δ, that has an associated exonuclease activity was found (75, 76, 23b). It has been purified to homogeneity (77, 23b), but its role in replication has not yet been determined.

A multistage selection mechanism that does not require 3'→5' exonuclease activity and that involves activated enzyme-nucleotide complexes (E*·dNDP and E*·dNMP) as intermediates prior to incorporation has been proposed (78). It is a dNTPase-like activity that activates the precursors for covalent bond formation with primer ends. The formation of dNDP during synthesis is consistent with this model (78).

Further studies with T4 DNA polymerase from gene 43 mutants with a mutator-antimutator phenotype show that this phenotype cannot be explained solely on the basis of an editing step. The exonucleases associated with the polymerases from strains L88 and CB120 do not discriminate between matched and mismatched nucleotides (67). However, the apparent K_m for the turnover reaction for mismatched bases for CB120 was higher than for a wild type, and for total utilization of the proper nucleotides (incorporation plus turnover) by L88, the K_m was very high (67). This suggests that CB120 is an antimutator because of the lower affinity of the wrong base for the primer-template-enzyme complex, and L88 is a mutator because the affinity of the normal base is lowered more than that of the wrong base by mutation (67). Thus these mutants are examples of changes in discrimination of nucleotides at stage (a).

No systematic studies of a similar nature appear to have been done with other polymerases. One of the temperature-sensitive mutants of T5 DNA polymerase, ts53, appears to have properties similar to those of CB120. The incorporation of dBrUTP by ts53 polymerase decreases with increase in temperature much faster than does dTTP incorporation, while both nucleotides were incorporated in the same amount independently of temperature with wild-type polymerase (10). There is no preferential turnover or hydrolysis of dBrUTP. This discrimination with respect to analog incorporation was observed also *in vivo* with ts53-infected cells (79).

Recent work on DNA glycosylases and apurinic endonucleases (80), and the discovery of an enzyme that inserts purines at apurinic sites (81), suggest the third stage (c), which may contribute to accuracy of replicated DNA. A newly replicated strand is probably undermethylated (82). If there are mismatched bases in that strand, they will be preferentially repaired, as suggested by studies on the repair of mismatched heteroduplexes (82). The repair may involve base excision by DNA glycosylase followed by cleavage by apurinic endonuclease, resynthesis by DNA polymerase, and relinking of the strand by ligase (80). Another possibility is base excision by a glycosylase followed by base insertion by an enzyme similar to the "purine inserting" enzyme (81). This process appears to be more efficient, but its accuracy is

TABLE III
ERROR RATE PER BASE-PAIR INCORPORATED[a]

	In vivo (70)
Escherichia coli	2×10^{-10}
Phase λ and T4	200×10^{-10}
Drosophila	0.5×10^{-10}
	In vitro
E. coli DNA polymerase I	130×10^{-7} (83,84)
T4 DNA polymerase	$10-100 \times 10^{-7}$ (85)
T4 DNA replication complex	$<5 \times 10^{-7}$ (86)
E. coli DNA polymerase III complex	5×10^{-7} (87)

[a] References are given in parentheses.

not known. In a similar vein, a pyrimidine insertion enzyme should be sought.

A combination of the above processes is expected to yield the accuracies of replication shown in Table III.

2. COMPARISON OF ACCURACY OF *in Vitro* AND *in Vivo* SYSTEMS

Error rates in *in vitro* systems are very difficult to determine; they vary from one in a few hundred base-pairs to one in 10^6 (70). Those shown in Table III are probably the best available. To determine such low frequencies of error, templates and precursor nucleotides must be very pure. In addition, the frequency of error depends upon the relative concentration of each precursor nucleotide, on the kinds and concentrations of divalent cations, and on the structure and nucleotide sequences of the template (84, 85, 88).

Recently, φX174 DNA was employed in an accuracy determination (88). The assay system used φX174 DNA with a DNA fragment, made by a restriction enzyme, annealed to a position just ahead of the am3 mutation site. After replication *in vitro*, the products were used to transfect spheroplasts, and the ratios of revertants to mutants were measured. *E. coli* polymerase I gave a value at the lower limit of the accuracy of replication, 1.3×10^{-4}; that is, no revertants were observed. The value shown in Table III is for poly(dA-dT), with well-characterized template, precursors, and product (83, 84). This value is also a minimal value for accuracy, because such a synthetic template is probably more tolerant of the incorporation of incorrect nucleotides than is natural DNA, and it allows mismatched nucleotides to loop out (88). T4 DNA polymerase, with poly(dC) as a template, incorporated dTTP into poly(dG) at a frequency of 10^{-5} to 10^{-6} (85). Thus the values shown in Table III for purified DNA polymerases are probably at the lower limit of accuracy.

The proteins in replication complexes contribute to the accuracy of DNA replication *in vitro*. Liu *et al.* (86) used randomly nicked φX-RF DNA (am3) with the T4 DNA replication complex, making 10 to 20 copies of the input. The products, mostly double-stranded DNA, were cut into unit lengths with restriction enzyme *Pst*I, which makes a single cut per φX genome. DNA was then treated with ligase and used for transfection, and the reversion frequency was measured. They observed no significant number of revertants and estimated the lower limit of accuracy to be less than 5×10^{-7} per base-pair. *In vivo*, specific reverse mutations range from about 10^{-4} to 5×10^{-11} and are strongly dependent on the DNA sequence. Therefore, the *in vitro* value was very close to the *in vivo* rate. Fersht (87) used the *E. coli* DNA polymerase III complex to determine the reversion frequency of φX-RF-I (am3) DNA. This system makes five to eight copies per input DNA, mostly single-stranded circular (+) DNA. The reversion frequencies increased with dGTP concentration. These rates were extrapolated back to physiological concentration of precursor nucleotides in *E. coli* and yielded an error rate of 5×10^{-7} per base-pair. This is considered very close to the *in vivo* rate.

These results show that the first two processes involving DNA polymerases mentioned previously (nucleotide selection and editing) contribute significantly to the overall accuracy of replicated DNA. For prokaryotic polymerases, $3' \rightarrow 5'$ exonuclease, in concert with polymerase, is believed to be involved in the editing step. Liu *et al.* (86) concluded that four proteins in the replication complex with T4 polymerase improve the efficiency of the editing step without affecting the nucleotide selection step. Since the replication complex systems of *E. coli* DNA polymerase III and T4 DNA polymerase give values very close to *in vivo* rate, any contribution by stage (c) (correction after replication) is minor, at least for these smaller DNAs. This type of correction may be more significant for maintenance of chromosomal DNA during the life cycle of cells.

3. SELECTION MECHANISMS

What are the selection mechanisms that enable DNA polymerases to achieve a high degree of accuracy in replication? The many hypotheses can be grouped, with modifications, into two fundamental groups. One, originally proposed by Watson and Crick, is that the strength of the hydrogen bonding between complementary base pairs is sufficiently greater than others to achieve the accuracy of replication (89). In this group belong proposals by Kornberg (29), Bessman (69), Hopfield (90), Ninio (91), and Branscomb and Galas (92). The other group states that specificity due to hydrogen bonding between complementary bases is not sufficient, and that there are allosteric sites on the enzymes that respond to substrates. Proposals by Speyer (93), Mildvan (94), and Loeb (95) belong to this group.

Recent models of the first group invoke coordinated action of base selection by Watson–Crick base-pair complementarity and editing by polymerase-associated exonuclease. This was first suggested by Kornberg (29). Hopfield (90) and Ninio (91) independently derived equations showing that accuracy is more than the sum of each step; it is amplified. Galas and Branscomb (92) derived specific equations and applied them to T4 DNA polymerase. They assumed that for both incorporation and editing, the only discriminatory factor is the difference in strength of the hydrogen bond between matched and mismatched base pairs. In polymerase function, the nucleotide that forms the most stable base-pair with a template base at the site adjacent to the primer end is favored for incorporation. In exonuclease function, the nucleotide at the primer end that is most weakly hydrogen-bonded is most likely to be hydrolyzed. They expressed differences in hydrogen-bond strength as differences in free energy (ΔG) between matched and mismatched base-pairs, and assumed that ΔG is the same for both processes. They further assumed that incorporation and editing occur in an ordered sequence at the 3' termini as proposed by Kornberg (5). However, whether a nucleotide at the 3'-OH end is hydrolyzed or acts as a primer for incorporation of the next nucleotide is stochastic in nature, since binding sites on the enzyme are nonspecific toward bases. There is probably one binding site for dNTP on an enzyme (96), and binding of dNTP involves Mg^{2+} and the β and γ phosphates (97). The primer end binds very close to it (98). The exonuclease active site is probably near the primer-end binding site, and its level of activity with respect to polymerase activity determines how effective it is in hydrolyzing nucleotides in "frayed" conformation. The hypothesis predicts that accuracy is dependent on the dNTP concentration. Increase in substrate concentration increases the error rates. The hypothesis also can predict turnover and incorporation rates if the ΔG's between a set of matched and mismatched bases are known or, alternatively, can determine the ΔG if turnover and incorporation rates are known. The theory also predicts consistency between error rate and $K_{m_{dNTP}}$.

Clayton et al. (99) used a set of mutator–antimutator mutants of T4 DNA polymerase to get experimental evidence for these predictions. They used L56, a mutator, L141, an antimutator, and the wild type. In previous studies, these strains showed most definite correlation between these phenotypes and level of exonuclease and polymerase ratio (68, 69). They studied misincorporation of 2-aminopurine deoxyribonucleotide (dAPMP) in place of dAMP using these polymerases. They showed that percentage of misincorporation is dependent on dNTP concentration for all three polymerases of T4 and that at lower concentration of dNTP all three polymerases have the same high accuracy of incorporation. Using Galas and Branscomb's model, a ΔG of the H-bond difference of 1.1 kcal/mol between

Ade·Thy and AP·Thy was determined from the concentration dependence of the incorporation and removal of aminopurine and adenine. It was also determined from the ratio of $K_{m_{AP}}$ and $K_{m_{Ade}}$ to be 1.1 kcal/mol. The difference in error frequency among the three polymerases (L56, L141, and the wild type) can be attributed to difference in the activity of their respective nucleases, not to altered selection or editing specificity. The value for ΔG of 1.1 kcal/mol is much less than previous calculations (94), yet the difference is sufficient for discrimination between matched and mismatched base-pairs.

Misincorporation frequency was also determined for *E. coli* DNA polymerase I, which has 3'-to-5' exonuclease activity that is less than 1% of that of the wild type T4, and for DNA polymerase α, which has no associated 3'→5' exonuclease. At high dNTP concentration, there is no detectable turnover of dATP or dAPTP with either enzyme. The misincorporation frequency of aminopurine was slightly higher for both enzymes than for L56, a T4 mutator, 12–15% and 10%, respectively. These values are expected for a ΔG (H-bond difference) of 1.1 kcal/mol in the case of very little or no 3'→5' exonuclease activity. Thus these results are consistent with the hypothesis.

With an *E. coli* polymerase III replication system using ϕX174 am3 DNA, the frequency of revertant increases with dGTP concentration (87). This is another independent confirmation of the proposition.

Experimental evidence for nucleotide incorporation and editing by a DNA polymerase at a given primer-template came from work with T5 DNA polymerase (100), which is suitable for this type of study because, in the absence of Mg^{2+}, it attaches preferentially to a primer end (101) to form a complex that can act either as polymerase or exonuclease (100). It is processive for both polymerization (25, 58) and exonuclease action (102). When Mg^{2+} was added to a complex of poly(dA)·oligo(dT) and polymerase, exonucleolytic action began. Denatured calf thymus DNA was present as a "challenger" polymer so that if the enzyme left the primer-template its chance of reattaching itself to the same primer-template would be negligible. During the course of the hydrolysis, dNTPs were added; this caused the polymerase to reverse its direction and to start nucleotide incorporation. The reverse of these processes is possible; that is, a polymerase moving in the 5'-to-3' direction during the chain elongation process could change its direction and act as a 3'→5' exonuclease or editing enzyme. This is evidence that a single enzyme attached to a primer end of a primer-template can function as both polymerase and 3'→5' exonuclease.

Among the second group of investigators, Loeb *et al.* have been the most active in trying to get evidence that the enzyme has allosteric sites that take part in the selection of nucleotides. Their studies showed that *E. coli* DNA polymerase I has many binding sites of varying affinity for divalent metal ions (103). These sites interact with nucleoside triphosphates, thus possibly altering the conformation of the active sites of the polymerase and thus influencing the base selectivity at the active site. However, their recent data for *E. coli* DNA polymerase I and DNA polymerase α and β are still inconclusive (*84, 104, 105*).

IV. Concluding Remarks on DNA Polymerase and the Replication Complex

As discussed, DNA polymerases have varied capabilities with respect to strand unwinding, processivity, and accuracy of replication. Each replicative polymerase is probably in a complex with other proteins at the replication fork in order to function efficiently in these processes. Each bacterial and eukaryotic replicative DNA polymerase is made up of many subunits (Table I), but phage T4 and T5 DNA polymerases are monomeric with quite different capabilities, making them ideal systems for comparative studies. Phage T7 DNA polymerase has two subunits (Table I), and it appears to require only two other proteins for the elongation process in a duplex DNA replication system (*37, 45*), but more may be necessary to enhance accuracy. Of these three phage enzymes, only the T4 DNA replication complex has been studied extensively in terms of function at a replication fork.

The T4 DNA replication complex appears to be made up of at least seven proteins (*86*). The function of five of these is probably to facilitate a translocation movement of the polymerase along the template, thus aiding the unwinding process and processiveness of the enzyme and increasing the accuracy of replication at the editing step (*86*). A detailed purification procedure for these proteins appeared recently, and thus more detailed work on these proteins and the replication complex can be carried out (*106–108*). On the basis of the hypothesis of Galas and Branscomb regarding base selection and editing, it appears that the slower the translocation rate the more efficient the editing process, and thus the more accurate the replication (*82*). However, when the translocation process is more processive, probably both elongation and editing processes will be faster, and accuracy may be improved. As stated previously, T5 DNA polymerase is processive in both polymerase (*25, 58*) and exonuclease function (*102*), and it is capable of switching direction during the translocation process (*100*). This may be true for other polymerases particularly in the replication complex. Properly functioning proteins in this complex are essential for accuracy, as seen by the effects on accuracy of mutations of genes for the proteins in the T4 DNA replication complex (*109, 110*).

T5 DNA polymerase probably exists also as a replication complex as judged by its inefficient replication of "nicked" linear duplex DNA (*40*).

Synthesis stops after a short time on such a primer-template, forming DNA with many branches. Recently, T5 DNA binding protein has been purified to homogeneity and found to have a property quite different from that of the T4 or *E. coli* helix-destabilizing protein (*111*). It binds cooperatively to double-stranded DNA, but only additively to single-stranded DNA. It is a product of gene D5 (*112*). Conditional lethal mutants of gene D5 are defective in DNA synthesis (*112*), yet D5 protein inhibits DNA synthesis on nicked circular DNA *in vitro* (Das and Fujimura, unpublished observation). Thus the role of D5 protein in DNA replication is different from that of the well-characterized helix-destabilizing protein of T4 or of *E. coli*. There are several amber mutants available (*8*) that affect DNA synthesis, but the functions of most of these gene products are not yet determined.

The *E. coli* DNA replication apparatus is very complex. The ingenious use of small circular DNAs and dna mutants has permitted many of the proteins in the complex to be purified and identified (*36, 65*). Judging from the great number of mutator genes isolated for *E. coli* DNA synthesis (*71*), and from the fact that the gene products of most of them are not known, many more proteins than those already identified may be involved in accurate replication of DNA.

Studies on replicative DNA polymerases and replication complexes at replication forks for eukaryotic systems seem difficult, owing to lack of mutants and the involvement of nucleosomes that form into higher-order structures. However, recent work suggests that nucleosomes are not present at replication forks (*113, 114*), and that higher-order structures of nucleosomes are formed very rapidly after replication (*113*). Thus questions arise as to how nucleosomes are removed prior to encountering replication forks and how they are reassembled after replication. These problems are outside the scope of this essay.

There are several DNA synthesis mutants available for studies in lower eukaryotes and animal DNA viruses. Thus, as in the case of *E. coli*, studies of chromosomal DNA replication through these systems may be worthwhile.

One of the promising lower eukaryotes is *Saccharomyces cerevisiae*, where several conditional lethal mutants of genes involved in DNA synthesis are available (*115*). A small circular DNA is present in yeast, replication of which requires several gene products involved in chromosomal DNA replication (*116*). This plasmid DNA has been isolated as a chromatin (*117*). Also, small fragments of chromosomal DNA containing an origin of replication, which are capable of autonomous replication as plasmids and are capable of transforming yeast cells, have been isolated as recombinants with *E. coli* plasmids, and are thus capable of replication in *E. coli* as well (*118, 119*). Two yeast DNA polymerases have been purified extensively, although it is not known which is the replicative polymerase (*120, 121*). Thus conditions are set for studies of replication and its control using the yeast *in vitro* system.

Mechanisms of animal virus DNA replication may be closer to that of higher eukaryotes. One most understood is adenovirus DNA replication. Adenovirus DNA is a linear duplex, but it is extracted from the virus in circular or oligomeric forms with a viral protein attached covalently to the 5' ends (*122*). One hypothesis says that this protein is involved in initiation of DNA replication from either end (*122*). There are three complementation groups of conditional lethal mutants for the viral DNA synthesis, and one of them codes for DNA binding protein, which is essential for initiation and elongation processes (*123*). Challberg and Kelly have developed an *in vitro* system consisting of an extract from the nuclei of HeLa cells infected with adenovirus 5 (Ad5), which carries out semiconservative replication specifically with exogenous Ad5 DNA (*124*). Maximal synthesis was observed with DNA–protein complex isolated from Ad5 virions. Replicating molecules, shown by electron microscopy, are linear duplex DNA containing one or two single-stranded branches. When similar extracts were prepared from cells infected with temperature-sensitive mutants of the DNA binding protein, the system could be inactivated at 37°C but completely complemented by addition of the DNA binding protein from the wild-type cells (*125*). Replication in such a system occurs at or near either end of a molecule (*125*), and thus all properties very closely resemble those *in vivo*.

The systems discussed above are examples of many ingenious approaches being used in eukaryotic systems. By proper focusing of problems and application of ingenious techniques, it is only a matter of time until the mechanisms involved in DNA replication will be understood even in higher eukaryotic systems.

REFERENCES

1. A. DeWaard, A. V. Paul, and I. R. Lehman, *PNAS* **54**, 1241 (1965).
2. P. DeLucia and J. Cairns, *Nature* **224**, 1164 (1969).
3. M. L. Gefter, Y. Hirota, T. Kornberg, J. A. Wechsler, and C. Barnoux, *PNAS* **68**, 3150 (1971).
4. A. Weissbach, *ARB* **46**, 25 (1977).
5. A. Kornberg, in "DNA Synthesis," p. 399, Freeman, San Francisco, 1974.
6. C. D. Steuart, S. R. Anand, and M. J. Bessman, *JBC* **243**, 5308 (1968).
7. C. D. Steuart, S. R. Anand, and M. J. Bessman, *JBC* **243**, 5319 (1968).
8. H. E. Hendrickson and D. J. McCorquodale, *J. Virol.* **9**, 981 (1972).
9. R. K. Fujimura and B. C. Roop, *JBC* **251**, 2168 (1976).
10. R. K. Fujimura and B. C. Roop, *Bchem.* **15**, 4403 (1976).
11. S. K. Das and R. K. Fujimura, *J. Virol.* **20**, 70 (1976).
12. D. M. Livingston, D. C. Hinkle, and C. C. Richardson, *JBC* **250**, 461 (1975).
13. D. M. Livingston and C. C. Richardson, *JBC* **250**, 470 (1975).
14. C. S. McHenry and W. Crow, *JBC* **254**, 1748 (1979).

15. S. Wickner, *PNAS* **73**, 3511 (1976).
17. I. R. Lehman, *Methods Enzymol.* **29 E**, 46 (1974).
18. P. Grippo and C. C. Richardson, *JBC* **246**, 6867 (1971).
19. P. Modrich and C. C. Richardson, *JBC* **250**, 5515 (1975).
19b. K. Hori, D. F. Mark, and C. C. Richardson, *JBC* **254**, 11591 (1979).
19c. S. Adler and P. Modrich, *JBC* **254**, 11605 (1979).
20. R. Kolodner, Y. Masamune, I. E. LeClerc, and C. C. Richardson, *JBC* **253**, 566 (1978).
21. F. J. Bollum, L. M. S. Chang, C. M. Tsiapalis, and J. W. Dorson, *Methods Enzymol.* **29 E**, 70 (1974).
22. P. A. Fisher and D. Korn, *JBC* **252**, 6528 (1977).
23. O. Fichot, M. Pascal, M. Mechali, and A-M. DeRecondo, *BBA* **561**, 29 (1979).
23b. Y-C. Chen, E. W. Bohn, S. R. Planck, and S. H. Wilson, *JBC* **254**, 11678 (1979).
24a. P. A. Fisher, T. S-F. Wang, and D. Korn, *JBC* **254**, 6128 (1979).
24b. P. A. Fisher and D. Korn, *JBC* **254**, 11033 (1979).
24c. P. A. Fisher and D. Korn, *JBC* **254**, 11040 (1979).
25. S. K. Das and R. K. Fujimura, *JBC* **254**, 1227 (1979).
26. R. L. Low, S. A. Rashbaum, and N. R. Cozzarelli, *JBC* **251**, 1311 (1976).
27. K. B. Gass and N. R. Cozzarelli, *JBC* **248**, 7688 (1973).
28. G. W. Bazill and J. D. Gross, *Nature NB* **240**, 82 (1972).
29. A. Kornberg, *Science* **163**, 1410 (1969).
30. I.-W. Knopf, *Cold Spring Harbor Meet. Herpes Viruses, 4th 1979*, (Abstr.), p. 116.
31. Y. Masamune and C. C. Richardson, *JBC* **246**, 2692 (1971).
32. P. Setlow, D. Brutlag, and A. Kornberg, *JBC* **247**, 224 (1972).
33. B. Alberts, J. Barry, M. Bittner, M. Davies, H. Hama-Inaba, C-C. Liu, D. Mace, L. Moran, C. F. Morris, J. Piperno, and N. K. Sinha, in "Nucleic Acid Protein Recognition" (H. J. Vogel, ed.), pp. 31–63. Academic Press, New York, 1977.
34. M. L. Gefter, *ARB* **44**, 45 (1975).
34b. J. Wolfson and D. Dressler, *JBC* **254**, 10490 (1979).
35. J. Tomizawa and G. Selzer, *ARB* **48**, 999 (1979).
36. R. Scheckman, A. Weiner, and A. Kornberg, *Science* **186**, 987 (1974).
37. E. Scherzinger, E. Lanka, G. Morelli, D. Seiffert, and A. Yuki, *EJB* **72**, 543 (1977).
37b. L. J. Romano and C. C. Richardson, *JBC* **254**, 10476 (1979).
37c. L. J. Romano and C. C. Richardson, *JBC* **254**, 10483 (1979).
38. S. Wickner, *PNAS* **74**, 2815 (1977).
39. L. Rowen and A. Kornberg, *JBC* **253**, 770 (1978).
40. R. K. Fujimura and D. P. Allison, *JBC* **251**, 2174 (1976).
41. N. G. Nossal, *JBC* **249**, 5668 (1974).
42. H. M. Sobel, E. D. Lozansky, and M. Lessen, *CSHS* **43**, 11 (1979).
43. N. G. Nossal and B. M. Peterlin, *JBC* **254**, 6032 (1979).
44. R. Kolodner and C. C. Richardson, *PNAS* **74**, 1525 (1977).
45. R. Kolodner and C. C. Richardson, *JBC* **253**, 574 (1978).
46. S. Eisenberg, J. F. Scott, and A. Kornberg, *PNAS* **73**, 3151 (1976).
47. J. F. Scott, S. Eisenberg, L. L. Bertsch, and A. Kornberg, *PNAS* **74**, 193 (1977).
48. G. T. Yarranton and M. L. Gefter, *PNAS* **76**, 1658 (1979).
49. S. Eisenberg, J. Griffith, and A. Kornberg, *PNAS* **74**, 3198 (1977).
50. M. Abdel-Monem, H. Dürwald, and H. Hoffmann-Berling, *EJB* **65**, 441 (1976).
51. M. Abdel-Monem, H. Dürwald, and H. Hoffmann-Berling, *EJB* **79**, 39 (1977).
52. M. Abdel-Monem, H. F. Lauppe, J. Kartenbeck, H. Dürwald, and H. Hoffmann-Berling, *JMB* **110**, 667 (1977).
52a. G. T. Yarranton, R. H. Das, and M. L. Geften, *JBC* **254**, 12002 (1979).

53. H. Krell, H. Dürwald, and H. Hoffmann-Berling, *EJB* **93**, 387 (1979).
54. M. Gellert, K. Mizuuchi, M. O'Dea, and H. A. Nash, *PNAS* **73**, 3872 (1976).
55. M. Gellert, M. H. O'Dea, T. Itoh, and J. Tomizawa, *PNAS* **73**, 4474 (1976).
56. T. Itoh and J. Tomizawa, *Nature* **270**, 78 (1977).
57. J. J. Champoux, *ARB* **47**, 449 (1978).
58. S. K. Das and R. K. Fujimura, *JBC* **252**, 8700 (1977).
59. S. K. Das and R. K. Fujimura, *JBC* **252**, 8708 (1977).
60. R. A. Bambara, D. Uyemura, and T. Choi, *JBC* **253**, 413 (1978).
61. W. R. McClure and T. M. Jovin, *JBC* **250**, 4073 (1975).
62. L. M. S. Chang, *JMB* **93**, 219 (1975).
63. L. A. Sherman and M. L. Gefter, *JMB* **103**, 61 (1976).
64. D. Uyemura, R. Bambara, and I. R. Lehman, *JBC* **250**, 8577 (1975).
65. S. H. Wickner, *ARB* **47**, 1163 (1978).
66. M. S. Hershfield, *JBC* **248**, 1417 (1973).
67. F. D. Gillin and N. G. Nossal, *JBC* **251**, 5219 (1976).
68. N. Muzyczka, R. L. Poland, and M. J. Bessman, *JBC* **247**, 7116 (1972).
69. M. J. Bessman, N. Muzyczka, M. F. Goodman, and R. L. Schnaar, *JMB* **88**, 409 (1974).
70. F. Bernardi and J. Ninio, *Biochimie* **60**, 1083 (1978).
71. E. C. Cox, *Annu. Rev. Genet.* **10**, 135 (1976).
72. G. Villani, S. Spadari, S. Boiteux, M. Defais, P. Caillet-Fanquet, and M. Radman, *Biochemie* **60**, 1145 (1978).
73. D. Brutlag and A. Kornberg, *JBC* **247**, 241 (1972).
74. R. M. Hall and W. J. Brammer, *Mol. Gen. Genet.* **121**, 271 (1973).
75. J. J. Byrnes, K. M. Downey, V. L. Black, and A. G. So, *Bchem* **15**, 2817 (1976).
76. J. J. Byrnes and V. L. Black, *Bchem* **17**, 4226 (1978).
77. M. Y. W. Lee, C-K. Tan, K. M. Downey, and A. G. So, *FP* **38**, 779 (1979).
78. J. Ninio, F. Bernardi, G. Brun, L. Assairi, M. Lauber, and F. Chapeville, *FEBS Lett.* **57**, 139 (1975).
79. R. K. Fujimura, *Bchem* **10**, 4386 (1971).
80. T. Lindahl, *This Series* **22**, 135 (1979).
81. W. A. Deutsch and S. Linn, *PNAS* **76**, 141 (1979).
82. R. Wagner, Jr. and M. Meselson, *PNAS* **73**, 4135 (1976).
83. T. A. Trautner, M. N. Swartz, and A. Kornberg, *PNAS* **48**, 449 (1962).
84. S. S. Agarwal, D. K. Dube, and L. A. Loeb, *PNAS* **254**, 101 (1979).
85. Z. Hall and I. R. Lehman, *JMB* **36**, 321 (1968).
86. C. C. Liu, R. L. Burke, U. Hibner, J. Barry, and B. Alberts, *CSHSQB* **43**, 469 (1979).
87. A. R. Fersht, *PNAS* **76**, 4946 (1979).
88. L. A. Weymouth, and L. A. Loeb, *PNAS* **75**, 1924 (1978).
89. J. D. Watson and F. H. C. Crick, *Nature* **171**, 964 (1953).
90. J. J. Hopfield, *PNAS* **71**, 4135 (1974).
91. J. Ninio, *Biochimie* **57**, 587 (1975).
92. D. J. Galas and E. W. Branscomb, *JMB* **124**, 653 (1978).
93. J. F. Speyer, J. D. Karam, and A. B. Lenny, *CSHSQB* **31**, 693 (1966).
94. A. S. Mildvan, *ARB* **43**, 357 (1974).
95. L. A. Loeb, *The Enzymes* **10**, 173 (1974).
96. P. T. Englund, J. A. Huberman, T. M. Jovin, and A. Kornberg, *JBC* **244**, 3038 (1969).
97. P. M. Burgers and F. Eckstein, *JBC* **254**, 6889 (1979).
98. T. R. Krugh, *Bchem* **10**, 2594 (1971).
99. L. K. Clayton, M. F. Goodman, E. W. Branscomb, and D. J. Galas, *JBC* **254**, 1902 (1979).

100. S. K. Das and R. K. Fujimura, *JBC*, in press (1980).
101. S. K. Das, *BBRC* **79**, 247 (1977).
102. S. K. Das and R. K. Fujimura, *NARes*, **8**, 657 (1980).
103. J. P. Slater, I. Tamir, L. A. Loeb, and A. S. Mildvan, *JBC* **247**, 6784 (1972).
104. M. D. Sirover, D. K. Dube, and L. A. Loeb, *JBC* **254**, 107 (1979).
105. G. Seal, C. W. Shearman, and L. A. Loeb, *JBC* **254**, 5229 (1979).
106. N. G. Nossal, *JBC* **254**, 6026 (1979).
107. C. F. Morris, H. Hama-Inaba, D. Mace, N. K. Sinha, and B. Alberts, *JBC* **254**, 6787 (1979).
108. C. F. Morris, C. A. Moran, and B. M. Alberts, *JBC* **254**, 6797 (1979).
109. S. M. Watanabe and M. F. Goodman, *J. Virol.* **25**, 73 (1978).
110. S. Muffti, *Virology* **94**, (1979).
111. A. C. Rice, T. A. Ficht, L. A. Holladay, and R. W. Moyer, *JBC* **254**, 8042 (1979).
112. G. Cinnadurai and D. J. McCorquodale, *Nature* **247**, 554 (1974).
113. H. Weintraub, *NARes* **7**, 781 (1979).
114. T. M. Herman, M. L. DePamphilis, and P. M. Wassarman, *Bchem* **18**, 4563 (1979).
115. L. H. Hartwell, *Bacteriol Rev.* **38**, 164 (1974).
116. D. M. Livingston and D. M. Kupfer, *JMB* **116**, 249 (1977).
117. D. M. Livingston and S. Hahne, *PNAS* **76**, 3727 (1979).
118. K. Struhl, D. T. Stinchcomb, S. Scherer, and R. W. Davis, *PNAS* **76**, 1035 (1979).
119. C.-L. Hsiao and J. Carbon, *PNAS* **76**, 3829 (1979).
120. E. Wintersberger, *EJB* **50**, 41 (1974).
121. L. M. S. Chang, *JBC* **252**, 1873 (1977).
122. D. M. K. Rekosh, W. C. Russell, A. J. D. Bellet, and A. J. Robinson, *Cell* **11**, 283 (1977).
123. E.-L. Winnacker, *Cell* **14**, 761 (1978).
124. M. D. Challberg and T. J. Kelly, Jr., *PNAS* **76**, 655 (1979).
125. L. M. Kaplan, H. Ariga, J. Hurwitz, and M. S. Horwitz, *PNAS* **76**, 5534 (1979).

Antibodies Specific for Modified Nucleosides: An Immunochemical Approach for the Isolation and Characterization of Nucleic Acids

THEODORE W. MUNNS AND
M. KATHRYN LISZEWSKI

Washington University
School of Medicine
Rheumatology Division
St. Louis, Missouri

I. Introduction	110
A. Antibodies Specific for Nucleic Acids	110
B. Antibodies Specific for Nucleosides	111
C. Complexes between Nucleic Acids and Antibodies	113
II. Nucleic Acid Methylation	114
A. Extent of Nucleic Acid Methylation	114
B. Methylated Nucleosides of Nucleic Acids	116
III. Immunochemical Procedures	124
A. Synthesis and Characterization of Nucleoside–Protein Conjugates	124
B. Immunization	127
C. Characterization of Antibodies Specific for Nucleosides by Radioimmunoassay	127
D. Purification of Antibodies Specific for Nucleosides	133
IV. Immunochemical Isolation of Oligonucleotides and Nucleic Acids Possessing Specific Modified Constituents	134
A. Isolation of Nucleotide Sequences Containing N^6-Methyladenine	135
B. Isolation of mRNAs Possessing a 5'-Terminal "Cap" Structure Containing 7-Methylguanine	140
V. Immunochemical Approaches for Assessing the Function of Modified Constituents	142
A. Antibodies as Site-Specific Probes	143
B. Antibodies Specific for 7-Methylguanine Inhibit mRNA Translation	145
VI. Immunoelectron Microscopy: The Mapping of Specific Determinants within Nucleic Acids	148
A. Antibodies Specific for Dinitrophenol	148
B. Antibodies Specific for 5-Methylcytosine	149
C. Antibodies Specific for N^6,N^6-Dimethyladenine	152
D. Antibodies Specific for N^6-Methyladenine	154
VII. Conclusion	156
References	158

I. Introduction

A. Antibodies Specific for Nucleic Acids

The employment of immunochemical procedures and reagents to obtain meaningful information regarding various aspects of molecular biology and nucleic acid biochemistry necessitates the availability of specific antibodies against nucleic acids. Only during the past few decades, however, have such antibodies become available (1). Although there is still some controversy as to the question, are nucleic acids antigens?, it is important to realize that their potential antigenicity is dependent upon a host of complex parameters, many of which have been the subject of detailed reviews (1-3). Such parameters include the following.

1. Physical and biological properties of the nucleic acids (i.e., natural or synthetic, single and/or double-stranded, base composition and/or content of unusual modified bases, nucleoprotein in nature)
2. Presence or absence of adjuvants and/or carrier proteins during immunization (2, 4)
3. Genetic predisposition of the host [e.g., certain strains of mice and patients with various autoimmune diseases spontaneously produce anti-nucleic acid and anti-(deoxy)ribonucleoprotein antibodies (5-9)]
4. The types of immunochemical procedures designed to evaluate such antibody preparations in terms of their specificity (1-3) (i.e., Are they specific for various proteins associated with nucleic acids and/or the nucleic acids themselves? What determinants are present in nucleic acid? Which are immunodominant?)

It has generally been accepted that free nucleic acids are only weakly immunogenic (if at all) in hosts other than those referred to above in item 3. However, when these macromolecules function as hapten components, i.e., when they are coupled to "carrier" proteins that by themselves are antigenic, the resulting immunogens in many instances elicit the production of anti-nucleic acid antibodies in the sera of immunized hosts (1, 2, 4). Although DNA and RNA can be covalently coupled to protein carriers, the simplest and perhaps most reliable immunogens are those in which a nucleic acid is mixed with an appropriate quantity of methylated bovine serum albumin (MeBSA) (4). These noncovalently linked conjugates (denatured DNA·MeBSA and RNA·MeBSA) have consistently yielded anti-DNA and anti-RNA populations (10-12).

In most instances, however, antibodies elicited in response to these immunogens possess broad ranges of specificity. For example, most anti-

DNAs[1] cross-react with denatured DNA preparations derived from a variety of sources (both pro- and eukaryotic), with numerous oligodeoxynucleotide sequences, and with individual deoxynucleotides and bases (1–4). Since the bases are common to both RNA and DNA, it is not surprising that many anti-DNAs cross-react with RNA-type substrates and vice versa. Although the above findings provide convincing evidence that nucleic acids can be rendered antigenic, they also emphasize the inability of these antibodies to interact selectively with specific DNA and RNA molecules that differ considerably in both sequence complexity and nucleotide composition.

Although some preparations of spontaneously occurring anti-DNAs appear to discriminate between single- and double-stranded DNA substrates (7–9), in general they seem to be devoid of unique specificities (1, 3). Various explanations can be advanced for the lack of specificity associated with both the spontaneous and experimentally induced antibodies. Perhaps most pertinent is the fact that nucleic acids are complex macromolecules displaying a large number of antigenic determinants (individual bases, nucleosides, nucleotides, oligonucleotide sequences, etc.). This implies that anti-DNAs possess a considerable degree of heterogeneity and accounts for their extensive cross-reactivity with a variety of nucleic acid substrates. Furthermore, since most specificity studies are evaluated with serum or unfractionated IgG preparations, the presence of unique specificities would often be overlooked or remain undetected if present as a minor fraction of the overall antibody population. It would be desirable, therefore, to reexamine the specificity of such antibodies after suitable fractionation and/or selection techniques.

B. Antibodies Specific for Nucleosides

In an attempt to restrict or minimize the multitude of determinants inherent in nucleic acid structures, attention first focused upon the use of individual bases, nucleosides, and nucleotides as simple hapten components (10–16, also see Table I). The impetus for this research came from the pioneering work of Landsteiner (17), who initially demonstrated that a multitude of organic compounds of low molecular weight could be rendered antigenic after coupling to suitable protein "carriers." Such investigations provided unequivocal evidence that nucleoside–protein conjugates can yield antibodies against nucleosides. Subsequent characterization of these antibodies revealed that they possess high degrees of specificity toward the nucleoside hapten, its homologous base and nucleotide components, yet minimal or no cross-reactivity with other bases, nucleosides, and nucleotides

---

[1]Anti-X is used as an abbreviated expression of "antibody specific for X," or "anti-X antibody."

TABLE I
NATURALLY OCCURRING METHYLATED NUCLEOSIDES EMPLOYED AS HAPTENS[a]

Nucleoside	Reference	Nucleoside	Reference
Adenosine	23–25	*Guanosine*	24,34,41,42
1-Methyl-	26	1-Methyl-	38
2-Methyl-	—	N^2-Methyl-	38,43
N^6-Methyl-	27,31	N^2,N^2-Dimethyl-	38,43
N^6,N^6-Dimethyl-	32,33	7-Methyl-	29–31,38,43–46
Cytidine	24,34–36	*Inosine*	47,48
3-Methyl-	37	1-Methyl-	26,37
5-Methyl-	38–40	7-Methyl-	26
2'-O-Methyl-(Cm)	—		
		Uridine	23–26
		3-Methyl-	37
		5-Methyl-	24, 34, 47, 48

[a] m^2A, Am, m^6Am, m^4C, Cm, m^4Cm, $m_3^{2,2,7}G$, Gm, Um, ψm are naturally occurring methylated nucleosides that have not yet been used as haptens.

of similar structure. Thus, the lack of specificity observed with anti-DNA (or RNA) antibodies could be overcome to a limited degree by the utilization of the appropriate anti-nucleoside; e.g., anti-thymidine cross-reacts with denatured DNA, but not with RNA (*24*).

However, most significant, and concurrent with this advancement, was an increasing awareness that nucleic acid populations possess a number of "minor" or modified nucleotides in addition to their four major nucleotides (*18–22*). Because of their variety (*20–22*), their limited occurrence in nucleic acids (hence the terms "minor" and "rare"), and their unique distribution within these molecules (*21, 22*), the feasibility of isolating and characterizing specific nucleic acids on the basis of their content of these unusual constituents became obvious.

As evidenced just by the number of antibodies specific for methylated nucleosides (*23–48*, see Table I), the attractiveness of this approach has not gone unnoticed. It is important to note that this list is by no means complete since other modified nucleosides, as well as nucleotides (*49–51*) and oligonucleotides (*52–54*), have also been employed with considerable success as haptens to prepare specific antibody populations. Such haptens include a variety of carcinogen-induced nucleoside adducts (*55–57a*), ultraviolet-induced photoproducts (*58–61*), and a host of other naturally occurring modified nucleosides present in RNA (*21, 22, 25, 62–68a*), e.g., N^6-(Δ^2-isopentenyl)adenosine (*62–64*), wyosine ("Y-nucleoside") (*65, 65b*), queuosine (*65a*), and pseudouridine (*19, 26, 66, 66a, 66b*).

C. Complexes between Nucleic Acids and Antibodies

Based upon the information presented above that (a) most nucleic acids contain a limited number of modified constituents and (b) antibodies specific for these constituents can be experimentally induced with the appropriate nucleoside–protein conjugate, one can envision various immunochemical approaches for the isolation and characterization of nucleic acids. Several of these approaches are listed in Fig. 1. It is important to note that each is dependent on the formation of a specific anti-nucleoside·nucleic-acid complex. Since such complexes can easily be isolated free of unreacted or nonantigenic material, the employment of specific antibody populations for isolating both oligonucleotides (I) and nucleic acids (II) possessing a specific hapten constituent should be quite useful. Further, the visualization of such complexes by electron microscopy would provide a simple, yet direct, approach for mapping or locating the position of modified constituents within nucleic acid or nucleoprotein structures (III). Last, information regarding the functional significance (IV) of the hapten component can be deduced. The latter approach assumes that if the activity of an enzyme is dependent upon the presence of a modified constituent, it would be inhibited by the masking effect of the antibody molecule.

Critical to the successful development of these immunochemical approaches is the production of very specific antibodies that possess high affinities for the hapten components. This characteristic is required to ensure that the interaction between the antibody and a hapten-containing nucleic acid component is both quantitative (high degree of affinity) and exclusive (specific). In considering specificity, it must be realized that modified con-

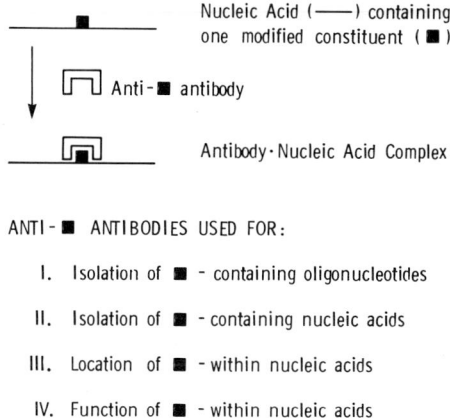

FIG. 1. A schematic representation of the immunochemical approaches for isolating and characterizing nucleic acids.

stituents (a) closely resemble in structure their parent, unmodified nucleotides; and (b) represent only a small percentage of the total nucleotide population within nucleic acids. It is mandatory, therefore, that the modification itself be immunodominant. If the criterion of specificity is not achieved, the usefulness of anti-nucleosides as reagents for the isolation and characterization of nucleic acids becomes rather limited.

The requirement of antibody specificity can be appreciated by examining the types and amounts of modified constituents present in nucleic acid. Such an examination also provides the necessary background to realize the potential of these and other immunochemical approaches.

II. Nucleic Acid Methylation

It is becoming increasingly apparent that most nucleic acids undergo a series of posttranscriptional modifications that result in the appearance of "minor" (rare) or modified nucleotide constituents in mature nucleic acids (20–22, 68a–70). Although numerous types of modification reactions exist, e.g., rearrangement (pseudouridine), reduction (dihydrouridine), deamination (inosine), thiolation (2-thiouridine), alkylation (isopentenyladenosine), etc., the last-named, in the form of methylation by S-adenosylmethionine and methyltransferases (68a–72) is undoubtedly the most common (19–22). While methylation of the ribose moieties is exclusively at the $2'$-position (e.g., $2'$-O-methyluridine), methylation of the bases occurs at several specific carbon, nitrogen, and oxygen atoms (68a). A limited number of nucleosides gain more than one methyl group, involving both the base (m_2^6A, m_2^2G, and $m_3^{2,2,7}G$) and/or ribose moiety (m^6Am, m^4Cm). A list of naturally occurring methylated nucleosides to which antibodies have been prepared appears in Table I.

A. Extent of Nucleic Acid Methylation

The degree to which nucleic acids become methylated varies considerably and appears to be most dependent upon the source and the type of nucleic acid in question, as becomes quite evident when comparing bacterial and mammalian RNAs. Such comparative data appear in Table II, together with data for nucleic acids of viral origin. Generally, the extent of nucleic acid methylation is directly related to the complexity of the biological system (mammalian > yeast > bacterial). Within a given system, the percentage of methylated constituents present in the three major classes of RNA increases from mRNA to tRNA (tRNA > rRNA > mRNA), bacterial mRNA not being methylated (73).

The extent of methylation of DNA appears to be more variable, especially in bacterial systems, where numerous modification and restriction enzyme systems exist (74) and account for the presence of methylated bases

TABLE II
EXTENT OF NUCLEIC ACID METHYLATION IN PROKARYOTIC AND EUKARYOTIC SYSTEMS

Nucleic acid[a]	Methyl groups per 1000 nucleotides	Methyl groups per nucleic acid molecule	References
Eukaryotes			
tRNA (unfx)	90–100	7	70,85–87
tRNA^{Ser_1}		9	85
tRNA^{Val_1}		6	85
rRNA (18 S)	20	40	88–90
rRNA (28 S)	14	70	88–90
mRNA (unfx)	3	6	69
DNA (cellular)	6–12	Variable	91,92
DNA (Ad-2, viral)	0.1–0.2	10–15	82
RNA (RSV, viral)	1.5–2.0	14–18	83,84,93
Prokaryotes			
tRNA (unfx)	30	2.5	30,85–87
tRNA^{Ser_1}		2	85
tRNA^{Val_1}		4	85
rRNA (16 S)	5	15	94,95
rRNA (23 S)	6	9	94,95
mRNA (unfx)	0	0	73
DNA (bacterial)	Variable	Variable	74,96,97
DNA (fd, bacteriophage)	0.3–0.6	2–4	75,76
DNA (ϕX174, bacteriophage)	0.2	1	77,78

[a] Unfx = unfractionated.

in bacteriophages (75–79). While the most common methylated bases in prokaryotic DNAs are 5MeCyt and 6MeAde (79, 80), 5MeCyt appears to be the major (if not exclusive) naturally occurring one in eukaryotic DNA (81). Viral DNA, from both bacteriophages and animal viruses, contain small amounts of 5MeCyt and/or 6MeAde (75–78, 82). In contrast, tumor virus RNAs possess a relatively large number of methylated constituents (83, 84), which approximate to a large extent those found in mammalian mRNA preparations (83).

It is important to note that the data in Table II reflect the average content of methylated constituents for unfractionated (bulk) tRNA and mRNA populations. Their actual number within individual species may vary considerably, as evidenced by the data presented for tRNA$_1^{Ser}$ and tRNA$_1^{Val}$ as well as by reviewing the composition of methylated nucleosides present in previously sequenced tRNAs (21, 22, 85).

B. Methylated Nucleosides of Nucleic Acids

In Table III are displayed the results of various assays, all based on chromatographic and electrophoretic techniques (69, 70, 87, 97), of the methylated nucleoside contents of the three major classes of RNAs from bacterial and mammalian sources. The data for 4 S (tRNAs) obviously reflect a composite of the values for the many discrete tRNA species. However, taken together with a knowledge of the *extent* of methylation (Table II), these data permit conclusions regarding the fraction of specific molecules

TABLE III
DISTRIBUTION (APPROXIMATE) AND SPECIES OF METHYLATED NUCLEOSIDES PRESENT IN BACTERIAL AND MAMMALIAN tRNA AND rRNA[a]

Methylated nucleosides	Percentage of total methylated nucleosides present					
	Bacterial			Mammalian		
	4 S	16 S	23 S	4 S	18 S	28 S
$m^1 A$				12		1.5
$m^2 A$	8.5		6.7			
$m^6 A$	1.5		13		2.5	1.5
$m^6_2 A$		22			5.0	
Am	1.5			3.0	27	30
$m^1 G$	6.0		6.7	8.0		
$m^2 G$		22	13	14		
$m^2_2 G$				8.0		
$m^7 G$	26	11	6.7	7.0	2.5	
Gm	13			9.0	17	25
$m^3 C$				3.0		
$m^4 Cm$		11				
$m^5 C$			13	17		1.5
Cm	2.5		20	2.0	17	22
$m^3 U$			6.7			1.5
$m^5 U$	38	22	6.7	13		
Um	2.5	11	6.7	4.0	25	15
ψm					2.5	1.5

[a] tRNA data are derived from references (69,70,86,87); rRNA data from (88,89,94). Some of the methylated nucleosides listed for 23 S rRNA are questionable (88). The percentages of methylated nucleosides of tRNA are derived from unfractionated populations, whereas those from rRNA are derived from discrete species. Slight differences due to experimental errors, techniques employed, and systems investigated dictate approximate distributions.

containing a particular methylated nucleoside. For example, since unfractionated ("bulk") *Escherichia coli* tRNA contains 2.5 methylnucleosides per tRNA (Table II), and knowing that 1.5 and 26% are m^6A and m^7G, respectively (Table III), the percentage of unfractionated tRNA containing these constituents are 2.5×0.015 or 3.7% for m^6A and 2.5×0.26 or 65% for m^7G. These values are consistent with sequence data (21, 22, 85), which reveal that many *E. coli* tRNAs contain m^7G, but only valine-1 tRNA has m^6A.

1. TRANSFER RNA

Compared to other nucleic acids, tRNAs contain both the greatest variety (Table III) and amount (Table II) of methylated nucleotides. Sequencing (21, 22, 85) reveals that many of the methylated constituents reside at specific locations within the tRNA structure. As illustrated by the composite tRNA molecule presented in Fig. 2, m^1G, m^2G, and m_2^2G appear consistently at the stem of the dihydrouridine (D) loop while other methylated nucleosides reside within that loop (Gm), the variable loop (m^5C, m^7G), and the TΨC loop (m^1A, m^5U). However, the region most modified in both pro- and eukaryotic tRNA is the anticodon loop. This region contains not only a variety of methylated constituents (m^2A, m^6A, Cm, m^3C, Gm, m^1G, m^1I, and Um), but the bulk of other types of modified constituents as well, e.g.,

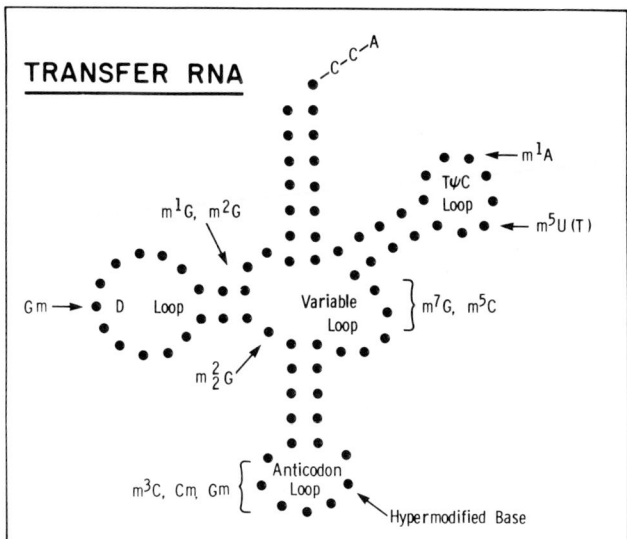

FIG. 2. A composite representation of a tRNA molecule in the two-dimensional "cloverleaf" conformation. As deduced from previously sequenced tRNAs (21, 22), specific locations possessing a high incidence of various modified constituents are presented.

N^6-isopentenyladenosine, inosine, N^6-(N-threonylcarbonyl)adenosine, wyosine (W), queuosine (Q) (21, 22, 65a, 98, 99).

From an immunochemical viewpoint, the importance of the modified nucleosides appearing in the anticodon region cannot be overstressed. In this location, the base moieties of these constituents are projected outward from the surface of the tRNA molecule and are therefore accessible to the combining site of an antibody molecule. However, inspection of the three-dimensional structure of tRNA (100–102) indicates that many of the other modified nucleosides (e.g., the variable loop m^7G) are confined to internal regions of the molecule. While structural considerations may limit the usefulness of specific antibodies to isolate the corresponding hapten-containing nucleic acids, one can still envision their usefulness as probes to evaluate secondary and tertiary structure.

The functional aspects of tRNA methylation are poorly understood, yet a hint of their importance followed the observation of Mandel and Borek (103) that tRNAs synthesized in a methionine auxotroph of E. coli (grown in the absence of methionine) are undermethylated. Subsequent investigations with these undermethylated tRNA species have suggested that certain translational functions are altered or absent as a result of this deficiency, e.g., amino-acid acceptance (104, 105), ribosomal binding (106, 107), and codon recognition and/or translation (108–110). Although these early studies imply impairment, specific functions could not be correlated with the absence of a particular methylated constituent.

Other investigations show that the modified constituents adjacent to the anticodon are essential for translation (111–114), and for the ability of a particular tRNA to recognize a specific codon. They also suggest other roles for modified constituents, such as the regulation of biosynthetic enzymes for specific amino acids (115, 116), the supression of amber mutations (117), the stabilization of the tertiary structure of tRNA (118–120), and the enhancement in the specificity of aminoacyl-tRNA synthetase recognition (121). These and other topics associated with tRNA synthesis, processing, and modification are the subjects of numerous reviews (21, 22, 68a, 98, 99, 122–124).

2. RIBOSOMAL RNA

The second most densely methylated RNA species, also containing a variety of methylated nucleosides (Table III), are the rRNAs of eukaryotes (18 S and 28 S) and prokaryotes (16 S and 23 S) [the low-molecular-weight rRNA species (5 S and 5.8 S) are described elsewhere (125–126a)].

The most prominent methylation characteristic of eukaryotic rRNAs is their high percentage of 2′-O-methylnucleosides (Nm, ca. 90%, see Table III) relative to those present in mRNAs (40–50%) and tRNAs (10–20%). Since numerous oligonucleotides from rRNAs possess two and sometimes three such constituents, they are in clusters at various unknown locations within the 18 S and 28 S rRNA molecules. Also clustered are the two m_2^6A modifications present in 18 S rRNA. Sequencing methods have established their presence at the apex of a proposed loop structure near the 3' terminus (127, 128). An almost identical m_2^6A-containing oligonucleotide sequence is present in bacterial 16 S rRNA (see Fig. 3).

Definitive data as to the functional role(s) of the methylated nucleosides of eukaryotic rRNA are lacking. Yet the importance of rRNA methylation is implied by the following observations. Methylation is restricted to those sequences conserved during the processing of the 45 S rRNA precursor (129) in which 45 S (14,000 nucleotides, 110 methylated nucleotides) is metabolized to the mature 28 S (5000 nucleotides, 70 methylated nucleotides) and 18 S (2000 nucleotides, 40 methylated nucleotides). Second, 45 S rRNA synthesized under methionine-starvation conditions (130), or in the presence of the methionine analog ethionine (131), is rapidly degraded. These results suggest that methylation is important in rRNA processing and/or ribosomal assembly. Related to the above is the hypothesis that the 100 or so 2'-O-methylnucleosides serve to protect rRNA from various nucleases whose mechanism of hydrolysis requires a free 2'-OH group (88). If this hypothesis is correct, it would be expected that the rRNA sequences containing the bulk of Nm constituents would be at the surface of the ribosomal subunits. While this information is not available, it is probable that the m_2^6A-m_2^6A-containing sequence is exposed on the ribosomal surface. The latter concept is based upon data (132) indicating that the methylation of these particular bases occurs during the late stages of 18 S rRNA maturation, i.e., when the 40 S ribosomal subunit is essentially assembled. Similarly, tRNA maturation is accompanied by early, intermediate, and late methylation events (133).

In contrast to eukaryotic rRNAs, the methylated nucleosides of prokaryotic rRNA are considerably less abundant and primarily of the base-methyl type. All but one (m^7G) of the nine methylated constituents of the 16 S RNA appear in the latter one-third of the sequence, from nucleotides 964 to 1541 (134, 135). By treating the 30 S ribosome with nucleases (136, 137) or with anti-m_2^6A antibodies (32, 138), it is possible to identify nucleotide sequences of 16 S rRNA that appear on the surface of the ribosomal subunit. Such investigations have revealed that the largest expanse of surface 16 S rRNA is a region of 175 nucleotides near the 3' end of the molecule (136, 137). This region contains over 50% of the methylated nucleotides present in 16 S rRNA, i.e., m^4Cm (nucleotide No. 1401), m^5C (No. 1405), Um (No. 1497), and m_2^6A-m_2^6A (Nos. 1516 and 1517).

Of particular interest is the sequence at the 3' terminus of 16 S rRNA

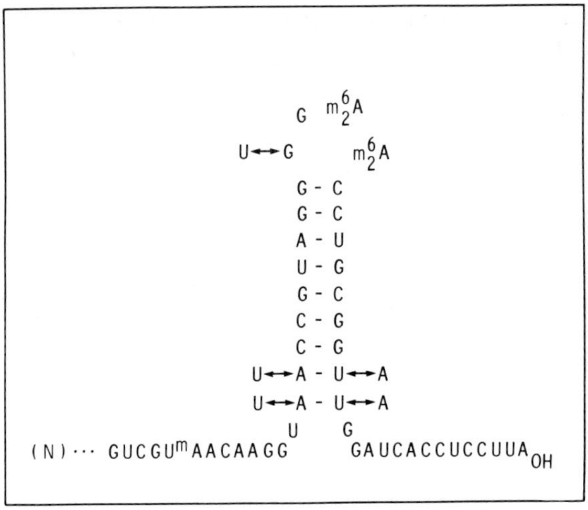

```
         G   m₂⁶A
    U↔G       m₂⁶A
       G - C
       G - C
       A - U
       U - G
       G - C
       C - G
       C - G
     U↔A - U↔A
     U↔A - U↔A
              U   G
(N)···GUCGUᵐAACAAGG   GAUCACCUCCUUA_OH
```

FIG. 3. Secondary structure for the 3' terminus of 16 S rRNA of *Escherichia coli*. The arrows indicate base changes between the *E. coli* (16 S) and rat liver (18 S) rRNAs (*139–141*).

(Fig. 3). Besides illustrating the presence of two adjacent m_2^6A residues, it appears that the oligo(Y) (Y = a pyrimidine nucleoside) sequence adjacent to the 3' terminus, ACCUCCUUA$_{OH}$, base-pairs with a complementary oligo(R) (R = a purine nucleoside) sequence that normally precedes the initiation codon (AUG) of viral and bacterial mRNAs (*139–141*). Accordingly, it has been proposed that the formation of base-pairs between a 5'-oligo(R) tract in the mRNA and a 3'-oligo(Y) tract in the 16 S rRNA aligns the mRNA on the ribosome and thereby discriminates against other internal AUG codons (*142*).

While the role of the dinucleotide m_2^6A-m_2^6A remains unresolved, its proximity to the 3'-terminal poly(pyrimidine nucleotide) sequence implies a translational function. Support for this assumption is provided by the finding that resistance to the antibiotic kasugamycin is associated with bacterial strains lacking these particular modifications (*143*). This indicates that the failure to methylate these particular bases prevents the binding of kasugamycin to the 30 S subunit. However, in bacterial strains sensitive to this antibiotic, protein biosynthesis is inhibited. Other changes in the base-methylation pattern of 23 S rRNA also appear to be associated with antibiotic resistance (*144, 145*).

3. MESSENGER RNA

The existence of methylated constituents in eukaryotic mRNAs (originally thought to be unmethylated) was initially demonstrated by Perry and Kelly (*146*) and by Desrosiers *et al.* (*69*). The success of these investigations was based upon an ability to isolate poly(A)-containing mRNAs free of rRNA contaminants. Employing an *in vitro* transcriptional system, Furuichi (*147*) also documented the methylation of cytoplasmic polyhedron virus mRNA and concluded that the initiation of synthesis of this viral mRNA is a methylation-coupled event. Since these landmark publications (*69, 146, 147*), a growing field of literature on mRNA methylation has accummulated (*148–154*).

As illustrated in Fig. 4, it appears that all poly(A)-containing mRNAs of eukaryotic cells possess a similar overall pattern of methylation: a 5'-terminal, m^7G-containing oligonucleotide, frequently termed "cap," and internal m^6A residues. An mRNA of average size (ca. 18 S, 2000 nucleotides) contains 2 to 6 methylated nucleosides. Generally, larger mRNAs have more internal m^6A residues (ca. 1 m^6A/1000 nucleotides), and smaller mRNAs can be completely devoid of this modification (*151, 152*).

The 5'-terminal cap structure is composed of an m^7G linked through its 5'-carbon by a triphosphate bridge (ppp) to the 5'-carbon of a 2'-O-methylnucleoside (the penultimate nucleoside), sometimes followed by a second Nm constituent (Fig. 4). Further fractionation of the terminal oligonucleotides has revealed considerable heterogeneity in their content of methylated bases (*155–157*); the penultimate nucleoside is usually Am, m^6Am, and Gm, and the adjacent 3'-nucleoside can be any one of the four Nm constituents.

The rapid elucidation of the functional aspects of mRNA methylation was made possible by the findings that many disrupted virions retain their ability to synthesize viral mRNA in both methylated and unmethylated form, the

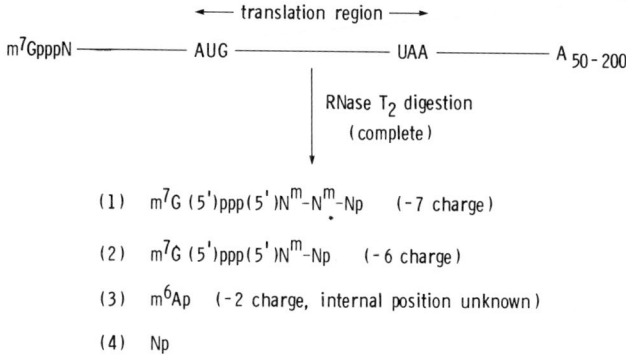

FIG. 4. A schematic representation of a eukaryotic mRNA molecule possessing both an m^7GpppN "cap" and poly(A) tract at the 5' and 3' terminus, respectively. The nucleotide products resulting from RNase T_2 digestion of mRNA are illustrated. The precise location of m^6A constituents is not known (*177–181*).

latter occurring in the presence of S-adenosylhomocysteine (*147, 158–160*).

In an *in vitro* translational system (wheat germ), the translation of capped but unmethylated viral mRNA is significantly impaired contrasted to control, m^{7}G-capped mRNA preparations (*161, 162*). Decreased translational efficiencies are also observed with other mRNA preparations chemically (*163, 164*) or enzymatically (*165*) treated for specific removal of the m^{7}G residues in their cap structures. The inability of these unmethylated or decapped mRNAs to interact with 40 S ribosomal subunits further suggests that 5'-terminal caps provide a recognition signal for subsequent binding of mRNA to the ribosome (*162*), presumably by interacting with one or more initiation factors (*166–168*).

Additional support for such a translational function is provided by the finding that pm^{7}G-containing cap analogs inhibit the translation of capped mRNAs, as well as their binding to the 40 S ribosomal subunit (*169, 170*). However, controversy persists as to the relative importance of the cap in translation. The degree of inhibition observed with uncapped mRNA or with cap analogs is influenced by [K$^+$] (*171*) as well as by the type of *in vitro* translational system employed (*172*). In view of their unusual structure, it is not surprising that cap structures protect mRNAs from 5'-exonucleolytic degradation (*173*). These and other investigations concerning the function, distribution, and metabolism of mRNA cap structures have been the subjects of extensive reviews (*148–150, 154*).

A second interesting feature of mRNA methylation is the appearance of one to four internal m^{6}A residues in most (but not all) mRNAs (*69, 151, 174–181*). Although not present in the 3' end of mRNA, including the poly (A) sequence, their precise location within the molecule remains to be established (*177–181*). Characterization of the m^{6}A-containing oligonucleotides derived from unfractionated mRNA from various cell lines (*174, 175, 181*), SV40 mRNA (*177*), and RSV RNA (*176*) has established a common nucleotide sequence adjacent to m^{6}A, i.e., G-m^{6}A-C and A-m^{6}A-C.

Most recently, Canaani *et al.* (*177*) identified and mapped three m^{6}A-containing sequences present in late SV40 16 S and 19 S mRNAs by hybridization of *methyl*-^{3}H-labeled mRNA to specific restriction fragments. Their results for 16 S mRNA reveal that while all three are present in the coding region, two appear to be clustered at the 5' end. Since the latter are located near the region of the mRNA in which an intervening sequence has been removed (*186, 187*), these observations provide supportive evidence for their potential role(s) in mRNA processing.

4. DNA

The occurrence and biological function of the methylated deoxynucleosides or methylated bases in bacterial DNAs (5MeCyt and 6MeAde) have been extensively studied (188, 189). Findings revealing that bacterial DNA methylases are species-specific (190) and appear to be elevated during phage infection (191) are illustrative of the concept of modification-restriction, i.e., the modification of specific adenine and cytosine DNA residues renders the adjacent sequence, normally sensitive to restriction endonucleases, immune to cleavage (188).

Although spectacular advances have been made in the understanding and exploitation of modification-restriction systems (188, 189) the functional significance of eukaryotic DNA methylation remains to be resolved. It is generally accepted that 5MeCyt is the sole methylated base in eukaryotic DNA (81), representing between 3 and 4% of the total cytosine content (91). Cytosines are methylated at the postreplicate level and invariably found in the sequence N-m^{5}C-G-N (where N is any one of the four deoxynucleosides) (192, 193).

On the basis of this recurring methylated sequence, a series of restriction enzymes have been selected to map the positions of 5MeCyt residues in eukaryotic DNA (194-197). Basically, the technique consists of selecting a pair of endonucleases that have identical sequence specificities (e.g., *Hpa*II and *Msp*I are specific for C-C-G-G), yet differ in their ability to cleave the methylated sequence (C-m^{5}C-G-G; *Hpa*II is refractory, *Msp*I is active). Examining sea urchin DNA in this manner, Bird *et al.* (198) determined that the bulk of its methylated DNA occupies only 40% of the genome. Further, relatively large tracts (15,000-50,000 bases) of both methylated and unmethylated DNA appeared to be specific subfractions of the genome, since enriched unique sequences of one fraction cross-reassociated poorly with the other.

Additionally, when transcriptionally active and repetitious DNA sequences were compared, only the latter contained appreciable quantities of 5MeCyt. These observations suggest (198) that transcriptionally active DNA sequences are not methylated, and are therefore accessible to proteins associated with transcription; transcriptionally inert sequences are inactive by virtue of being methylated. Consistent with this hypothesis are the recent findings (199) that nonsecreting tumor-cell lines induced by Herpesvirus saimiri contain viral DNA sequences that are extensively methylated, while the viral DNA present in other actively secreting cell lines is unmethylated. Although equivocal, these observations imply that methylation may inhibit the expression of specific genomic information required for complete viral assembly and secretion (199).

It is anticipated that the continued employment of such restriction enzymes as well as the potential use of anti-m^{5}C[1] to map accurately the position of 5MeCyt residues in DNA should provide interesting and relevant information regarding the biological function of eukaryotic DNA methylation.

III. Immunochemical Procedures

The major steps involved in the production of anti-nucleoside antibodies are presented in Fig. 5. As illustrated by the methylated nucleoside m^6A, they consist of (I) the synthesis of the immunogen, (II) the induction of the anti-nucleoside antibody, (III) the characterization of the antiserum by radioimmunoassay to evaluate antibody concentration, specificity, and affinity, and (IV) the purification and, in certain instances, the immobilization of the antibody on Sepharose adsorbents.

The purpose of this section is to describe the critical aspects of the steps outlined above. While the selection and employment of a particular antibody is dependent on the background, interest, and imagination of each investigator, the descriptions provided herein will assist in the preparation of antibody populations that possess the necessary prerequisites of specificity, affinity, and purity.

A. Synthesis and Characterization of Nucleoside–Protein Conjugates

The chemistry involved in the coupling of various low-molecular-weight organic compounds to protein carriers has been reviewed (200). Although there are numerous synthetic schemes for preparing nucleoside- and nucleotide–protein conjugates, the majority of these types of immunogens have been prepared by periodate oxidation techniques (12, 47). The almost universal acceptance of this procedure, as originally described by Erlanger and Beiser (12), is based upon (a) the simplicity of the reactions involved; (b)

I. SYNTHESIS OF ANTIGEN
m^6A + BSA $\longrightarrow$ m^6A-BSA

II. PRODUCTION OF ANTIBODY
m^6A-BSA + Rabbit $\longrightarrow$ anti-m^6A antisera

III. CHARACTERIZATION OF ANTIBODY (RIA)
anti-m^6A antisera + $[^3H]$ m^6A $\longrightarrow$ anti-m^6A antibody·$[^3H]$ m^6A

IV. PURIFICATION AND IMMOBILIZATION OF ANTIBODY
A. anti-m^6A antisera + m^6A-BSA/Sepharose $\longrightarrow$
anti-m^6A antibody·m^6A-BSA/Sepharose

B. anti-m^6A antibody + CNBr/Sepharose $\longrightarrow$
anti-m^6A antibody/Sepharose

FIG. 5. A schematic outline of conventional immunochemical protocols employed to obtain anti-nucleoside antibodies as represented by the nucleoside m^6A. BSA, bovine serum albumin; RIA, radioimmunoassay.

the variety of nucleoside and nucleotide constituents amenable to this form of conjugation; (c) the degree to which the various haptens can be conjugated per molecule of carrier protein; and (d) the lack of significant degradation or modification of the base moiety during conjugation.

The major features of the periodate coupling reaction are presented in Fig. 6. The reaction takes advantage of the vicinal hydroxyl groups contained in the ribose moiety of both nucleosides and nucleotides. Oxidation of these hydroxyl groups with periodate (IO_4^-) yields a nucleoside-dialdehyde structure that readily undergoes condensation with primary amino groups, e.g., the ϵ-NH_2 group of lysine, present in the carrier protein. The resulting Schiff-type bases are reduced into more stable compounds by treatment with borohydride. Alternatively, direct reduction of the nucleoside dialdehyde with [^{3}H]borohydride results in the production of a ^{3}H-labeled nucleoside trialcohol derivative (86). These latter constituents serve as useful ^{3}H-labeled haptens in sensitive radioimmunoassays and are discussed below in connection with the specificity of selected anti-nucleoside antibodies.

The extent to which a particular nucleoside can be conjugated in the

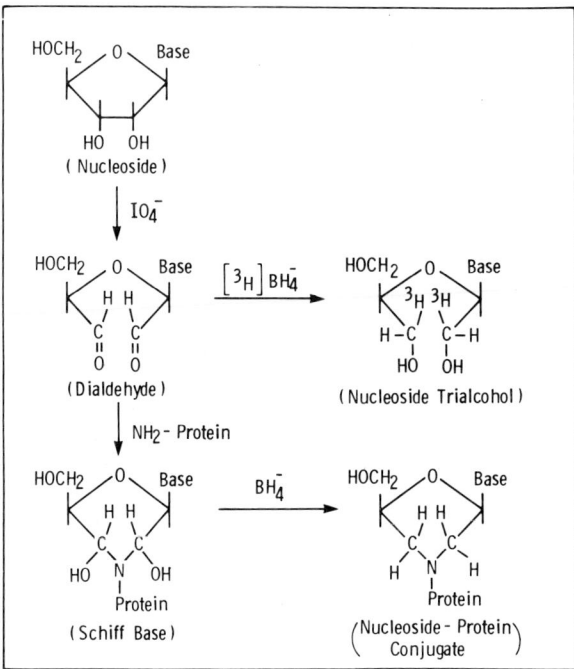

FIG. 6. A schematic representation of the conjugation reaction whereby nucleosides are coupled to protein carriers (12). The dialdehyde can be directly reduced with [^{3}H]BH_4^- to the [^{3}H]nucleoside trialcohol (86) for use in sensitive radioimmunoassays.

above reaction scheme is dependent upon the concentrations of nucleoside and carrier protein employed (12, 38, 68). Carrier proteins such as bovine serum albumin make excellent immunogens since they are very antigenic by themselves and contain a considerable number of lysine residues (bovine serum albumin has 56). Keyhole limpet hemocyanin and the serum albumins derived from human, bovine, and rat all have been used with considerable success (12, 44, 68). Our experience, however, has been confined to the bovine albumin, which generally incorporates 15 to 35 hapten residues when equal quantities (masses) of nucleoside and albumin are employed (28). Higher levels of conjugation can be expected by varying the amount of nucleoside (increase) and/or protein (decrease). At the conclusion of the conjugation reaction (borohydride stabilization), the protein conjugates are dialyzed overnight against distilled water, lyophilized, and stored (desiccated) at 4°C. They have remained stable under these storage conditions for over three years.

The importance of characterizing both the reactants (nucleoside and carrier protein) and product (nucleoside–protein conjugate) by spectral analysis is emphasized because immunization schedules are rather lengthy and the immunogens themselves are sometimes employed to determine antibody concentration and specificity (microquantitative precipitin assays). Coupled with a knowledge of the appropriate extinction coefficients (201), such analyses provide the necessary information to determine the degree of nucleoside conjugation as well as whether the base portion of the molecule has become altered during the coupling reaction (28, 45, 68). While the importance of the latter parameter is self-evident, the former provides some assurance as to the reliability of the antibody titer, as protein carriers possessing low levels of hapten (e.g., less than 10 nucleosides per albumin) are usually poor immunogens (low quantities of antibody, i.e., 100 μg per milliliter of antiserum). In addition, low levels of conjugation can imply that one of the reagents (borohydride, periodate) is of poor quality, or that an error in the synthesis protocol has been made.

In view of the above, and the instability of some nucleosides at various pHs (m^1A, m^6A, m^7G), we routinely determine the spectra of all nucleosides and nucleoside–albumin conjugates we prepare (28). Our results indicate that the periodate conjugation conditions described (12) cause no detectable alteration in the base moieties of m^6A, m^1I, m^1G, m^2G, m^3C, m^5C, or m^5U. However, m^1A undergoes rearrangement to m^6A (202). Recently, Meredith and Erlanger described a modified coupling procedure that appears to be more amenable to such pH-sensitive nucleosides (45). They also recommend the use of nucleoside 5'-phosphates, especially pm^7G, on the basis of stability at higher pH (45, 204).

B. Immunization

There are as many, or perhaps more, immunization protocols as there are investigators in the field. An excellent description of general immunization procedures and the logic for them have been reviewed (203). Below is a brief description of the protocol that has provided us with very satisfactory antibody titers.

Nucleoside–albumin conjugates are dissolved in phosphate-buffered saline (P·/NaCl) (150 mM NaCl, 10 mM phosphate, pH 7.4) to a final concentration of 5 mg/ml. This antigen is prepared for injection in the form of a stable emulsion by mixing equal volumes of complete Freund's adjuvant and conjugate. New Zealand White rabbits (female) are inoculated twice (at day 1 and day 8) with an equivalent of 1–2.5 mg of antigen per animal at multiple subcutaneous sites. One week after the second inoculation (day 15) a small sample of blood (2–5 ml) is withdrawn from each animal and processed for the presence of specific antibodies by microquantitative precipitin analysis (28) and/or by radioimmunoassay (see below). Thereafter, rabbits possessing antibody are boosted at intervals of 6–8 weeks with a single dose of antigen prepared as before, but emulsified in incomplete Freund's adjuvant and injected at multiple subcutaneous sites. About 40 ml of blood are withdrawn (cardiac puncture) 7–10 days after each booster immunization, and the resulting antiserum is stored at −20°C until processed.

Using this protocol, we obtain yields of specific antibody in the range of 0.5 to 2 mg/ml of antiserum. It is important to note that continued boosting at regular intervals of 6–8 weeks promotes significant increases in antibody affinity (205, 206). In most instances, three rabbits are used per antigen. Those yielding unsatisfactory responses after initial immunization are discarded.

C. Characterization of Antibodies Specific for Nucleosides by Radioimmunoassay

The three most common procedures for characterizing anti-nucleosides[1] consist of complement-fixation techniques (17, 27), quantitative precipitin analyses (17, 28, 207), and, more recently, radioimmunoassays (43, 48). While each has its advocates, radioimmunoassays (RIA), because of their sensitivity, speed, and reliability, are being used with increasing frequency to characterize anti-nucleosides. The delay in the introduction of this technique can be attributed primarily to the lack of techniques for obtaining radioactively labeled methylated nucleosides of high specific activity. Several of the methylated nucleosides (m_2^2G and m^7G) have been prepared by tritium substitution of their 8-bromo derivatives (43) and employed in sensitive radioimmunoassays to evaluate the specificity of anti-m$_2^2$G and -m^{7}G antisera preparations (43). Similar procedures have been described recently for other nucleosides (32, 44, 45) and various carcinogen-induced nucleoside adducts (55–57).

Perhaps as convenient a method as any for preparing tritiated derivatives of nucleosides or nucleotides that undergo periodate oxidation is the [^{3}H]borohydride procedures introduced by Randerath (86) for determining the composition of methylated base constituents of various RNA preparations. The method is convenient because [^{3}H]trialcohol derivatives can be prepared simultaneously with the nucleoside–protein conjugates (see Fig. 6) by excluding the carrier protein and modifying (slightly) the reaction conditions described by Randerath (86). We have used this procedure with considerable success, having prepared the [^{3}H]nucleoside trialcohols of m^{5}C, m^{7}G, and m^{6}A with specific activities ranging from 1 to 2 Ci/mmol (i.e., 5 to 10 × 10^3 dpm/ng), and with purities judged to be greater than 95% as evaluated by thin-layer chromatographic techniques (208). To distinguish such trialcohol (ta) derivatives from their original nucleosides, they are referred to as ta[^{3}H]m^{6}A, ta[^{3}H]m^{7}G, etc., throughout the remainder of the text.

A number of procedures are available for determining the binding of ta[^{3}H]nucleosides to their respective antibodies, the only prerequisite being the ability to separate the ta[^{3}H]nucleoside anti-nucleoside complex from unbound ta[^{3}H]nucleoside. Procedures to accomplish this include filtering on filters that selectively retain protein (209), equilibrium dialysis (210), and precipitation with ammonium sulfate (211). In our laboratory, comparable results have been obtained with each procedure when identical incubation conditions are employed. For the most part, however, we have selected the ammonium sulfate precipitation procedure in view of its ease when handling a large number of samples. A brief summary of this procedure appears in the legend of Fig. 7.

The ability of the above procedures to discriminate between bound and free hapten permits one to determine not only the concentration and specificity of the antibody, but its affinity for the hapten constituent as well.

1. Antibody Concentration

Measurements of antibody concentration are performed by determining the point at which increasing amounts of [^{3}H]hapten do not increase its binding to the antibody, i.e., the point at which the antibody combining sites are saturated with respect to hapten. This condition can be met by incubating a small amount of antibody with increasing amounts of [^{3}H]hapten and measuring the amount of bound hapten after the reactants and products have achieved equilibrium. Equilibrium in these systems is rapid, and is usually attained within an hour at either 4° or 37° (*48*).

The data obtained from such experiments are graphically presented in Fig. 7 in which the amount of bound hapten is plotted against the amounts of hapten used in each assay. Thus, as shown in Fig. 7A, the amount of ta[^{3}H]m^{6}A bound to anti-m^{6}A continues to increase until a plateau value is reached (0.45 ng or 1.6 pmol). This plateau value represents the point at which all of the anti-m^{6}A combining sites are saturated with ta[^{3}H]m^{6}A, since a fivefold increase in the amount of ta[^{3}H]m^{6}A (10 ng vs 50 ng) does not increase its binding to the antibody. Because each antibody molecule has *two* combining sites and a molecular weight of 150,000, the above data indicate the presence of 0.8 pmol of antibody per 0.10 μl of antiserum (or

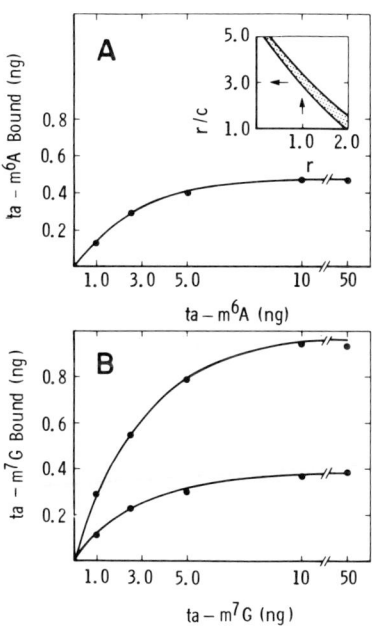

FIG. 7. The determination of antibody concentration and affinity by radioimmunoassay. Reaction conditions: in panel A, the equivalent of 0.1 μl of anti-m^{6}A antiserum was incubated (1 hour at 37°C, overnight at 4°C) in the presence of increasing quantities of ta[^{3}H]m^{6}A (2200 cpm/ng), in a total volume of 0.3 ml in 150 mM NaCl, 10 mM phosphate, pH 7.4; in panel B, same as above with 0.1 and 0.25 μl of anti-m^{7}G antiserum and ta[^{3}H]m^{7}G (3000 cpm/ng). The extent of [^{3}H]hapten binding is measured by selectively precipitating the [^{3}H]hapten·antibody complex with (NH$_4$)$_2$SO$_4$ (*211*). Graphs of the amount of [^{3}H]hapten incubated (abscissa) vs the amount bound (ordinate) yield both antibody concentration and affinity. A Scatchard plot of this and other anti-m^{6}A data is presented in the inset of panel A (for further details refer to discussion in Section III,C); ta = trialcohol.

1.20 mg/ml antiserum). A comparable value has been obtained by using the same antiserum in microquantitative precipitin analysis, i.e., 1.25 mg/ml (data not shown). Besides measuring the concentration of anti-m^7G, Fig. 7B illustrates that the maximum binding of ta[^{3}H]m^7G to anti-m^7G is directly proportional to the concentration of antibody. In this particular study, 0.1 and 0.25 µl aliquots of anti-m^7G antiserum bound 0.38 and 0.95 ng of ta[^{3}H]m^7G, respectively.

2. Antibody Affinity

By measuring the degree of hapten binding to a constant amount of antibody over a wide range of hapten concentrations (see Fig. 7), the antibody's affinity for the hapten component can be determined. This is accomplished by arranging the data in the form of a Scatchard plot (212, 213) derived from the equation

$$r/c = nK - rK$$

where r is the moles of bound hapten per mole of antibody, c is the concentration of free hapten, n is the number of antibody combining sites, and K is the affinity of the antibody for its hapten constituent. A plot of r/c versus r using the data presented in Fig. 7 for anti-m^6A antibodies results in a graph of slope K and intercept n (see inset, Fig. 7). This yields both antibody affinity and the maximum number of antibody combining sites (i.e., n, the intercept = 2.0). The slope is usually not a straight line because antisera generally contain heterogeneous populations of antibodies of different affinities, thus necessitating terms such as "average" or "approximate" affinity. Since the antibody has two combining sites, the above equation reduces to $r/c = 2K - rK$; and when $r = 1.0$, then $K = 1/c$.

Thus, the affinity of anti-m^6A can be readily estimated by extrapolation of the Scatchard plot that is presented in the inset of Fig. 7A, i.e., $K = 4 \times 10^7$ M^{-1}. The latter equation ($K = 1/c$, when $r = 1.0$) also provides one with a definition of antibody affinity, the reciprocal of the free hapten concentration when half the antibody combining sites are occupied. Therefore, an approximation of affinity without Scatchard plot analysis can be obtained from the equation $K = 1/c$ and data such as those presented in Fig. 7. When half of the anti-m^7G sites are occupied with ta[^{3}H]m^7G, the concentration of the free hapten can be determined and yields an affinity of 5.0×10^7 M^{-1} for this particular anti-m^7G population.

3. Antibody Specificity

Another important aspect of these assays is their usefulness in determining the specificity of the anti-nucleoside antibodies. Specificity assays are performed simply by measuring the ability of various unlabeled nucleoside (nucleotide) constituents to compete with, and thus inhibit, the binding

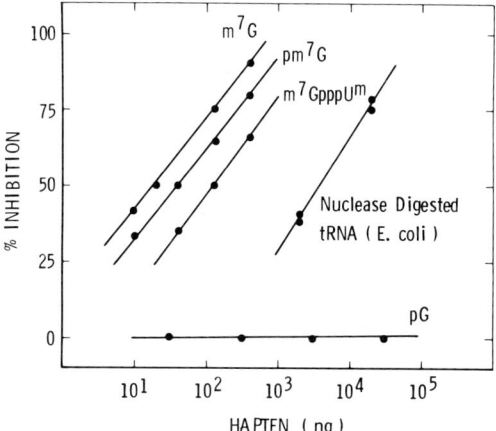

FIG. 8. Anti-m⁷G specificity, evaluated by the ability of selected unlabeled haptens to inhibit the binding of ta[^{3}H]m⁷G to anti-m⁷G antibodies. The reaction conditions were identical to those described in Fig. 7, except that 1.0 μl of anti-m⁷G antisera was incubated with 10 ng of ta[^{3}H]m⁷G (30,000 cpm). Control studies (without competing haptens) revealed that 20.5% (6150 cpm) of the ta[^{3}H]m⁷G was precipitated with the antibody, i.e., 0% inhibition. The types and amounts of competing haptens employed in these assays are presented in the figure. *Escherichia coli* tRNA was digested into 5'-mononucleotides by nuclease P$_1$ prior to addition (28).

between the (labeled) hapten and its corresponding antibody. The reaction conditions employed for these assays are described in the legend of Fig. 8, which presents the results of competition assays between various unlabeled hapten-containing constituents and ta[^{3}H]m⁷G.

These data demonstrate that m⁷G, its 5'-phosphate (pm⁷G), and other m⁷G-containing oligonucleotides (mRNA "cap" structures) effectively compete with ta[^{3}H]m⁷G for anti-m⁷G, while pG is ineffective in this regard. The actual amount of competing hapten required to inhibit the binding of ta[^{3}H]m⁷G by 50% is presented in Table IV, and again reveals that m⁷G-containing constituents are highly competitive haptens. The small differences in inhibition observed between [ta]m⁷G and other haptens containing m⁷G (2- to 5-fold) are attributed to small structural differences in the ribose moieties of the latter. These differences are not considered significant in view of the high affinity of the anti-m⁷G. Most significant, however, is the lack of inhibition observed with 10,000-fold molar excesses of nonmethylated nucleotides. These results indicate that this anti-m⁷G preparation does not cross-react to any detectable extent with the bulk of unmodified constituents present in RNA. This observation is of utmost importance when attempting to form a specific complex between an anti-m⁷G and an m⁷G-containing RNA in which the latter may contain as many as 10^3 to 10^4 G residues. Similar

TABLE IV
INHIBITION OF BINDING OF THE TRIALCOHOL DERIVATIVE
OF 7-METHYLGUANINE TO 7-METHYLGUANINE ANTISERUM
BY UNLABELED COMPETING HAPTENS[a]

Competing haptens	Amount required for 50% inhibition (pmol)[b]	Amount relative to ta[^{3}H]m^{7}G (mol/mol)
[ta]m^{7}G	32	1.0
m^{7}G	70	2.2
pm^{7}G	109	3.4
m^{7}GpppU^m	134	4.2
m^{7}GpppG^m	126	3.9
GpppG	[c]	$>10^4$
tRNA[d]	179	5.6
pA	[c]	$>10^4$
pC	[c]	$>10^4$
pG	[c]	$>10^4$
pU	[c]	$>10^4$

[a] Reaction conditions were those described in the legend of Fig. 8.

[b] 32 pmol of ta-m^{7}G (the trialcohol of m^{7}G) = 10 ng.

[c] No inhibition detected at $>10^4$ molar excess of competing hapten.

[d] tRNA from *Escherichia coli* H previously digested to 5′-mononucleotides with nuclease P$_1$. Picomoles of tRNA required for 50% inhibition were calculated from a molecular weight estimate of intact tRNA, i.e., 25,000. Similar results have been obtained with anti-m^{6}A antiserum.

results have been obtained with m^{6}A-containing haptens and anti-m^{6}A antiserum preparations. Most important was the finding that a 10^4 molar excess of pA relative to [ta]m^{6}A was ineffective as a competing hapten.

The inability of unmethylated nucleotide constituents to inhibit these binding assays implies that the inhibition observed with P$_1$ nuclease-digested tRNA (Fig. 8, Table IV) reflects its content of pm^{7}G. By comparing the amount of digested tRNA (based upon the molecular weight of intact tRNA, 25,000) required to achieve 50% inhibition (179 pmol) with that of the pm^{7}G standard (109 pmol), one can determine that approximately 61% (109/179) of these tRNAs contain pm^{7}G. This value is in excellent agreement with that reported (65%) in connection with the data presented in Tables II and III. These observations illustrate the well-known fact that RIAs are not restricted to assessing the concentration, affinity, and specificity of the antibody molecule.

D. Purification of Antibodies Specific for Nucleosides

Although specificity and affinity are the most significant prerequisites when employing anti-nucleoside antibodies to isolate and characterize nucleic acids, they become meaningless when antibody preparations are contaminated with serum nucleases. Furthermore, since the concentration of a particular anti-nucleoside may represent as little as 0.2% (usually 0.5 to 2%) of the total serum protein, a number of purification schemes must be adopted for the isolation of either a total IgG fraction, or if need be, an individual antibody population. Conventional procedures for the isolation of unfractionated IgG have been described in detail and include ammonium sulfate precipitation (214), DEAE-cellulose chromatography (215), and, more recently, affinity chromatographic techniques employing protein A-coupled Sepharose adsorbents (216). The latter are based upon the ability of protein A to bind selectively to the F_c region of the IgG molecules obtained from a variety of sources.

Affinity chromatography for the isolation of a specific antibody population employs an adsorbent to which the homologous antigen has been immobilized. We have frequently used this procedure for the isolation of anti-m^6A and anti-m^7G with m^6A- and m^7G-albumin/Sepharose (28). Antibody populations retained by such antigen–Sepharose adsorbents can be dissociated by a number of reagents including acetic acid (0.1 to 1.0 M), urea (4 to 8 M), and various chaotropic ions such as SCN^-, ClO_4^-, and I^- (217). In many instances, both IgG preparations and purified antibodies are in turn coupled to Sepharose, and these immunoadsorbents are utilized for the isolation of antigens and/or hapten-containing constituents (30, 63, 65). The use of such antibody-coupled adsorbents for the isolation of hapten-containing oligonucleotides and various RNA molecules is presented in the next section.

Although the affinity chromatographic techniques described above should (in theory) remove nuclease contaminants, it is recommended that additional steps be taken for their inactivation or complete removal. We have had considerable success in this regard by adopting the combined DEAE- and CM-cellulose technique of Palacios et al. (218). It should be noted, however, that complete removal of detectable nuclease activity is accompanied by a 20–40% loss in the antibody population. Other procedures reported include the use of charcoal (219) and alum (45) to bind residual nucleases present in antibody preparations. Whatever the method employed, all solutions and glassware should be autoclaved or dry-heat sterilized (200°C) to prevent introducing exogenous nucleases.

IV. Immunochemical Isolation of Oligonucleotides and Nucleic Acids Possessing Specific Modified Constituents

The use of immobilized anti-nucleoside antibodies as a means of fractionating RNA populations on the basis of their composition of modified nucleotides has been documented (see Table V). Noting that only a small fraction of *E. coli* tRNAs contain inosine (I), Inouye *et al.* (220) immobilized anti-I antibodies to Sepharose and isolated two I-containing tRNA species

TABLE V
IMMUNOCHEMICAL ISOLATION OF tRNA[a]

Antibody	tRNAs isolated	References
Anti-nucleoside(tide)		
A. Anti-I	*E. coli* tRNA$_1^{Arg}$	68,220
	E. coli tRNA^{Pro} (minor species)	
B. Anti-W[b]	Yeast tRNA^{Phe}	65,65b,221
	Rat liver tRNA^{Phe}	
C. Anti-m^{6}A	*E. coli* tRNA$_1^{Val}$	28,30
D. Anti-i^{6}A	*E. coli* tRNA^{Ser}	62,63
	E. coli tRNA^{Leu}	
	E. coli tRNA^{Phe}	
	E. coli tRNA^{Trp}	
	E. coli tRNA^{Tyr}	
Anti-tRNA		
E. Spontaneous[c] NZB/NZW mice	*E. coli* tRNA unfx[e] Rat liver tRNA unfx	219,222
F. Anti-tRNA[d]	*E. coli* tRNA unfx Yeast tRNA unfx	223

[a] Immunochemical isolation of tRNAs by anti-nucleoside (A–D) and anti-tRNA (E,F) antibodies.

[b] Anti wye-antibodies were experimentally induced with yeast tRNA^{Phe} conjugated to bovine γ-globulin, the main determinant being directed toward the highly modified wybutine.

[c] Spontaneously occurring antibodies present in the serum of NZB/NZW mice.

[d] Anti-tRNA antibody experimentally induced by *Escherichia coli* tRNA–γ-globulin conjugate.

[e] unfx = unfractionated.

($tRNA_1^{Arg}$ and a minor $tRNA^{Pro}$ species) by affinity chromatographic techniques. Based upon this success, antibodies that recognize such minor constituents as yW (wybutosine), m^6A, and i^6A have been prepared and employed to isolate specific tRNAs that possess the corresponding hapten constituent (Table V). Besides recognizing their homologous hapten, both anti-yW and -i^6A antibodies cross-react to a considerable extent with structurally related derivatives of these constituents (63, 65, 65b, 221), i.e., peroxywybutosine and the 2-methylthio derivative of i^6A (ms^2i^6A). As a result, anti-yW antibodies are capable of recognizing both yeast $tRNA^{Phe}$ (wybutine) and rat liver $tRNA^{Phe}$ (peroxywybutine) isoaccepting species, while anti-i^6A antibodies have been employed to characterize tRNA populations containing either i^6A or ms^2i^6A (63). Contributing to the successful immunochemical isolation of these tRNAs is the collective fact that each of the hapten constituents is located in the anticodon loop.

A distinctly different type of anti-tRNA antibody can be experimentally induced with tRNA–protein conjugates (223). These and similar anti-tRNA antibodies present in the serum of mice with autoimmune traits (219), appear to recognize various conformational features of tRNA. This observation is based upon the findings that RNase digests of tRNA and/or nucleosides themselves are not effective as competing haptens (222, 223). Further, denatured tRNA substrates are only poorly recognized while other RNAs possessing secondary structures (based-paired regions) appear to cross-react to variable extents with these anti-tRNA antibodies. Although such antibodies are of potential interest in examining RNA conformation, they are not suitable for the isolation of specific tRNA populations.

The information presented below will serve to illustrate various methodologies and techniques for examining the ability of anti-nucleoside antibodies to isolate specific hapten-containing oligonucleotodes and RNAs.

A. Isolation of Nucleotide Sequences Containing N^6-Methyladenine

Anti-m^6A antibodies, selected on the basis of their high specificity and affinity (see preceding section, Table IV, Figs. 7 and 8), were purified by affinity chromatography using both albumin Sepharose (to remove anti-albumin antibodies) and m^6A-albumin/Sepharose adsorbents. Purified anti-m^6A was, in turn, coupled to Sepharose, and this immunoadsorbent was assessed for its ability to retain specific hapten-containing oligonucleotides and tRNAs. The latter were derived from unfractionated preparations of labeled *E. coli* tRNAs. These tRNAs were chosen for studies because they (a) are readily isolated free of other RNA populations; (b) can be labeled *in vivo* to high specific activities with [$methyl$-^{3}H]methionine; and (c) possess the labeled hapten under consideration, i.e., [Me-^{3}H]m^6A.

1. AFFINITY CHROMATOGRAPHY

Initially, our strategy was to evaluate the effectiveness of anti-m⁶A/Sepharose adsorbents to retain increasingly larger m^6A-containing nucleotide sequences. To accomplish this, aliquots of unfractionated tRNA were digested with nuclease P_1 (mononucleotides) or RNase-T_1 (oligonucleotides). These digests, together with intact tRNA, were incubated with anti-m⁶A/Sepharose (30 minutes at 24°C). The adsorbent was then washed repeatedly before the immunospecifically retained radioactivity was quantitatively eluted with 7 M urea. (for details, see legend of Table VI). The results are presented in Table VI, and indicate that approximately 1.5, 1.5, and 5.2% of *methyl*-³H-labeled mononucleotides, oligonucleotides, and tRNA were retained by anti-m⁶A/Sepharose.

To determine whether the bound nucleotide sequences were immunospecifically adsorbed, identical studies were conducted using the competing nucleotide haptens pA and pm⁶A. These results, also presented in Table VI, revealed that 10 μg of pm⁶A completely abolished the ability of the immunoadsorbent to retain the radioactivity. However, even a 50-fold molar excess of pA (relative to that of pm⁶A) had no effect in this regard. These

TABLE VI
IMMUNOSPECIFIC RETENTION OF MONONUCLEOTIDES, OLIGONUCLEOTIDES, AND tRNAs WITH ANTI-m⁶ A/SEPHAROSE ADSORBENT[a]

[*methyl*-³H]Methylated constituents	Incubated with adsorbent (cpm)	Retained by Adsorbent	
		Cpm	%
Mononucleotides, pN			
pN	88,000	1230	1.4
pN + pA (500 μg)	88,000	1310	1.5
pN + pm⁶A (10 μg)	88,000	<50	<0.1
Oligonucleotides, p(N)x			
p(N)x	96,000	1560	1.6
p(N)x + pA (500 μg)	96,000	1350	1.4
p(N)x + pm⁶A (10 μg)	96,000	<50	<0.1
tRNA			
tRNA	120,000	5920	4.9
tRNA + pA (500 μg)	120,000	6250	5.2
tRNA + pm⁶A (10 μg)	60,000	<50	<0.1

[a] Details have been described elsewhere (28–30). These data are presented with the permission of the American Chemical Society and the American Society of Biological Chemistry.

findings imply that the binding of methylated nucleotides, oligonucleotides, and tRNA is immunospecific and in accord with the specificity data obtained for anti-m^6A as derived from radioimmunoassays.

2. ANALYSIS OF METHYLATED NUCLEOSIDES

To provide unequivocal evidence that anti-m^6A/Sepharose adsorbents were retaining m^6A-nucleotide sequences, both the nonretained and retained fractions were processed (desalted and enzymatically degraded to nucleosides, which were separated by chromatographic techniques; see 28–30) to determine both the types and amounts of methylnucleosides present. These data are presented in Table VII and show that while $[Me\text{-}^3H]m^6A$ represents 1.3% of the total labeled nucleosides present in tRNA, their absence in the appropriate nonretained fractions suggests that those mono- and oligonucleotides possessing m^6A were quantitatively retained. This was demonstrated directly by the finding that more than 95% of the radioactivity retained by the immunoadsorbent was $[Me\text{-}^3H]m^6A$.

Similar analysis of the tRNA fractions again revealed the absence of

TABLE VII
IMMUNOSPECIFIC RETENTION OF METHYLATED-NUCLEOTIDES, -OLIGONUCLEOTIDES, AND -tRNA BY ANTI-m^6 A/SEPHAROSE ADSORBENT. DISTRIBUTION AND IDENTIFICATION OF METHYLATED NUCLEOSIDES[a]

Methylated nucleosides	Unfx[b] tRNA	Percentage of methylated nucleosides in					
		Mononucleotides		Oligonucleotides		tRNA	
		NR	R	NR	R	NR	R
m^2A	8.5	8.3	ND	8.5	ND	9.0	ND
m^6A	1.3	ND	>95	ND	>95	ND	27
m^1G	6.0	6.2	ND	5.9	ND	6.4	ND
m^7G	26	26	ND	26	ND	27	23
m^5U	39	41	ND	40	ND	39	25
N^m	19	19	ND	20	ND	19	24 (Gm)

[a] Nonretained (NR) and immunospecifically retained (R) fractions obtained from anti-m^6 A/Sepharose adsorption of [methyl-^{3}H]methionine-labeled tRNA, of nuclease P_1 (mononucleotides) and RNase T_1 (oligonucleotides) digests of tRNA processed for identification of [methyl-^{3}H]methylnucleosides as previously described (28–30). Parts of these data are presented with the permission of the American Chemical Society and the American Society of Biological Chemistry.

[b] Unfx = unfractionated. ND represents not detected, or less than 0.2% of the applied radioactivity. Nm represents radioactivity incorporated into 2'-O-methylnucleosides.

[Me-^{3}H]m^6A in the nonretained fraction and the presence of equimolar amounts of [Me-^{3}H]m^6A, -m^7G, -m^5U, and -Gm (Table VII). These results imply that only those tRNA species possessing m^6A are retained by anti-m^6A/Sepharose. The almost equimolar quantities of the four methylated constituents present in this retained fraction further suggest the presence of a single tRNA species.

In contrast, our inability to isolate m^7G-containing tRNAs with anti-m^7G/Sepharose adsorbents (data not shown) suggests that this hapten constituent is inaccessible to anti-m^7G antibodies. Examination of the three-dimensional structure of tRNA (102, 226) supports this observation since the m^7G (nucleoside No. 46) of tRNA^{Phe} is in a triplet with nucleosides 13 (C) and 22 (G). Thus, not only can selected antibodies be employed for the isolation of specific tRNA species, but also they can confirm and extend other findings regarding various aspects of tRNA secondary and tertiary structure.

3. Characterization of N^6-Methyladenine-Containing Oligonucleotides and tRNA

On the basis of the data presented in Table VI, it was concluded that anti-m^6A/Sepharose quantitatively and exclusively retained nucleotide sequences containing m^6A. However, these data provide no information as to the nature of the oligonucleotide(s) and tRNA(s) retained, or the possibility that significant degradation had occurred during the adsorption process. To this end, studies were initiated to characterize both the oligonucleotide(s) (DEAE-cellulose chromatography) and tRNA(s) (two-dimensional polyacrylamide gels) retained by the immunoadsorbents.

The results obtained by chromatographing both an unfractionated and the [Me-^{3}H]m^6A-containing oligonucleotide(s) are presented in Fig. 9. Although five peaks of radioactivity were observed in the unfractionated preparation, only the late-eluting peak was retained by anti-m^6A/Sepharose. Subsequent analysis of the methylnucleoside composition of these peaks of radioactivity indicated that the labeled m^6A was exclusively present in fractions 40 and 41 (Fig. 9). These results confirmed the ability of anti-m^6A/Sepharose to select only those oligonucleotides possessing m^6A, while further indicating the lack of significant degradation during the adsorption process. In addition, the isolation of a single m^6A-containing oligonucleotide provides additional evidence that only one m^6A-containing tRNA is present in these unfractionated preparations.

To examine this possibility further, tRNA obtained from $E.$ $coli$ cells previously labeled with ^{32}P were subjected to the immunospecific adsorption procedures described above and subsequently characterized by two-dimensional gel electrophoreses. Reproduced in Fig. 10 are autoradiographs of these gels. Whereas unfractionated tRNA could be resolved into a number

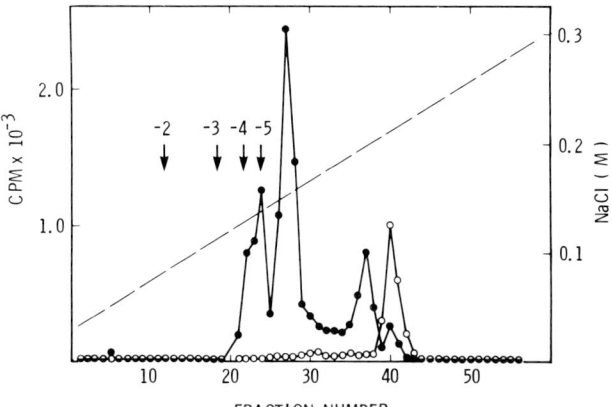

FIG. 9. DEAE-cellulose characterization of unfractionated (●———●) and anti-m^6A/Sepharose adsorbed (○———○) preparations of [$methyl$-^{3}H]oligonucleotides derived from an RNase T_1 digest of *Escherichia coli* tRNA (unfractionated). Arrows denote the elution of mono- (-2), di- (-3), tri- (-4), and tetra- (-5) phosphates of adenosine and are included for reference. Details have been published elsewhere (29). Presented by permission of the American Society of Biological Chemistry.

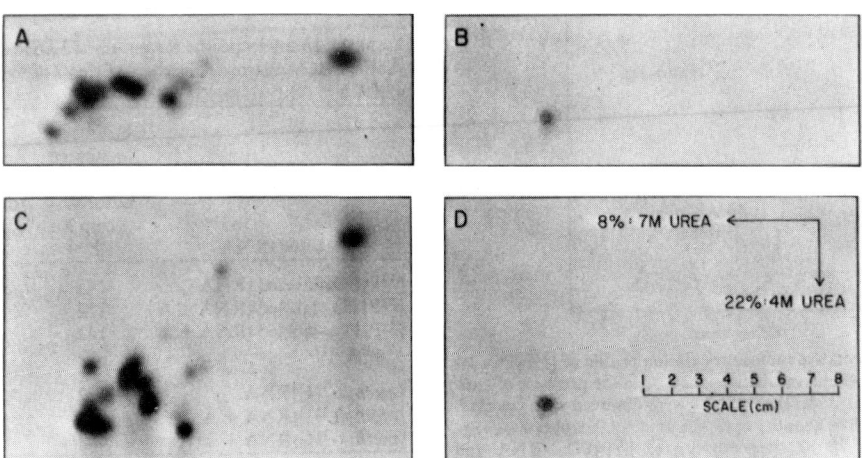

FIG. 10. Two-dimensional acrylamide gel characterization of unfractionated bulk (panels A and B) and anti-m^6A/Sepharose adsorbed (panels C and D) preparations of *Escherichia coli* [^{32}P]tRNA. Details regarding these data have been published elsewhere (30). Presented by permission of the American Chemical Society.

of species (Fig. 10A and 10C), only a single, discrete tRNA was observed in the immunospecifically retained fraction (Fig. 10B and 10D). These results confirmed the above speculation that a single m^6A-containing tRNA is present in *E. coli* tRNA and that there was no significant degradation of tRNA during the immunospecific adsorption. Additional aminoacylation studies (*30*) revealed the identity of this tRNA, namely tRNA$_1^{Val}$.

B. Isolation of mRNAs Possessing a 5'-Terminal "Cap" Structure Containing 7-Methylguanine

The unique poly(A) sequence at the 3' terminus of most eukaryotic mRNAs provides a means of isolating these RNAs by affinity chromatography on oligo(dT)-cellulose (*227*) or poly(U)-Sepharose (*228*) adsorbents. This is accomplished simply by passing unfractionated RNA preparations through these adsorbents under conditions (moderate to high ionic strength buffers, e.g., 0.1 to 0.5 M NaCl) that promote hybridization, thereby selecting for poly(A)-containing mRNAs while excluding rRNA and tRNA populations. Although histone mRNA appears to be the only species purified thus far that has no poly(A) sequence (*229*), other mRNAs lacking a poly(A) tract have been characterized in HeLa cells (*230*) and sea urchin embryos (*231*). Thus, although some mRNAs are not amenable to isolation by affinity chromatography, others that are selected could be devoid of 5'-terminal modifications as a result of partial degradation (either *in vivo* and/or during their isolation). It would be desirable, therefore, to select for mRNA populations not only on the basis of their 3'-terminal poly(A) tract, but on the basis of their 5'-terminal modification as well.

In view of the above and other findings that document the ability of anti-m^7G antibodies to specifically recognize m^7G-containing caps (Table IV and Fig. 8), Morrow *et al.* (*232, 233*) conducted a series of filter-binding assays to determine whether these antibodies would retain [5-^{3}H]uridine-labeled chorion mRNAs previously isolated by oligo(dT)-cellulose. This particular assay is based upon selected filters (HAWP) that retain protein and protein–nucleic acid complexes, but exclude unbound nucleic acids (*209*). The results of such experiments, presented in Table VIII, indicate that approximately 75% of the chorion mRNAs previously incubated with anti-m^7G antibody was retained by the filter. By including pm^7G and pG into these experiments, it was also demonstrated that mRNA retention was immunospecific, i.e., the presence of pm^7G (2.7 nmol) completely inhibited the interaction between antibody (0.3 nmol) and mRNA (3.3 pmol), while pG (14 nmol) was without effect. Most significant was the finding that more than 90% of the mRNA recovered from the filter, i.e., poly(A)-selected, m^7G-selected, could be retained by a second incubation in the presence of antibody. This observation implies that the binding between anti-m^7G antibody

TABLE VIII
THE BINDING OF m⁷G-CONTAINING CHORION mRNA TO ANTI-m⁷G ANTIBODIES[a]

Incubation mixtures	mRNA input (cpm)	mRNA retained cpm	%
Poly(A)-selected mRNA			
mRNA	2210	75	3.4
mRNA	2210	55	2.5
mRNA + antibody	2210	1720	78
mRNA + antibody	2210	1630	74
mRNA + antibody + pm⁷G	2210	45	2.0
mRNA + antibody + pm⁷G	2210	55	2.5
mRNA + antibody + pG	2210	1590	72
mRNA + antibody + pG	2210	1700	77
Poly(A)-selected, m⁷G-selected mRNA[b]			
mRNA	1270	60	4.7
mRNA	1270	45	3.6
mRNA + antibody	1270	1170	92
mRNA + antibody	1270	1250	98

[a] The isolation of [5-^{3}H]uridine-labeled, poly(A)-containing chorion mRNA and the purification of nuclease-free anti-m⁷G antibodies have been described (242). Chorion mRNA (0.5 μg/assay) was incubated alone, with antibody (45 μg), and with antibody plus either pm⁷G (10 μg) or pG (50 μg) for 2 hours at 0°C in 100 μl of 15 mM NaCl and 10 mM phosphate, pH 7.5. Each assay mixture was filtered through HAWP filters (prewashed) and subsequently washed 5 times with 1 ml of NaCl/phosphate buffer, dried and processed for retained radioactivity.

[b] Poly(A)-selected, m⁷G-selected mRNA is mRNA recovered from the filter and reevaluated for its ability to bind with anti-m⁷G antibodies as described above. Data are presented with the permission of C. S. Morrow (232,233).

(300 pmol) and chorion mRNA (3.3 pmol) is quantitative and does not result in significant degradation of RNA. The latter was confirmed by dodecyl sulfate/sucrose gradient analysis of chorion mRNA before and after its immunospecific retention (data not shown).

The results presented in Table VII further suggest the potential of anti-m⁷G antibodies as specific reagents for selecting mRNA on the basis of their 5'-terminal caps. However, it is equally important to note that the above investigation was conducted with poly(A)-containing mRNAs in which the m⁷G-containing tRNAs and 18 S rRNA were removed. Thus, additional studies are required to determine whether these RNAs are also retained by anti-m⁷G antibodies (bacterial m⁷G-containing tRNA do not react with anti-m⁷G antibodies; see preceding section). Interestingly, both tRNA and

18 S rRNA might be removed from unfractionated RNA populations with anti-tRNA and -m$_2^6$A antibodies, respectively, thereby providing one approach for selecting nonpolyadenylated, yet capped, mRNAs and hnRNAs by anti-m^{7}G antibodies.

V. Immunochemical Approaches for Assessing the Function of Modified Constituents

Both anti-nucleic acid and anti-nucleoside antibodies have been employed to inhibit a variety of functions associated with nucleic acids. For example, cellular processes such as sea urchin development (234), bacterial transformation (235, 236), and bacteriophage infectivity (237) appear to be inhibited in the presence of antisera containing antibodies that react with nucleic acids. Similarly, in vitro experiments with antibodies specific for DNA (238) or for G or T (239) reveal their effectiveness to inhibit both DNA- and RNA-polymerase activities. While a number of specific conclusions can be derived from these early investigations, collectively they serve to illustrate the effectiveness of these antibodies to compete for nucleotide sequences with a host of proteins associated with nucleic acid metabolism. In most instances, however, this competition is indiscriminate, because the antibody will react with and mask numerous sequences of which only one will be recognized by a particular protein or enzyme. Thus, even though antibody-dependent inhibition of normal processes is observed, it may be quite difficult to separate primary from secondary effects.

While the experimental induction of antibodies that recognize specific nucleotide sequences such as AUG is desirable, investigations of this type have had only limited success (240, 241). The major difficulty encountered here appears to be the extensive cross-reactivity observed with one or more of the nucleotides present in a given sequence. For example, it has been reported that anti-AUG antibodies are quantitatively precipitated in the presence of G-albumin conjugates as well as with AUG-albumin (241).

In contrast to the above approaches, antibodies that recognize specific modified nucleosides would have the distinct advantage of binding to only a limited number of hapten-containing nucleotide sequences. In this regard, if one assumes that various cellular processes (replication, transcription, post-transcriptional processing and translation) are dependent upon the presence of modified constituents in nucleic acids, it would seem logical to conclude that the interaction between these constituents and their corresponding antibodies would inhibit or block such processes. The data presented below test these assumptions by (a) examining the nature of the anti-m^{7}G·chorion mRNA complex and (b) assessing the ability of these antibodies to inhibit the *in vitro* translation of chorion mRNA in a wheat germ translational system.

A. Antibodies as Site-Specific Probes

In the preceding section, it was shown that anti-m^{7}G antibodies quantitatively bind to chorion mRNA and that this binding is dependent upon the presence of an m^{7}G-containing cap structure (Table VIII). To provide additional evidence for immunospecific binding, [*Me*-^{3}H]methionine-labeled chorion mRNA was digested with nuclease P_1, and this digest was incubated in the presence of anti-m^{7}G adsorbent. The resulting unadsorbed and immunospecifically retained radioactive fractions were subsequently characterized by DEAE-cellulose chromatography together with an unfractionated mRNA digest. Consistent with the specificity of nuclease P_1 (i.e., m^{7}GpppN-N-N... → m^{7}GpppN + p[N]$_x$), the unfractionated mRNA digest (Fig. 11A) exhibited two peaks of radioactivity that were subsequently iden-

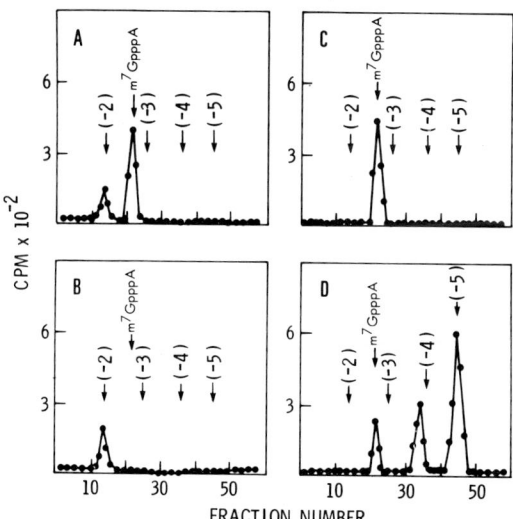

FIG. 11. DEAE-cellulose characterization of the *methyl*-^{3}H-labeled caps and mononucleotides of chorion mRNA previously digested with nuclease P_1 and fractionated by immunospecific adsorption to anti-m^{7}G/Sepharose. Panels represent the resulting radioactivity profiles of unfractionated (A), unadsorbed (B), and immunospecifically adsorbed (C) digests. Panel D represents the elution of [*methyl*-^{3}H]m^{7}GpppN(pN)$_x$ caps of chorion mRNA that were digested *after* their adsorption to anti-m^{7}G/Sepharose (see Fig. 12 for protocol). Arrows reflect the elution of pA (-2), ppA (-3), pppA (-4), and ppppA (-5) standards relative to the m^{7}GpppA cap standard. Details have been published elsewhere (*242*). Presented by permission of the American Chemical Society.

tified as p[Me-^{3}H]Nm (fractions 12–15) and [Me-^{3}H]m^{7}GpppNm (fractions 20–23) by thin-layer chromatography (232). Identical characterization of the unabsorbed and immunospecifically retained fraction revealed that anti-m^{7}G/Sepharose adsorbent quantitatively retained only the [Me-^{3}H]m^{7}GpppNm caps (Fig. 11C) while excluding the [Me-^{3}H]Nm constituents (Fig. 11B). These results, together with the data presented in Table VIII, indicate that the interaction between anti-m^{7}G and chorion mRNA occurs at the m^{7}G-containing, 5′-terminal cap (242).

The distinct possibility existed, however, that antibody binding to mRNA masked additional nucleotide sequences adjacent to the cap. To examine the potential of this masking effect, chorion mRNA was digested with nuclease P_1 *before* (predigested) and *after* (postdigested) its adsorption to anti-m^{7}G/Sepharose. This protocol (illustrated in Fig. 12) assumes that if significant antibody-dependent masking is occurring, the complex formed between immobilized antibody and mRNA (mRNA·anti-m^{7}G/Sepharose) would inhibit or obviate the hydrolytic activity of nuclease P_1 at or near the 5′-cap of mRNA. Furthermore, if masking did exist, the cap obtained by immunospecific adsorption of the predigested mRNA (m^{7}GpppN, see Fig. 11C) would be considerably smaller in size than the cap(s) derived by digesting the mRNA after its adsorption to anti-m^{7}G/Sepharose, i.e., m^{7}GpppN-N . . . .

However, as illustrated in Fig. 11D, characterization of a postdigested preparation of chorion mRNA by DEAE-cellulose chromatography revealed only three small cap structures: m^{7}GpppN (fractions 21–23), m^{7}GpppN-N (fractions 32–35), and m^{7}GpppN-N-N (fractions 43–47). The assignment of these structures was based upon their elution relative to adenosine nucleotide and cap standards, as well as a knowledge of the specificity of nuclease P_1. Although not unequivocal, these latter results (Fig. 12) provide

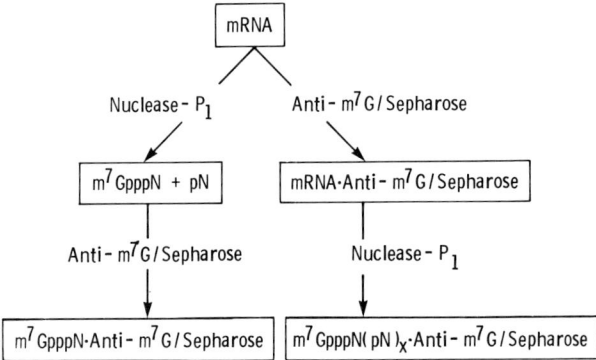

Fig. 12. Schematic presentation of the protocol employed to evaluate the nature of the anti-m^{7}G antibody·chorion mRNA complex. For details see Munns *et al.* (242).

evidence that anti-m⁷G antibodies were behaving as site-specific probes, i.e., their ability to mask nucleotide sequences adjacent to the cap (m⁷GpppN) appeared to be limited to either a mono- or dinucleotide adjacent to the penultimate nucleoside. An almost identical distribution of methyl-labeled caps as those observed in Fig. 11D has been obtained with postdigested preparations of [Me-³H]methionine-labeled adenovirus mRNA (205).

B. Antibodies Specific for 7-Methylguanine Inhibit mRNA Translation

In view of the above results, as well as the functional importance of the 5'-terminal cap in translation (148), it was anticipated that the presence of anti-m⁷G antibodies would inhibit the *in vitro* translation of chorion mRNA. However, in assessing antibody-dependent inhibition, experiments had to be designed to accommodate other m⁷G-containing RNAs endogenous to the wheat germ system, i.e., tRNA and rRNA. To circumvent the potential binding of antibodies to these RNAs and to enhance antibody interaction with chorion mRNA, antibodies were first incubated with chorion mRNA for 2 hours as well as added directly to the wheat germ system at the onset of translation, i.e., with no prior incubation. Similar assays employing identical incubations, but with chorion mRNA alone, with m⁷G, or with anti-m⁷G plus m⁷G, provided the appropriate controls.

The results of these experiments are presented in Table IX and indicate that the translation of chorion mRNA was inhibited significantly by anti-m⁷G antibody only after a preliminary 2-hour incubation; i.e., translation was inhibited 50-55% with this prior incubation, but only 10-15% with no prior incubation. Whereas the inhibition observed in the latter instance could be attributed to the interaction of anti-m⁷G with any m⁷G-containing RNAs in the wheat germ system (including both endogenous tRNAs and rRNA and the exogenous chorion mRNA), the *difference* in inhibition observed between the 2 and 0 hours incubation (approximately 40%) was attributed to the specific interaction of anti-m⁷G antibody with chorion mRNA.

Similar incubation periods of mRNA with m⁷G or with anti-m⁷G plus m⁷G (1 μg/assay) had little or no effect on subsequent translation. These results were expected in view of the findings that m⁷G does not inhibit the *in vitro* translation of globin mRNA (169). However, the introduction of this competing hapten (m⁷G) into these experiments serves several purposes. First, it confirms the absence of nuclease contamination in these antibody preparations; and, second, it provides additional evidence that the antibody-dependent inhibition of chorion mRNA translation is immunospecific.

Additional experiments with anti-m⁷G antibody were conducted to assess

TABLE IX
Inhibition of Translation of Chorion mRNA by Anti-m^{7}G Antibody.
Effects of Prior Incubation of Antibody with mRNA.
Effects of the Competing Hapten m^{7}G[a]

Prior incubation mixture	Prior incubation time (hours)	Acid-precipitable radioactivity (% of control)		Inhibition (%)	
		Expt 1	Expt 2	Expt 1	Expt 2
mRNA (control)	0	(100)	(100)		
mRNA	2	99	98	<5	<5
mRNA + antibody	0	86	89	15	11
mRNA + antibody	2	44	51	56	49
mRNA (control)	2	(100)	(100)		
mRNA + m^{7}G	2	95	102	<5	<5
mRNA + m^{7}G + antibody	2	98	97	<5	<5
mRNA + antibody	2	52	46	48	54

[a] Chorion mRNA (1.5 μg per assay) was incubated alone, with anti-m^{7}G (7.5 μg), and with anti-m^{7}G plus m^{7}G (1.0 μg) for 0 and 2 hours at 0°C prior to addition to a wheat germ translation system. Incubations were conducted at 25°C for 60 minutes (50 μl per assay), and aliquots (5 μl) were withdrawn at selected time intervals and processed for determination of trichloroacetic acid precipitable radioactivity. Values presented were derived from two independent experiments; each value represents the mean of duplicate determinations after a 60-minute translation period. These data have been published (242) and are presented with the permission of the American Chemical Society.

both kinetic and dose-response parameters. The data are presented in Fig. 13 and demonstrate that a preliminary incubation (2 hours) of anti-m^{7}G (1.5 μM) with mRNA (0.2 μg) resulted in 93, 90, and 89% inhibition of translation after 10-, 20-, and 30-minute incubation periods, respectively (Fig. 13A). By use of increasing amounts of anti-m^{7}G (0.50–1.5 μM) in the prior incubation, it was further observed that the inhibition of translation of chorion mRNA was dose-dependent. These results are presented in Fig. 13B and indicate that antibody concentrations of 0.5, 1.0, and 1.5 μM inhibited translation to an extent of 67, 80, and 90%, respectively. Additional control experiments indicated that translation of chorion mRNA (0.2 μg) was inhibited to an extent of 10% or less when equivalent amounts of antibody (1.4–4.5 μg) and 1 μg of m^{7}G were added at the onset of translation. Since the average molecular weight of chorion mRNAs has been estimated to be approximately 150,000 (243), the data in Fig. 13B suggest that approximately six antibody molecules were required per mRNA to inhibit the translation by 50%, i.e., a 6:1 molar ratio.

While these studies only confirm the well established observation of the

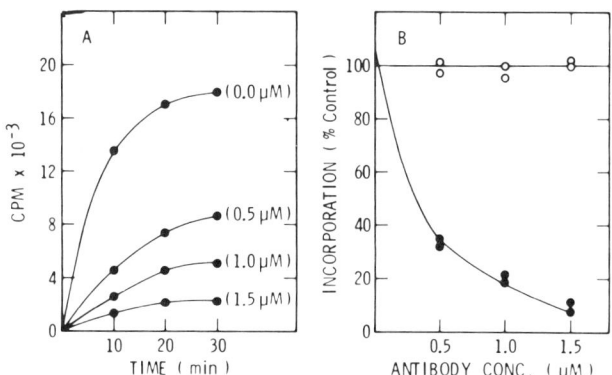

FIG. 13. Effects of anti-m⁷G antibodies on the *in vitro* translation of chorion mRNA. Kinetic and dose-response parameters. Chorion mRNA (0.2 μg) was incubated in the presence of increasing amounts of antibody (1.5, 3.0, and 4.5 μg/20 μl or 0.5, 1.0, and 1.5 μM) for 2 hours at 0°C prior to addition into the wheat germ translational system. The kinetics of incorporation of [^{3}H]amino acids into Cl$_3$AcOH-precipitable protein is presented in panel A. Panel B depicts the ability of increasing concentrations of antibody to inhibit the incorporation of labeled amino acids into protein after a 10-minute incubation. Open circles represent incorporation in the presence of 1 μg of competing nucleoside hapten (m⁷G). Details have been published elsewhere (*242*). Presented by permission of the American Chemical Society.

importance of the 5'-terminal cap in the initiation of translation of mRNA, they serve to demonstrate the utility of antibodies specific for modified nucleosides to assess the functional significance of the hapten component. One of the more intriguing aspects of nucleic acid metabolism is the post-transcriptional conversion ("processing") of large precursor mRNAs into their smaller functional counterparts. As reviewed in Section II, the internal m⁶A modifications present in most hnRNA and mRNA may serve a critical role in mRNA processing, yet definitive evidence for this function is lacking. However, coupled with the recent observations (*244, 245*) that an early adenovirus precursor mRNA can be processed to a mature size mRNA in various *in vitro* systems, it would be interesting to reevaluate its processing in the presence of anti-m⁶A antibodies.

Additionally, the functional aspects of m⁶A and other modified nucleosides might be clarified by determining their precise location in selected nucleic acid populations, e.g., the location of m⁶A in mRNAs, hnRNAs, and RNA-tumor virus genomes. For example, the realization that m⁷GpppN cap structures of eukaryotic mRNAs are located at the 5'-terminus of these molecules implies their involvement in the initiation of translation. In view of the ability of various antibodies specific for nucleosides to form specific complexes with nucleic acids (at the site of the hapten constituent), it would appear that the visualization of such complexes by immunoelectron microscopy should provide a rapid and direct approach for locating modified constituents within a given nucleic acid.

VI. Immunoelectron Microscopy: The Mapping of Specific Determinants within Nucleic Acids

The direct visualization of protein·nucleic acid complexes by electron microscopy has provided an enormous quantity of information about nucleic acids and nucleoprotein structure. Examples include the mode and direction of DNA replication (246), the repeating substructure of chromatin (247), and the topography of the ribosome and its subunits (248). The same techniques applied to complexes of protein and nucleic acid have established new insights into the abundance and location of specific binding sites within both DNA and RNA. Representative of the latter are the location and number of promoter sequences whereby RNA polymerase attaches to DNA prior to the onset of transcription (249). In view of the above and the findings presented herein, it becomes evident that the visualization of complexes of antinucleoside antibody and nucleic acid via electron microscopy will provide the necessary information to map the position(s) of specific modified nucleosides or other haptens associated with nucleic acid structures.

A. Antibodies Specific for Dinitrophenol

The introduction of the antibody molecule into the electron microscopy technology of nucleic acid was originally designed to enhance the mass of proteins specifically bound to DNA. This becomes especially apparent when cytochrome c or "Kleinschmidt" techniques (249-250) are employed. While the envelope of denatured cytochrome c that surrounds the polynucleotide sequences provides the necessary "bulk" to visualize nucleic acids, it also masks the presence and, hence, location of other specifically attached proteins (polymerase, repressors, endonucleases, etc.). Although protein-free spreading techniques offer high resolution, many proteins of low or intermediate size (i.e., 50,000 and below) associated with nucleic acid are still too small to be observed.

To circumvent this difficulty, a number of investigators (251-253, also reviews 247, 254) have increased the mass surrounding these proteins with appropriate antibodies prior to cytochrome c spreading. For example, Reed et al. (251) established that a protein required in the initiation of SV40 replication (T antigen) was bound at the origin of replication. To visualize this site, the mass surrounding the T antigen was initially enhanced with hamster anti-T antibody and ferriten-conjugated goat anti-hamster IgG. However, this procedure requires the availability of an antibody that recognizes the protein being examined (anti-T antibody), which, in many instances, is either unavailable or extremely difficult to obtain.

A very appealing alternative to the above approach is a technique recently described by Wu and Davidson (252, 253) in which various protein·nucleic acid complexes are treated with fluorodinitrobenzene (N_2ph-F). Under appropriate conditions, the covalent coupling of the dinitrophenyl (N_2ph) moiety to protein can be maximized, yet its attachment to nucleic acid is reduced to an insignificant level (252). Subsequently, the complex of N_2ph-protein and nucleate can be incubated in the presence of rabbit anti-N_2ph antibodies, and this complex, containing one or several anti-N_2ph antibodier per protein, visualized in conventional cytochrome c spreads (250). If necessary, the size of the anti-N_2ph·N_2ph-nucleate complex can be enlarged with goat anti-rabbit IgG (both antibodies are commercially available). Employing these procedures, Wu and Davidson (252, 253) illustrated the versatility of their method by mapping the positions of a host of proteins that recognize specific domains (sequences and/or conformations) within selected nucleic acids (e.g., *E. coli* polymerase/T7 DNA, Ad-2-terminal protein/Ad-2 DNA, T antigen/SV40, and VPg protein/polioviron RNA). Presented in Fig. 14 are the results of their investigations with the complex between Ad-2 terminal protein and Ad-2 DNA. While this particular protein (molecular weight 55,000) is not readily visible by protein-free or conventional cytochrome c spreading techniques, it is readily detected after treatment with fluorodinitrobenzene and the corresponding antibody (252). It has been suggested that this terminal protein binds to the 5′ ends of the Ad-2 genome, thereby facilitating replication of the free 3′-OH termini (249).

Other investigations, using a distinctly different format, have utilized anti-N_2ph antibodies to establish the location of the peptidyl-tRNA binding site (255) and the 3′ terminus of 16 S rRNA in 30 S ribosomal subunit (256). Whereas the former investigation employed N_2ph-valyl-tRNA$_1^{Val}$ as the initial probe, the latter specifically incorporated N_2ph into the 3′-terminal nucleoside (periodate-oxidized) of 16 S rRNA in the 30 S ribosomal subunit.

B. Antibodies Specific for 5-Methylcytosine

The potential of antibodies specific for ribonucleosides (anti-A, anti-C, anti-G, anti-T, and anti-m^{5}C) to bind to their respective bases in the DNA of metaphase chromosomes was initially realized by Erlanger and Beiser and colleagues (35, 39, 40, 257–263). Because these antibodies appear to recognize purine and pyrimidine determinants only in single-stranded DNA regions, chromosomal preparations were initially treated with a variety of DNA denaturing agents. The results obtained by incubating fluorescein-conjugated antibodies with these chromosomal preparations were striking, yielding distinctive chromosomal or fluorescent banding patterns reminis-

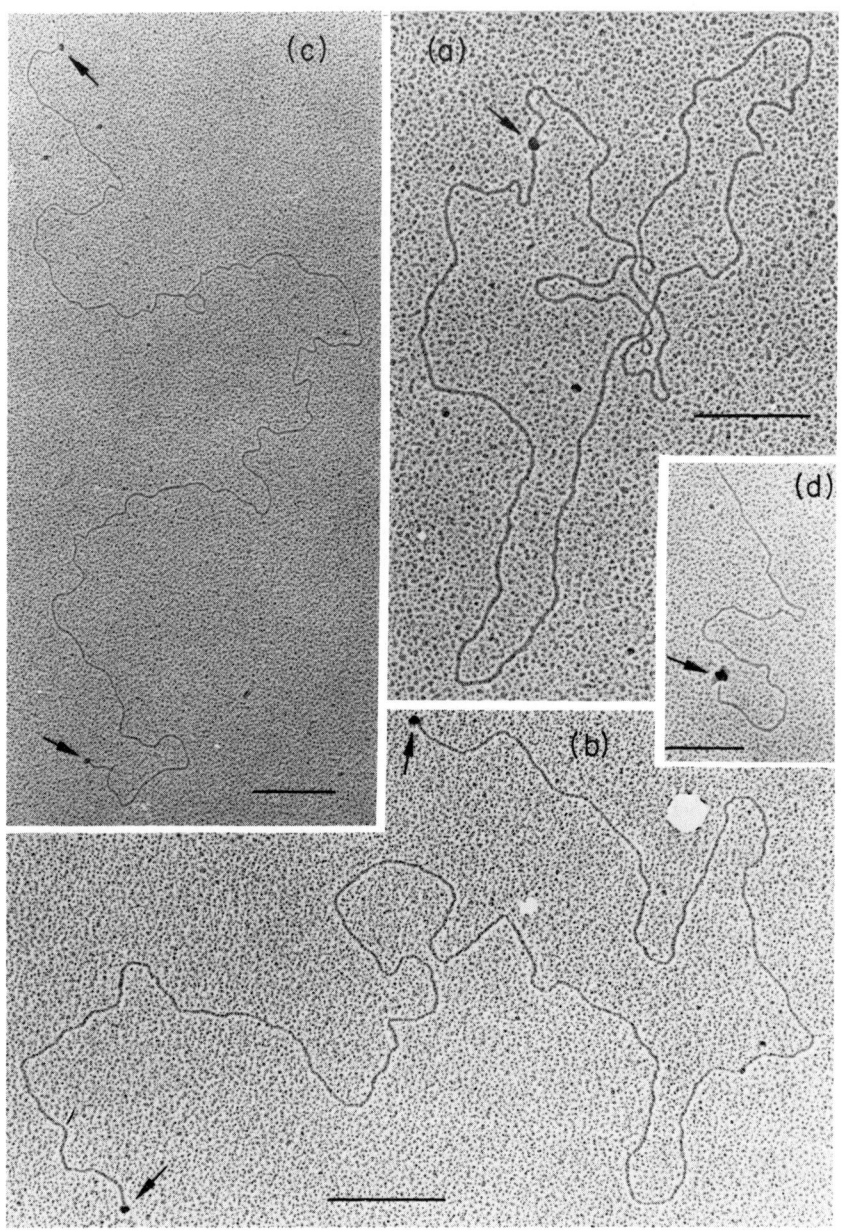

cent of those obtained preparations stained with quinacrine or Giemsa. For example, preparations of anti-A and anti-T appeared to react with localized regions rich in dA and dT in chromosomal preparations previously irradiated with UV-light to promote denaturation (257–259). This conclusion was based upon the specificity of these antibodies and the similarity in banding patterns achieved by quinacrine-staining techniques, reportedly specific for adenine and thymine residues in DNA (264). An additional feature of these studies was the appearance of intense fluorescence observed in the paracentromeric region of human chromosomes 1, 9, and 16 (260), regions known to contain (dA-dT)-rich sequences (satellite DNA) and relatively high levels of 5-methylcytosine (40, 265–267).

Because of the latter finding, Lubit et al. (39) employed anti-m^{5}C to examine the abundance and location of 5-methylcytosine residues in human metaphase chromosomes by both light- and immunoelectron microscopy techniques. In these studies, the visualization of anti-m^{5}C binding was aided by an indirect immunoperoxidase assay using sheep anti-rabbit IgG previously conjugated with horseradish peroxidase (268). A light micrograph of UV-irradiated chromosomes subjected to the above procedures is presented in Fig. 15A and illustrates the intense staining (anti-m^{5}C binding) observed in the centromeric heterochromatin regions of specific chromosomes, particularly chromosomes 1, 9, and 16. Additional electron micrographs of individual chromosomes (such as chromosome 15 in Fig. 15B) permitted isolated foci of 5-methylcytosine residues to be observed throughout the arms of the chromosomes. The inability to stain identical chromosomal preparations in the presence of m^{5}C (in molar excess relative to anti-m^{5}C) further implied the absence of nonspecific binding.

From an assumed base-stacking distance of 3.5 Å, Lubit et al. (39) concluded that a 100 Å antibody molecule would overlap approximately 30 bases. Thus, a single electron-dense aggregate (Fig. 15B) could represent between 1 and 30 5MeCyt residues per 30 bases. Their estimate of 3000 aggregates appearing on chromosome 1 would indicate, therefore, the presence of 3000–90,000 5MeCyt residues. While this extreme range is consistent with the low levels of 5MeCyt in human DNA, i.e., 0.4%, it precludes the use of these immunochemical procedures to quantitate such constituents. However, at the chromosomal level, these findings are significant

---

FIG. 14. Electron micrographs of Ad-2 DNA·Protein complexes labeled by treatment with N$_2$ph-F and rabbit anti-N$_2$ph IgG and spread by the cytochrome c method. Arrows point to the several amplified protein dots. The bar represents 0.5 μm. (a) A circular DNA molecule with one cluster of anti-N$_2$ph IgG. (b) and (c) Linear molecules with both ends clearly labeled with anti-N$_2$ph IgG clusters. (d) One end of a linear molecule labeled with anti-N$_2$ph IgG and goat anti-rabbit IgG. This figure is reproduced by permission of N. Davidson (252).

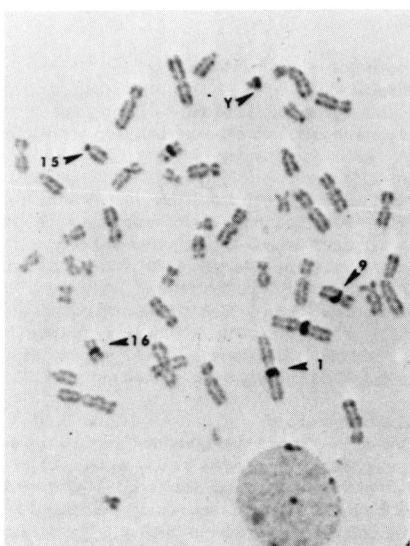
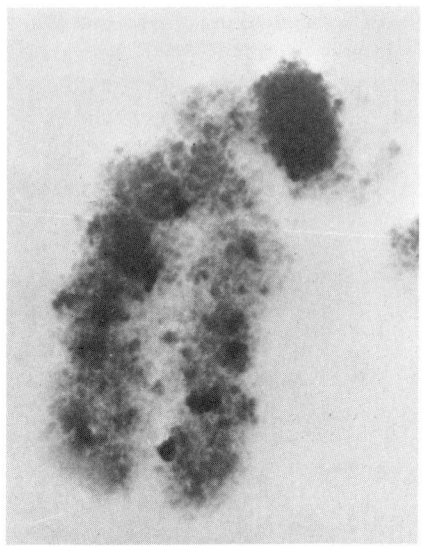

FIG. 15. Anti-m⁵C·chromosome (human metaphase) complexes. The left panel illustrates UV-irradiated chromosomes stained by an indirect immunoperoxidase procedure (268) and observed by light microscopy. The right panel presents an electron micrograph of chromosome 15 (upper left in left panel) with foci of electron-dense aggregates along long arms and with intensely stained regions on the short arm nearest the centromeric constriction (× 26,000). For additional details and micrographs of other chromosomes, see Lubit et al. (39). This figure is reproduced by permission of B. F. Erlanger and the MIT Press.

in terms of consistently identifying DNA sequences containing high densities of 5MeCyt residues within specific chromosome regions (paracentromeric), and further support the concept that clusters of this base exist along the arms of condensed chromosomes. Such investigations have been useful in examining and defining abnormalities associated with chromosomal structure (60, 264, 269).

C. Antibodies Specific for N^6,N^6-Dimethyladenine

While the nucleotide sequence of bacterial 16 S rRNA has been established (134, 135), only a limited amount of information is available regarding its three-dimensional structure or function within the 30 S ribosomal subunit. One approach in examining nucleotide sequences that appear on the surface of ribosomal subunits is to exploit the limited number of different types of methylated nucleosides present in rRNA (Table III). Although some of these constituents could be "buried" within the ribosomal subunit, others appearing on the surface should provide a unique set of determinants capable of interacting with their corresponding anti-nucleoside antibodies.

The occurrence of two successive m_2^6A residues located near the 3' terminus of bacterial 16 S rRNA (see Fig. 3) prompted (32) an investigation of the potential of anti-m_2^6A as an immunochemical probe to map the position of these hapten constituents on the 30 S ribosomal subunit. On the basis of a double antibody precipitation technique (goat anti-rabbit IgG), approximately 90% of a ^{32}P-labeled 30 S ribosomal subunit preparation was precipitated by anti-m_2^6A. Sucrose density gradient centrifugation revealed that 30 S subunits previously incubated in the presence of antibody sedimented more rapidly than unincubated controls. Additional control studies with ribosomal subunits derived from a kasugamycin-resistant *E. coli* strain lacking the m_2^6A modifications (32, 143), demonstrated that these precipitation and sedimentation studies were consistent with the formation of specific complexes between anti-m_2^6A and m_2^6A-containing ribosome subunits.

Examination of the 30 S ribosomal subunits by electron microscopy revealed images of these particles similar to those described earlier (248, 270). However, preparations previously incubated with anti-m_2^6A consistently (90%) appeared as dimers in which the antibody combining site was localized adjacent to the antibody cleft. A schematic representation of this subunit (Fig. 16) illustrates the cleft and platform sites together with the postulated antibody combining sites (hatched areas). The interpretation of the data (32) supports previous knowledge regarding the 30 S subunit topology and function, which includes (a) an involvement of the 3' terminus of 16 S rRNA with mRNA binding to the ribosomes (139–141); (b) the chemical cross-linking of an initiation factor (IF-3) with the 3' terminus of 16 S rRNA (271) and with other ribosomal proteins that have been localized within the platform structure (270); (c) the localization of the decoding region (255) and 3' end of 16 S

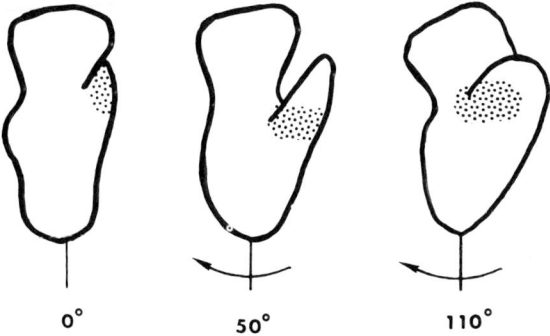

FIG. 16. Localization of m_2^6A in the small ribosomal subunit of *Escherichia coli*. The shaded area of the model adapted from Lake (270) indicates the binding site of antidimethyladenosine, as determined by electron microscopy. For details, see Politz and Glitz (32). This schematic is reproduced by permission of D. G. Glitz.

rRNA (256) at the platform site of the 30 S ribosomal subunit; and (d) the anti-m$_2^6$A-dependent inhibition of IF-3 binding to the 30 S ribosomal subunit (138).

In view of these observations, it is interesting to note that the largest expanse of surface 16 S rRNA sequences appears to reside in a region of 175 nucleotides near the 3' terminus (136, 137), a region occupied by four distinctly different methylated nucleosides, namely m^{4}Cm, m^{5}C, Um, and m$_2^6$A (134, 135).

D. Antibodies Specific for N^6-Methyladenine

Although the above studies (32, 39) readily document the use of specific antibodies for evaluating the abundance and location of methylated bases in nucleoprotein structures, no information is available regarding the use of such antibodies to localize specific methylated constituents within "naked" nucleic acids. This is somewhat surprising in light of the documented affinities and specificities that can be achieved with these antibody preparations.

In view of the wealth of information that can be derived from such investigations, an examination of the interaction between anti-m^{6}A and the DNA of bacteriophage fd was undertaken. This particular nucleic acid was selected because it is a well defined, single-stranded, circular DNA (272) that reportedly contains up to four 6MeAde residues per DNA molecule (272-275). Initial studies concentrated on determining the number of 6MeAde residues in a commercial fd preparation, which yielded 1.7 ± 0.4 6MeAde/fd DNA by (a) radioimmunoassay (see Section III), and (b) an antibody-binding assay that measured the maximum extent of anti-m^{6}A binding to a constant amount of fd DNA. The latter assay consisted of incubating 10 μg quantities of fd DNA with increasing amounts of [^{3}H]uridine-conjugated antibody, and determining the amount of radioactive antibody coeluting with the DNA after Sepharose-4B chromatography. This chromatographic step readily separates the larger antibody·fd complexes (> 2.2 × 10^6) from unreacted antibody (150,000).

Similar chromatographic steps were used throughout this investigation to prepare anti-m^{6}A·fd DNA and complexes of ferriten-conjugated goat anti-rabbit antibody with anti-m^{6}A·fd DNA. Examination of the antibody·(antibody·fd DNA) complexes (see Fig. 17) as visualized in conventional cytochrome c spreads revealed the presence of two distinct anti-m^{6}A antibodies

---

FIG. 17. Electron micrographs of anti-m^{6}A·fd DNA prepared by the cytochrome c spreading method (273). Arrows denote the sites where anti-m^{6}A binds to the circular DNA as amplified by ferritin-conjugated goat anti-rabbit IgG antibodies (× 132,700).

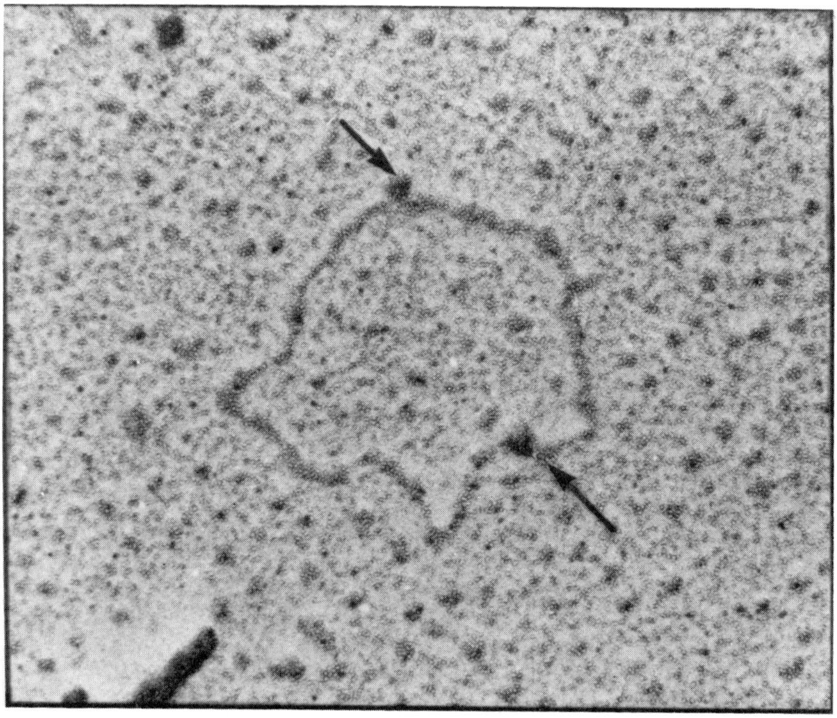

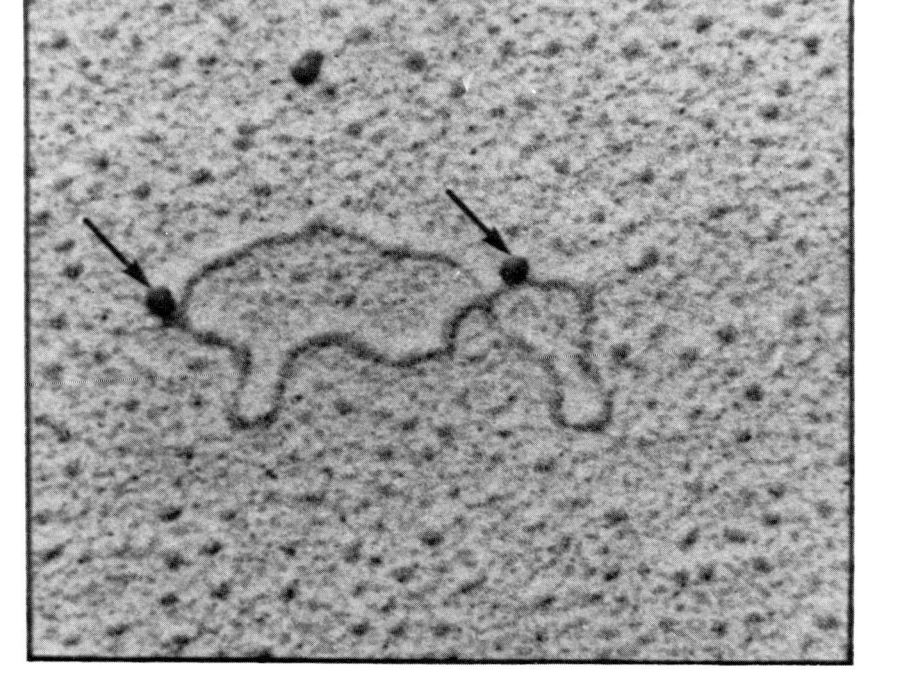

attached to the circular DNA and spaced along the nucleic acid approximately 0.6 map unit apart (the entire molecule represents 1.0 unit). The number of antibodies attached (two) and their consistent spacing presents two arguments against random or nonspecific attachment; another is the inability of anti-m^{6}A to bind to ϕX174 DNA, which is devoid of 6MeAde residues (77, 78).

Although preliminary, these results together with those presented in connection with anti-m^{5}C (39) and anti-m$_2^6$A (32) suggest the utility of this particular immunochemical approach for accurately mapping the position of modified constituents within nucleic acids. Particularly amenable to analysis are single-stranded nucleic acids, such as the m^{6}A-containing mRNA and hnRNA populations (177–181), as well as the genomes of various RNA-tumor viruses (83, 84). Since each of these RNA species possesses an m^{7}G-cap structure at their 5′ terminus, it is conceivable that a combination of anti-m^{7}G and -m^{6}A will provide additional resolution in terms of mapping internal m^{6}A moieties relative to the 5′ end of these molecules.

VII. Conclusion

The rationale of an immunochemical approach for the isolation and characterization of nucleic acids is based upon two important observations. First, most nucleic acids contain a limited number of modified nucleosides, which are uniquely distributed throughout their structures (Section II). Second, antibodies that possess high degrees of affinity and specificity toward these modified nucleosides can be prepared, with minimal or no significant cross-reactivity with the bulk of unmodified constituents present in nucleic acids (Section III). From the information presented in these sections, the feasibility of an immunochemical approach is demonstrated by using specific anti-nucleoside antibodies (a) to isolate hapten-containing oligonucleotides and RNAs (Section IV), (b) to assess the functional significance of the hapten component (Section V); and (c) to map the position of the hapten within selected nucleic acid structures (Section VI).

It is anticipated that these and related immunochemical techniques will continue to provide new and/or alternative approaches for examining nucleic acid and/or nucleoprotein structures, as well as the functional significance of the modified nucleosides present in their sequences. This becomes evident when reviewing the various approaches employed to locate the internal m^{6}A nucleosides present in various hnRNA (180) and mRNA (84, 175, 177) populations. They include (a) determination of the content of m^{6}A in poly(A)- and m^{7}GpppN-containing nucleotide sequences (84, 175, 180) derived from limited digests of mRNA (isolated by oligo(dT)-cellulose and borate gel chromatographic techniques); (b) comparison of the m^{6}A content in the RNA genome of RSV-tumor virus (12 m^{6}A) and a transformation-defective deletion mutant of RSV (7 m^{6}A) that lacks the *src* gene *(84);* and *(c)* hybridization of m^{6}A-containing nucleotide sequences of the 16 S and 19 S mRNAs of SV40 to specific restriction fragments of the SV40 genome *(177).*

The data obtained from each of the above investigations imply that many of the m^{6}A nucleosides are present in the translatable region of mRNA, but their exact location remains to be resolved (particularly with respect to the splicing junction(s) of these molecules). Further, both approaches *(b)* and *(c)* are restricted to mRNAs derived from well characterized genomes.

In contrast to the above, the utilization of anti-m^{6}A antibodies as probes to assess m^{6}A function and location appears promising, particularly with regard to *(a)* isolating and characterizing (sequencing) m^{6}A-containing sequences of mRNAs with anti-m^{6}A/Sepharose adsorbents; *(b)* assessing the ability of anti-m^{6}A antibodies to inhibit the processing of large precursor mRNAs into their smaller structural counterparts; and *(c)* mapping the position of m^{6}A in mRNAs and hnRNAs via immunoelectron microscopy techniques.

While the above illustrate the use of anti-m^{6}A antibodies to investigate a single aspect of nucleic acid metabolism, these and other immunochemical approaches appear to be limited only by the number of unique determinants in various nucleic acid structures, e.g., 5MeCyt (eukaryotic DNA), m$_2^2$G, m^{7}G, and/or m$_3^{2,2,7}$G (low-molecular-weight nuclear RNAs and ribonucleoprotein structures) (see Tables I–III).

Last, the employment of the anti-nucleoside antibody is by no means restricted to naturally occurring modified constituents of nucleic acids. Equally or perhaps more significant, are their application in the detection, quantitation, isolation, and location of aberrant nucleoside adducts that appear in the nucleic acids of organisms exposed to various forms of radiation (e.g., pyrimidine dimers, *58–61)* and a host of chemical carcinogens *(55–57, 276–280).*

In regard to the latter, many environmental carcinogens have been grouped into three major classes (see reviews *57a, 276, 277),* the polyaromatic hydrocarbons, aromatic amines, and alkylating (predominantly methylating) agents. While the O^6-methylguanine adduct produced by methylating carcinogens is frequently mispaired and results in base-substitution mutations, frame-shift or deletion mutations are associated with those carcinogens that introduce a bulky substituent into the purine ring system *(57a)* such as dimethylbenzanthracene and acetamidofluorene.

Investigations with these carcinogens reveal that the extent of carcinogen-induced modifications in DNA approximates 0.1 to 0.001%, i.e., 1 nucleotide adduct per 10^3 to 10^5 residues *(276).* Modifications induced by the polycyclics and aromatic amines (0.001%) are usually one to two orders of magnitude less than that observed with alkylating agents (0.1%). Although this level of modification appears to be quite small in physical terms, it is significant biologically. For example, based upon the number of nucleotide pairs present in the DNA of a mammalian cell, between 10^4 and 10^6 nucleotides could become modified by the appropriate carcinogen. Thus, depending upon the efficiency and capacity of the repair systems present, the potential for extensive mutational events to occur is evident, particularly in cells receiving large doses and/or lengthy exposures to carcinogens.

During the past several years, antibodies specific for O^6-methylguanine (56), the acetamidofluorene-guanine adduct (57, 279, 280), and pyrimidine dimers (58–61) have resulted in the development of sensitive radioimmunoassays for the detection and quantitation of these adducts in nucleic acids. It is probable, therefore, that these immunochemical reagents will be useful in monitoring carcinogen- and radiation-induced adduct formation, persistence, and repair as well as in providing information concerning which nucleotide sequences are most sensitive to modification (57a, 279, 280).

Acknowledgments

We are grateful to B. F. Erlanger, D. G. Glitz, and N. Davidson for providing us with manuscripts and data. We express our gratitude to B. Goewert, T. Rucinsky, C. Morrow, R. Oberst, C. Graves, G. Grabau, and H. Sims for their participation in some of the experiments described herein. Research originating in the authors' laboratory was supported by grants CA-17715, CA-27801, and HL 15619 from the National Institutes of Health.

References

1. B. D. Stollar, in "The Antigens" (M. Sela, ed.). Academic Press, New York, 1973.
2. O. J. Plescia and W. Braun, *Adv. Immunol.* **6**, 231 (1967).
3. L. Levine and B. D. Stollar, *Progr. Allergy* **12**, 161 (1968).
4. O. J. Plescia, in "Methods of Enzymology" (K. Moldave and L. Grossman, eds.), Vol. 12B, p. 893. Academic Press, New York, 1968.
5. D. D. Notman, N. Kurata, and E. M. Tan, *Ann. Intern. Med.* **83**, 464 (1975).
6. T. T. Provost, *J. Invest. Dermatol.* **72**, 110 (1979).
7. R. Arana and M. Seligmann, *J. Clin. Invest.* **46**, 1867 (1967).
8. E. M. Tan and P. G. Natali, *J. Immunol.* **104**, 902 (1970).
9. R. J. Samaha and W. S. Irvin, *J. Clin. Invest.* **56**, 446 (1975).
10. V. P. Butler, S. M. Bieser, B. F. Erlanger, S. W. Tanenbaum, S. A. Cohen, and A. Bendich, *PNAS* **48**, 1597 (1962).
11. S. W. Tanenbaum and S. M. Beiser, *PNAS* **49**, 662 (1963).
12. B. F. Erlanger and S.M. Beiser, *PNAS* **52**, 68 (1964).
13. M. J. Halloran and C. W. Parker, *J. Lab. Clin. Med.* **64**, 865 (1964).
14. M. J. Halloran and C. W. Parker, *J. Immunol.* **96**, 373 (1966).
15. M. Sela, H. Ungar-Waron, and Y. Scheckter, *PNAS* **52**, 285 (1964).
16. J. P. Coat, S. David, and J. C. Fisher, *Bull. Soc. Chim. Fr.* **170**, 2489 (1966).
17. K. Landsteiner, "The Specificity of Serological Reactions." Harvard Univ. Press, Cambridge, Ma., 1946.
18. W. E. Cohn, *JBC* **235**, 1488 (1960).

19. J. L. Starr and B. H. Sells, *Physiol. Rev.* **49**, 623 (1969).
20. R. H. Hall "The Modified Nucleosides in Nucleic Acids." Columbia Univ. Press, New York, 1972.
21. R. P. Singhal and P. A. M. Fallis, *This Series* **23**, 227 (1979).
22. D. B. Dunn and R. H. Hall, *in* "Handbook of Biochemistry and Molecular Biology, 3rd edition" (G. D. Fasman, ed.), Vol. I, pp. 65–250. CRC Press, Cleveland, 1975; also M. A. Sodd, ibid., Vol. II, pp. 423–456.
23. S. Estrada-Parra, A. Schmill, and R. M. Martinez, *Nature* **208**, 1010 (1965).
24. H. Ungar-Waron, E. Hurwitz, J. C. Jaton, and M. Sela, *BBA* **138**, 513 (1967).
25. M. H. Karol and S. W. Tanenbaum, *PNAS* **57**, 713 (1967).
26. S. Estrada-Parra, E. Garcia-Ortigoza, and F. Quesada-Pascual, *Immunochemistry* **15**, 791 (1978).
27. D. L. Sawicki, B. F. Erlanger, and S. M. Beiser, *Science* **174**, 70 (1971).
28. T. W. Munns, M. K. Liszewski, and H. F. Sims, *Bchem* **16**, 2163 (1977).
29. T. W. Munns, H. F. Sims, and M. K. Liszewski, *JBC* **252**, 3102 (1977).
30. T. W. Munns, M. K. Liszewski, R. J. Oberst, and H. F. Sims, *Bchem* **17**, 2573 (1978).
31. T. W. Munns, R. J. Oberst, H. F. Sims, and M. K. Liszewski, *JBC* **254**, 4327 (1979).
32. S. M. Politz and D. G. Glitz, *PNAS* **74**, 1468 (1977).
33. P. Thammana and C. R. Cantor, *NAR* **5**, 805 (1978).
34. S. M. Beiser, S. W. Tanenbaum, and B. F. Erlanger, *in* "Methods of Enzymology" (K. Moldave and L. Grossman, eds.), Vol. 12B, p. 889. Academic Press, New York, 1968.
35. B. Gutter, Y. Nishioka, W. T. Speck, H. S. Rosenkranz, B. Lubit, and B. F. Erlanger, *Exp. Cell Res.* **102**, 413 (1976).
36. S. Estrada-Parra, A. Schmill, and R. A. Martinez, *Nature* **208**, 1010 (1965).
37. T. W. Munns and M. K. Liszewski, unpublished observations.
38. L. Levine, H. Van Vunakis, and R. C. Gallo, *Bchem.* **10**, 2009 (1971).
39. B. W. Lubit, T. D. Pham, O. J. Miller, and B. F. Erlanger, *Cell* **9**, 503 (1976).
40. O. J. Miller, W. Schnedl, J. Allen, and B. F. Erlanger, *Nature* **251**, 636 (1974).
41. A. L. Sandberg and B. D. Stollar, *J. Immunol.* **96**, 764 (1966).
42. S. S. Wallace, B. F. Erlanger, and S. M. Beiser, *JMB* **43**, 41 (1969).
43. L. Levine and H. Gjika, *ABB* **164**, 583 (1974).
44. L. Rainen and B. D. Stollar, *NAR* **5**, 4877 (1978).
45. R. D. Meredith and B. F. Erlanger, *NAR* **6**, 2179 (1979).
46. D. L. Sawicki, S. M. Beiser, D. Srinivassan, and P. Srinivassan, *ABB* **176**, 457 (1976).
47. A. J. Garro, B. F. Erlanger, and S. M. Beiser, *J. Immunol.* **106**, 442 (1971).
48. W. L. Highes, M. Christine, and B. D. Stollar, *Anal. Biochem.* **55**, 468 (1973).
49. A. L. Steiner, D. M. Kipnis, R. Utiger, and C. Parker, *PNAS* **64**, 367 (1969).
50. A. L. Steiner, C. W. Parker, and D. M. Kipnis, *JBC* **247**, 1106 (72).
51. H. Van Vunakis, E. Seaman, and L. Levine, *in* "Nucleic Acids in Immunology" (O. J. Plescia and W. Braun, eds.), p. 58. Springer-Verlag, Berlin and New York, 1968.
52. B. Bonavida, S. Fuchs, M. Sela, P. W. Roddy, and H. A. Sober, *EJB* **31**, 155 (1972).
53. S. M. Beiser and B. F. Erlanger, *Cancer Res.* **26**, 2012 (1966).
54. S. S. Wallace, B. F. Erlanger, and S. M. Beiser, *Bchem* **10**, 679 (1971).
55. F. L. Moolten, N. J. Capparell, E. Boger, and P. Mahathalang, *Nature* **272**, 614 (1978).
56. W. T. Briscoe, J. Spizizen, and E. M. Tan, *Bchem* **17**, 1896 (1978).
57. M. C. Poirier, S. H. Yuspa, I. B. Weinstein, and S. Blobstein, *Nature* **270**, 186 (1977).
57a. D. Grunberger and I. B. Weinstein, *This Series*, **23**, 105 (1979).
58. E. Seaman, H. Van Vunakis, and L. Levine, *JBC* **247**, 5709 (1972).
59. J. J. Cornelis, J. Rommelaere, and M. Errera, *Photochem. Photobiol.* **26**, 241 (1977).
60. H. L. Lewis, D. R. Muhleman, J. F. Ward, *Radiat. Res.* **75**, 305 (1978).

61. P. G. Natali and E. M. Tan, *Radiat. Res.* **46**, 506 (1971).
62. B. Hacker, H. Van Vunakis, and L. Levine, *J. Immunol.* **108**, 1726 (1972).
63. D. S. Milstone, B. S. Vold, D. G. Glits, and N. Shutt, *NAR* **5**, 3439 (1978).
64. B. S. Vold, J. M. Lazar, and A. M. Gray, *JBC* **254**, 7262 (1979).
65. H. Inouye, S. Fuchs, M. Sela, and U. Z. Littauer, *BBA* **240**, 549 (1971).
65a. F. Harada and S. Nishimura, *Bchem* **11**, 301 (1972).
65b. S. Fuchs, A. Aharonov, M. Sela, S. Von Der Haar, and J. F. Cramer, *PNAS* **71**, 2800 (1974).
66. S. W. Tanenbaum and S. M. Beiser, *PNAS* **49**, 662 (1963).
66a. E. Goldwasser and R. L. Heinrikson, *This Series* **5** (1966).
66b. R. W. Chambers, *This Series* **5** (1966).
67. B. J. Underwood and H. Eisen, *J. Immunol.* **106**, 1431 (1971).
68. H. Inouye, S. Fuchs, M. Sela, and U. Z. Littauer, *Isr. J. Med. Sci.* **6**, 442 (1970).
68a. B. Singer and M. Kröger, *This Series* **23**, 151 (1979).
69. R. Desrosiers, K. Friderici, and F. Rottman, *PNAS* **71**, 3971 (1976).
70. T. W. Munns and P. A. Katzman, *Bchem* **13**, 4409 (1974).
71. L. R. Mandel and E. Borek, *Bchem* **2**, 555 (1963).
72. E. Borek and S. J. Kerr, *Adv. Cancer Res.* **15**, 163 (1972).
73. P. B. Moore, *JMB* **18**, 38 (1966).
74. J. D. Smith, W. Arber, and U. Kühnlein, *JMB* **63**, 1 (1972).
75. D. Nathans and H. O. Smith, *ARB* **44**, 273 (1975).
76. U. Kühnlein and W. Arber, *JMB* **63**, 9 (1972).
77. A. Razin, J. W. Sedat, and R. S. Sinsheimer, *JMB* **53**, 251 (1970).
78. A. S. Lee and R. L. Sinsheimer, *J. Virol.* **14**, 872 (1974).
79. M. Gold and J. Hurwitz, *PNAS* **50**, 161 (1963).
80. D. Fujimoto, P. R. Srinivassan, and E. Borek, *Bchem* **4**, 2849 (1965).
81. P. D. Lawley, A. R. Crathorn, S. A. Shah, and B. D. Smith, *BJ* **128**, 133 (1972).
82. U. Günthert, M. Schweiger, M. Stupp, and W. Doerfler, *PNAS* **73**, 3923 (1976).
83. K. Dimock and C. M. Stoltzfus, *Bchem* **16**, 471 (1977).
84. K. L. Beemon and J. M. Keith, *JMB* **113**, 165 (1977).
85. M. O. Dayhoff and P. L. McLaughlin, in "Atlas of Protein Sequence and Structure" (M. O. Dayhoff, ed.). National Biomedical Research Foundation, Washington, D.C., 1976.
86. K. Randerath and E. Randerath, *Methods Cancer Res.* **9**, 3 (1973).
87. T. W. Munns and H. F. Sims, *Anal. Bchem* **64**, 537 (1975).
88. B. E. H. Maden and M. Salim, *JMB* **88**, 133 (1974).
89. M. S. N. Khan, M. Salim, and B. E. H. Maden, *BJ* **169**, 531 (1978).
90. B. E. H. Maden and M. S. N. Khan, *BJ* **167**, 211 (1977).
91. J. W. Kappler, *J. Cell. Physiol.* **78**, 33 (1971).
92. M. J. Brown and R. H. Burdon, *NAR* **4**, 1025 (1977).
93. C. M. Stoltzfus and K. Dimock, *J. Virol.* **18**, 586 (1976).
94. P. Fellner and F. Sanger, *Nature* **219**, 236 (1968).
95. P. Fellner, *EJB* **11**, 12 (1969).
96. D. B. Dunn and J. D. Smith, *BJ* **68**, 627 (1958).
97. G. G. Brownlee, in "Determination of Sequences in RNA" (T. S. Work and E. Work, eds.). North-Holland Publ., Amsterdam.
98. S. Nishimura, *This Series* **12**, 49 (1972).
99. J. A. McCloskey and S. Nishimura, *Acc. Chem. Res.* **10**, 403 (1977).
100. S. H. Kim, J. L. Sussman, F. L. Suddath, G. J. Quigley, A. McPherson, A. H. J. Wang, N. C. Seeman, and A. Rich, *PNAS* **71**, 4970 (1974).
101. J. D. Robertus, J. E. Ladner, J. T. Finch, D. Rhodes, R. S. Brown, B. F. C. Clark, and A. Klug, *Nature* **250**, 546 (1974).

102. S. H. Kim, in "Transfer RNA" (S. Altman, ed.), p. 248. MIT Press, Cambridge, Ma., 1978; also in Vol. 17 of this series.
103. L. R. Mandel and E. Borek, *BBRC* **4**, 14 (1961).
104. L. Shugart, G. D. Novelli, and M. P. Stulberg, *BBA* **157**, 83 (1968).
105. A. Peterkofsky, M. Litwack, and J. Marmor, *Cancer Res.* **31**, 675 (1971).
106. E. Fleissner, *Bchem* **6**, 621 (1967).
107. R. Stern and U. Z. Littauer, *Bchem* **9**, 10 (1970).
108. J. D. Capra and A. Peterkofsky, *JMB* **33**, 591 (1968).
109. J. D. Capra and A. Peterkofsky, *JMB* **21**, 455 (1966).
110. M. Revel and U. Z. Littauer, *JMB* **15**, 389 (1966).
111. M. L. Gefter and R. L. Russell, *JMB* **39**, 145 (1969).
112. F. Fittler and R. H. Hall, *BBRC* **25**, 441 (1966).
113. R. Thiebe and H. G. Zachau, *EJB* **5**, 546 (1968).
114. J. P. Miller, Z. Hussoin, and M. P. Schweizer, *NAR* **3**, 1185 (1976).
115. C. E. Singer, G. R. Smith, R. Cortese, and B. N. Ames, *Nature NB* **238**, 72 (1972).
116. H. S. Allaudeen, S. K. Yang, and D. Söll, *FEBS Lett.* **28**, 205 (1972).
117. M. G. Marinus, N. R. Morris, D. Söll, and T. C. Kwong, *J. Bact.* **122**, 257 (1975).
118. K. Watanabe, T. Oshima, and S. Nishimura, *NAR* **3**, 1703 (1976).
119. T. Igo-Kelmers and H. G. Zachau, *EJB* **18**, 292 (1971).
120. K. Watanabe, T. Oshima, and S. Nishimura, *FEBS Lett.* **43**, 59 (1974).
121. B. Roe, M. Michael, and B. Dudock, *Nature NB* **246**, 135 (1973).
122. S. Altman, in "Transfer RNA" (S. Altman, ed.), p. 48. MIT Press, Cambridge, Ma., 1978.
123. R. LaRossa and D. Soll, in "Transfer RNA" (S. Altman, ed.), p. 136. MIT Press, Cambridge, Ma., 1978.
124. S. Nishimura, in "Transfer RNA" (S. Altman, ed.), p. 168. MIT Press, Cambridge, Ma., 1978.
125. T. A. Walker, J. S. Betz, and J. Olah, *FEBS Lett.* **54**, 241 (1975).
126. B. E. H. Maden and J. S. Robertson, *JMB* **87**, 227 (1974).
126a. V. A. Erdmann, *This Series* **18** (1976).
127. O. Hagenbüchle, M. Santer, J. Steitz, and R. J. Mans, *Cell* **13**, 551 (1978).
128. M. Santer, S. C. Chung, G. Harmon, M. Estner, J. P. Hendrick, J. Hopper, C. Brecht, and P. Pandhi, *J. Bact.* **140**, 131 (1979).
129. E. K. Wagner, S. Penman, and V. Ingram, *JMB* **29**, 371 (1967).
130. M. H. Vaughan, R. Soeiro, J. R. Warner, and J. E. Darnell, *PNAS* **58**, 1527 (1967).
131. S. F. Wolf and D. Schlessinger, *Bchem* **16**, 2783 (1977).
132. M. Salim and B. E. H. Maden, *Nature NB* **244**, 334 (1973).
133. T. W. Munns and H. F. Sims, *JBC* **250**, 2143 (1975).
134. J. Brosius, M. L. Palmer, P. J. Kennedy, and H. F. Noller, *PNAS* **75**, 4801 (1978).
135. P. Carbon, C. Ehresmann, B. Ehresmann, and J. P. Ebel, *FEBS Lett.* **94**, 152 (1978).
136. M. Santer and U. Santer, *J. Bact.* **116**, 1304 (1973).
137. M. Santer and M. Shane, *J. Bact.* **130**, 900 (1977).
138. P. Thammana and C. R. Cantor, *NAR* **5**, 805 (1978).
139. J. Shine and L. Dalgarno, *Nature NB* **254**, 34 (1975).
140. J. Shine and L. Dalgarno, *EJB* **57**, 221 (1975).
141. J. A. Steitz and K. Jakes, *PNAS* **72**, 4734 (1975).
142. L. Dalgarno and J. Shine, in "The Ribonucleic Acids" (P. R. Stewart and D. S. Letham, eds.), p. 195. Springer-Verlag, Berlin and New York, 1977.
143. T. L. Helser, J. E. Davies, and J. E. Dahlberg *Nature NB* **233**, 12 (1971).
144. C. J. Lai, B. Weisblum, S. R. Fahnestock, and M. Nomura, *JMB* **74**, 67 (1973).
145. C. J. Lai and B. Weisblum, *PNAS* **68**, 856 (1971).
146. R. P. Perry and D. E. Kelley, *Cell* **1**, 37 (1974).

147. Y. Furuichi, *NAR* **1**, 809 (1974).
148. A. Shatkin, *Cell* **9**, 645 (1976).
149. R. P. Perry, *ARB* **45**, 605 (1976).
150. Y. Furuichi, S. Muthukrishnan, J. Tomasz, and A. J. Shatkin, *This Series* **19**, 3 (1976); also, *in* the same volume, papers by Rottman *et al.*, Busch *et al.*, Moss *et al.* (*154*).
151. T. W. Munns, C. S. Morrow, J. R. Hunsley, R. J. Oberst, and M. K. Liszewski, *Bchem* **18**, 3804 (1979).
152. R. P. Perry and K. Scherrer, *FEBS Lett.* **57**, 73 (1975).
153. M. J. Ensinger, S. A. Martin, E. Paoletti, and B. Moss, *PNAS* **72**, 2525 (1975).
154. B. Moss, S. A. Ma:rtin, M. J. Ensinger, R. F. Boone, and C-M. Wei, *This Series* **19**, 63 (1976).
155. K. Friderici, M. Kaehler, and F. Rottman, *Bchem* **15**, 5234 (1976).
156. S. Corey and J. M. Adams, *JMB* **99**, 519 (1975).
157. C. M. Wei, A. Gershowitz, and B. Moss, *Nature NB* **257**, 251 (1975).
158. D. P. Rhodes, S. A. Moyer, and A. K. Banerjee, *Cell* **3**, 327 (1974).
159. T. Urushibara, Y. Furuichi, C. Nishimura, and K. Miura, *FEBS Lett.* **49**, 385 (1975).
160. C. M. Wei and B. Moss, *PNAS* **72**, 318 (1975).
161. G. W. Both, A. K. Banerjee, and A. J. Shatkin, *PNAS* **72**, 1189 (1975).
162. G. W. Both, Y. Furuichi, S. Muthukrishnan, and A. J. Shatkin, *Cell* **6**, 185 (1975).
163. S. Muthukrisnan, G. W. Both, Y. Furuichi, and A. J. Shatkin, *Nature NB* **255**, 33 (1975).
164. B. Kemper, *Nature NB* **262**, 321 (1976).
165. K. Shimotohno, Y. Kodama, J. Hashimoto, and K. Miura, *PNAS* **74**, 2734 (1977).
166. N. Sonenberg, K. M. Rupprecht, S. M. Hecht, and A. J. Shatkin, *PNAS* **76**, 4345 (1979).
167. J. E. Bergman, H. Trachsel, N. Sonenberg, A. J. Shatkin, and H. F. Lodish, *JBC* **254**, 1440 (1979).
168. N. Sonenberg, M. A. Morgan, W. C. Merrick, and A. J. Shatkin, *PNAS* **75**, 4843 (1978).
169. E. D. Hickey, L. A. Weber, and C. Baglioni, *PNAS* **73**, 19 (1976).
170. L. A. Weber, E. D. Hickey, and C. Baglioni, *JBC* **253**, 178 (1978).
171. L. Chu and R. E. Rhodes, *Bchem* **17**, 2450 (1978).
172. H. F. Lodish and J. K. Rose, *JBC* **252**, 1181 (1977).
173. Y. Furuichi, A. LaFiandra, and A. J. Shatkin, *Nature NB* **266**, 235 (1977).
174. C. M. Wei and B. Moss, *Bchem* **16**, 1672 (1976).
175. C. M. Wei, A. Gershowitz, and B. Moss, *Bchem* **15**, 397 (1977).
176. C. M. Stoltzfus and K. Dimock, *J. Virol.* **18**, 586 (1976).
177. D. Canaani, C. Kahana, S. Lavi, and Y. Groner, *NAR* **6**, 2879 (1979).
178. M. Revel and Y. Groner, *ARB* **47**, 1079 (1978).
179. Y. Furuichi, M. Morgan, A. J. Shatkin, W. Jelinek, M. Salditt-Georgeiff, and J. E. Darnell, *PNAS* **72**, 1902 (1975).
180. M. Salditt-Georgieff, W. Jelinek, J. E. Darnell, Y. Furuichi, M. Morgan, and A. J. Shatkin, *Cell* **7**, 227 (1976).
181. M. Faust, K. E. M. Hastings, and S. Millward, *NAR* **2**, 1329 (1975).
182. C. M. Wei, and B. Moss, *PNAS* **72**, 318 (1975).
183. G. Abraham, D. P. Rhodes, and A. K. Banerjee, *Nature NB* **255**, 37 (1975).
184. R. P. Perry, D. E. Kelley, K. Friderici, and F. Rottman, *Cell* **6**, 13 (1975).
185. J. Abelson, *ARB* **48**, 1035 (1979).
186. Y. Aloni, R. Dhar, O. Laub, M. Horowitz, and G. Khoury, *PNAS* **74**, 3686 (1977).
187. M-T. Hsu and J. Ford, *PNAS* **74**, 4982 (1977).
188. M. Meselson, R. Yuan, and J. Heywood, *ARB* **41**, 447 (1972).
189. D. Nathans and H. O. Smith, *ARB* **44**, 273 (1975).
190. M. Gold, J. Hurwitz, and M. Anders, *PNAS* **50**, 164 (1963).

191. D. Fujimoto, P. R. Srinivasan, and E. Borek, *Bchem* **4**, 2849 (1965).
192. J. Doskocil and F. Sarm, *BDA* **55**, 953 (1962).
193. P. Grippo, M. Iaccarino, E. Parisi, and E. Scarano, *JMB* **36**, 195 (1968).
194. A. P. Bird and E. M. Southern, *JMB* **118**, 27 (1978).
195. F. Gautier, H. Bunemann, and L. Grotjahn, *EJB* **80**, 175 (1977).
196. C. Waalwijk and R. A. Flavell, *NAR* **5**, 4631 (1978).
197. J. Singer, J. Roberts-Ems, and A. D. Riggs, *Science* **203**, 1019 (1979).
198. A. P. Bird, M. H. Taggart, and B. A. Smith, *Cell* **17**, 889 (1979).
199. R. C. Desrosiers, C. Mulder, and B. Fleckenstein, *PNAS* **76**, 3839 (1979).
200. M. Becker, *in* "Immunoadsorbents in Protein Purification" (E. Ruoslahti, ed.), *Scand. J. Immunol. Suppl.* **3**, 11 (1978).
201. R. H. Symons, *in* "Data for Biochemical Research" (R. M. C. Dawson, D. E. Elliott, W. H. Elliott and K. M. Jones, eds.), p. 145. Clarendon, Oxford, 1974.
202. P. Brookes and P. D. Lawley, *JCS* **1960**, 539.
203. W. J. Herbert, *in* "Handbook of Experimental Immunology" (D. M. Weir, ed.), p. A3.1. Alden Press, Oxford, 1978.
204. R. D. Meredith and B. F. Erlanger, *FP* **37**, 1503 (1978).
205. M. K. Liszewski, unpublished observations.
206. H. N. Eisen and G. W. Siskind, *Bchem* **3**, 996 (1964).
207. E. A. Kabat, *in* "Experimental Immunochemistry" (E. A. Kabat and H. Meyer, eds.). Thomas, Springfield, Ill., 1961.
208. T. W. Munns and H. F. Sims, *J. Chromatogr.* **111**, 403 (1975).
209. M. Z. Humayun and T. M. Jacobs, *BBA* **331**, 41 (1973).
210. R. N. Pinckard, *in* "Handbook of Experimental Immunology" (D. M. Weir, ed.), p. 17.1. Alden Press, Oxford, 1978.
211. K. Heide and H. G. Schwick, *in* "Handbook of Experimental Immunology" (D. M. Weir, ed.), p. 7.1. Alden Press, Oxford, 1978.
212. M. J. Taussig, *in* "The Immune System" (M. J. Hobart and I. McConnell, eds.), p. 42. Alden Press, Oxford, 1975.
213. F. Karush, *Ann. N.Y. Acad. Sci.* **169**, 56 (1970).
214. P. Minden and R. S. Farr, *in* "Handbook of Experimental Immunology" (D. M. Weir, ed.), p. 13.1. Alden Press, Oxford, 1978.
215. J. L. Fahey and E. W. Terry, *in* "Handbook of Experimental Immunology" (D. M. Weir, ed.), p. 8.1. Alden Press, Oxford, 1978.
216. H. Hjelm and J. Sjöquist, *in* "Immunoadsorbents in Protein Purification" (E. Ruoslahti, ed.), *Scand. J. Immunol. Suppl.* **3**, 51 (1978).
217. E. Ruoslahti, *in* "Immunoadsorbents in Protein Purification" (E. Ruoslahti, ed.), *Scand. J. Immunol. Suppl.* **3**, 3 (1978).
218. R. Palacios, R. D. Palmiter, and R. T. Schimke, *JBC* **247**, 2316 (1972).
219. D. Eilat, P. Natale, A. D. Steinberg, and A. N. Schecter, *J. Immunol.* **118**, 1016 (1977).
220. H. Inouye, S. Fuchs, M. Sela, and U. Z. Littauer, *JBC* **248**, 8125 (1973).
221. R. Salomon, S. Fuchs, A. Aharonov, D. Giveon, and U. Z. Littauer, *Bchem* **14**, 4046 (1975).
222. E. Eilat, A. N. Schecter, and A. D. Steinberg, *Nature NB* **259**, 5539 (1976).
223. A. Aharonov, S. Fuchs, B. D. Stollar, and M. Sela, *EJB* **42**, 73 (1974).
224. T. W. Munns, K. C. Podratz, and P. A. Katzman, *J. Chromatogr.* **76**, 401 (1973).
225. T. W. Munns, K. C. Podratz, and P. A. Katzman, *Bchem* **13**, 4409 (1974).
226. B. F. C. Clark, *in* "Transfer RNA" (S. Altman, ed.), p. 14. MIT Press, Cambridge, Ma., 1978.
227. H. Aviv and P. Leder, *PNAS* **69**, 1408 (1972).

228. U. Lindberg and T. Persson, *EJB* **31**, 246 (1972).
229. M. Grunstein and P. Schedl, *JMB* **104**, 323 (1976).
230. C. Milcarek, R. Price, and S. Penman, *Cell* **3**, 1 (1974).
231. M. Nemer, M. Graham, and L. M. Dubroff, *JMB* **89**, 435 (1974).
232. C. S. Morrow, St. Louis Univ, Ph.D. Thesis, in preparation.
233. C. S. Morrow, J. R. Hunsley, M. K. Liszewski, and T. W. Munns, submitted.
234. H. S. Rosenkranz, B. F. Erlanger, S. W. Tanenbaum, and S. M. Beiser, *Science* **145**, 282 (1964).
235. R. Erickson, W. Braun, O. J. Plescia, and Z. Kwiatkowski, *in* "Nucleic Acids in Immunology" (O. J. Plescia and W. Braun, eds.), p. 201. Springer-Verlag, Berlin and New York, 1968.
236. S. W. Tanenbaum and M. H. Karol, *in* "Nucleic Acids in Immunology" (O. J. Plescia and W. Braun, eds.), p. 222. Springer-Verlag, Berlin and New York, 1968.
237. B. U. Bowman and R. A. Patnode, *PSEBM* **115**, 338 (1964).
238. A. M. Williams and F. J. Bollum, *PSEBM* **112**, 701 (1963).
239. S. S. Wallace, B. F. Erlanger, and S. M. Beiser, *JMB* **43**, 41 (1969).
240. R. D'Alisa and B. F. Erlanger, *J. Immunol.* **116**, 1629 (1976).
241. R. D'Alisa and B. F. Erlanger, *Bchem* **13**, 3575 (1974).
242. T. W. Munns, C. S. Morrow, J. R. Hunsley, R. J. Oberst, and M. K. Liszewski, *Bchem* **18**, 3804 (1979).
243. R. E. Gelinas and F. C. Kafatos, *PNAS* **70**, 3764 (1973).
244. J. M. Blanchard, J. Weber, W. Jelinek, and J. E. Darnell, *PNAS* **75**, 5344 (1978).
245. C. J. Goldenberg and H. J. Raskas, *Cell* **16**, 131 (1979).
246. H. Kasamatsu and J. Vinograd, *ARB* **43**, 695 (1974).
247. J. D. Griffith and G. Christiansen, *Annu. Rev. Biophys. Bioeng.* **7**, 19 (1978).
248. C. G. Kurland, *ARB* **46**, 173 (1977).
249. D. M. K. Rekosh, W. G. Russel, A. J. D. Bellett, and A. J. Robinson, *Cell* **11**, 283 (1977).
250. R. Davis, M. Simon, and N. Davidson, *in* "Methods of Enzymology" (K. Moldave and L. Grossman, eds.), Vol. 21, p. 413. Academic Press, New York, 1971.
251. S. I. Reed, J. Ferguson, R. W. Davis, and G. R. Stark, *PNAS* **72**, 1605 (1975).
252. M. Wu and N. Davidson, *NAR* **5**, 2825 (1978).
253. M. Wu, N. Davidson, and E. Wimmer, *NAR* **5**, 4711 (1978).
254. H. W. Fisher and R. C. Williams, *ARB* **48**, 649 (1979).
255. M. Keren-Zur, M. Boublik, and J. Ofengand, *PNAS* **76**, 1054 (1979).
256. H. M. Olson and D. G. Glitz, *PNAS* **76**, 3769 (1979).
257. O. J. Miller, R. R. Schrick, S. M. Beiser, and B. F. Erlanger, *Nobel* **23**, 43 (1973).
258. V. G. Dev, D. Warburton, O. J. Miller, D. A. Miller, B. F. Erlanger, and S. M. Beiser, *Exp. Cell Res.* **74**, 288 (1972).
259. R. R. Schrick, D. Warburton, O. J. Miller, S. M. Beiser, and B. F. Erlanger, *PNAS* **70**, 804 (1973).
260. R. R. Schreck, V. G. Dev, B. F. Erlanger, O. J. Miller, *Chromosoma* **62**, 337 (1977).
261. R. R. Schreck, B. F. Erlanger, and O. J. Miller, *Exp. Cell Res.* **88**, 31 (1974).
262. W. J. Klein, S. M. Beiser, and B. F. Erlanger, *J. Exp. Med.* **125**, 61 (1967).
263. D. Liebeskind, K. C. Hsu, B. F. Erlanger, and R. Bases, *Exp. Cell Res.* **83**, 399 (1974).
264. R. R. Schreck, W. R. Berg, B. F. Erlanger, and O. J. Miller, *Hum. Genet.* **36**, 1 (1977).
265. K. Harbers, B. Harbers, and J. H. Spencer, *BBRC* **66**, 738 (1975).
266. J. E. Gill, J. A. Mazrimas, and C. C. Bishop, *BBA* **335**, 330 (1974).
267. R. Gantt and V. J. Evans, *In Vitro* **8**, 288 (1973).
268. B. W. Lubit, R. R. Schreck, O. J. Miller, and B. F. Erlanger, *Exp. Cell Res.* **89**, 426 (1974).

269. G. C. Koo, S. S. Wachtel, K. Krupen-Brown, L. R. Mittl, W. R. Berg, M. Genel, I. M. Rosenthal, D. S. Borgaonkar, D. A. Miller, R. Tantravahi, R. R. Schreck, B. F. Erlanger, and O. J. Miller, *Science* **198**, 940 (1977).
270. J. A. Lake, *JMB* **105**, 131 (1976).
271. J. Van Duin, C. G. Kurland, J. Dondon, and M. Grunberg-Manago, *FEBS Lett.* **59**, 287 (1975).
272. E. Beck, R. Sommer, E. A. Auerswald, Ch. Kruz, B. Zink, G. Osterberg, and H. Schaller, *NAR* **5**, 4495 (1978).
273. T. W. Munns, R. Goewert, R. J. Oberst, T. Rucinsky, and M. K. Liszewski, in preparation.
274. J. D. Smith, W. Arber, and U. Kühnlein, *JMB* **63**, 1 (1972).
275. G. E. Geier, and P. Modrich, *JBC* **254**, 1408 (1979).
276. C. Heidelberger, *ARB* **44**, 79 (1975).
277. E. K. Weisburger, *Annu. Rev. Pharmacol. Toxicol.* **18**, 395 (1978).
278. R. L. Taylor, H. B. Gjika, and H. Van Vunakis, *BBRC* **80**, 213 (1978).
279. M. Guigues and M. Leng, *NAR* **6**, 733 (1979).
280. G. DeMurcia, M-C. E. Lang, A-M. Freund, R. P. P. Fuchs, M. P. Daune, E. Sage, and M. Leng, *PNAS* **76**, 6076 (1979).

DNA Structure and Gene Regulation

R. D. WELLS,
T. C. GOODMAN, W. HILLEN,
G. T. HORN, R. D. KLEIN,
J. E. LARSON, U. R. MÜLLER,
S. K. NEUENDORF,
N. PANAYOTATOS, AND
S. M. STIRDIVANT

University of Wisconsin
Department of Biochemistry
College of Agricultural and Life Sciences
Madison, Wisconsin

I. Introduction	168
A. Overview	168
B. Potential Recognition Sites on DNA for Regulatory Proteins	169
II. DNA Structure	172
A. Duplex DNA Conformations	172
B. Temperature-Dependent Conformational Changes	182
C. Long-Range Interactions in DNA	194
D. Cruciforms	199
III. Preparation of Large Amounts of DNA Restriction Fragments	200
IV. Transcription Recognition Sites	203
A. Bacterial Promoters	203
B. Effect of Supercoiling on Transcription	209
C. T7 Late Promoters	214
D. Terminators	217
V. Recognition at the Origins of RNA-Primed DNA Replication	218
A. Location of Origins	218
B. Primase	218
C. Auxiliary Proteins	219
VI. Effect of Protein Binding on DNA Conformations	224
A. General Considerations	224
B. Amino Acids and Polypeptides	226
C. Helix-Destabilizing Proteins	227
D. *E. coli* Lactose Repressor	233
E. *E. coli* RNA Polymerase	235
F. Other Proteins That Alter DNA Conformation	236
G. Conclusions: Protein Binding and DNA Conformations	237
VII. Structure in Single-Stranded Viral DNAs	238
A. General Considerations	238
B. Restriction Endonuclease Susceptibility	239
C. Single-Strand-Specific Nuclease Reactions	240
D. DNA Polymerase Reactions *in Vitro*	241
E. Electron Microscopy	243
F. Biological Importance	243

VIII. DNA Secondary Structures in Intercistronic Regions 244
    A. General Considerations and Definitions 244
    B. Single-Stranded DNA Phages 245
    C. Phage Lambda 252
    D. Summary: Secondary Structure in Intercistronic Regions 253
IX. Conclusions and Prospects for the Future 253
    References 255
    Note Added in Proof 267

> "... the idea of DNA as an inert repository of genetic information seems really to have breathed its last."
> 1979 *Nature* "News and Views"

I. Introduction

A. Overview

Our purpose here is to assess the role of DNA structure in gene regulation. We consider the properties and conformations of various DNAs and review static structure as well as dynamic transitions. Moreover, both duplex DNA and single-stranded viral genomes are examined.

Some of the goals of current research in this area are the following: determination of the properties of regions of DNA along the high-molecular-weight chromosomes; identification of the interactions between neighboring regions of DNA; determination of how the properties of DNA influence the specificity or affinity of regulatory proteins that interact with specific regions of DNA; identification of how the interaction of regulatory proteins with DNA modifies the properties of the DNA target site; investigation of the presumed correlation between the physical properties of a region of DNA and its genetic function.

Progress in this area is accelerating but continues to be hampered by the lack of suitable systems for study. A number of physical and biochemical investigations have been performed on regulatory sites contained in large viral DNAs. However, extremely sensitive probes with highly defined mechanisms must be utilized in order to have the necessary specificity and sensitivity, since the target site comprises a very small percentage of the genome. Alternatively, at the other end of the spectrum, a number of relatively small synthetic DNAs, prepared in order to answer specific conformational questions, have been investigated. Thus, in the latter case, it was possible to have an interesting structural feature in sufficiently high "concentration" to obtain rigorous answers to questions relating to their conformation and properties. However, there is good reason for optimism in the near future, since it is now possible to isolate quantities of pure DNA restriction fragments (see Section III) as well as highly purified regulatory proteins with defined genetic functions.

The concept underlying this review is that DNA properties and conformation have an intimate role in gene regulation and that DNA does not have the same conformation throughout its entire length. Instead, DNA has interesting structural features at certain loci. A corollary to this notion is that DNA structure is not unalterable but may be perturbed when certain proteins are bound. Here it should be emphasized that rigorous proof for this notion is only beginning to emerge in a number of systems. Recent data indicate that chromosomal DNA does not have the same static conformation throughout its entire length and suggest that neighboring nucleotide sequences can mutually influence the dynamic properties and conformations of each other. Thorough evaluation of these concepts must await further studies, but they are of fundamental importance for our eventual understanding of the details of genetic expression.

Of necessity, there is a transition in this review from studies on DNA properties to biochemical and genetic studies on regulatory sites such as promoters, origins of replication, operators, and intercistronic regions. Whenever possible, emphasis is placed on investigations of the properties and conformations of these regions and correlations with genetic function.

Most of the studies described were performed on prokaryotic systems because of the relative wealth of information on specific regulatory proteins that bind to defined DNA sites. Similar progress in eukaryotic systems has been hampered by a lack of this type of information. Of necessity, these investigations cut across the boundaries of different genetic systems, since work is being carried out in a number of laboratories on a variety of viral and cellular protein–DNA complexes. Thus, it is necessary to describe a number of types of interactions between specific regulatory proteins and DNA target sites.

This review is not intended to be comprehensive, but rather to be a critical evaluation of the contemporary status of the field with emphasis on developments in the last few years. It is necessarily speculative on some topics; however, it is hoped that the speculation will provoke revealing experiments.

B. Potential Recognition Sites on DNA for Regulatory Proteins

Before describing specific aspects of DNA structure and of regulatory proteins, it may be useful to identify the general types of DNA structural features that may be recognized by regulatory proteins. Several previous reviews have touched on this subject from a somewhat different standpoint (1-4).

1. Duplex DNA

a. Single-Stranded Regions. Many biological processes involve DNA in its single-stranded form. This is true for transcription and recombination as well as for the initiation of DNA replication (at least in some systems); hence there is good reason to consider that this form of DNA can provide important recognition sites. This is quite significant from the structural standpoint since a single-stranded polynucleotide, which may have a partially disordered structure, contains an abundance of specific sites for the recognition by regulatory proteins. These sites include the heterocyclic bases as well as the phosphate and sugar moieties. The structural features of these components have been reviewed previously (*1, 4*) and are also briefly discussed below.

Segments of normally duplex DNA may be partially nonhelical owing to decreased stability (Fig. 1A). In addition, certain nonhelical regions may be

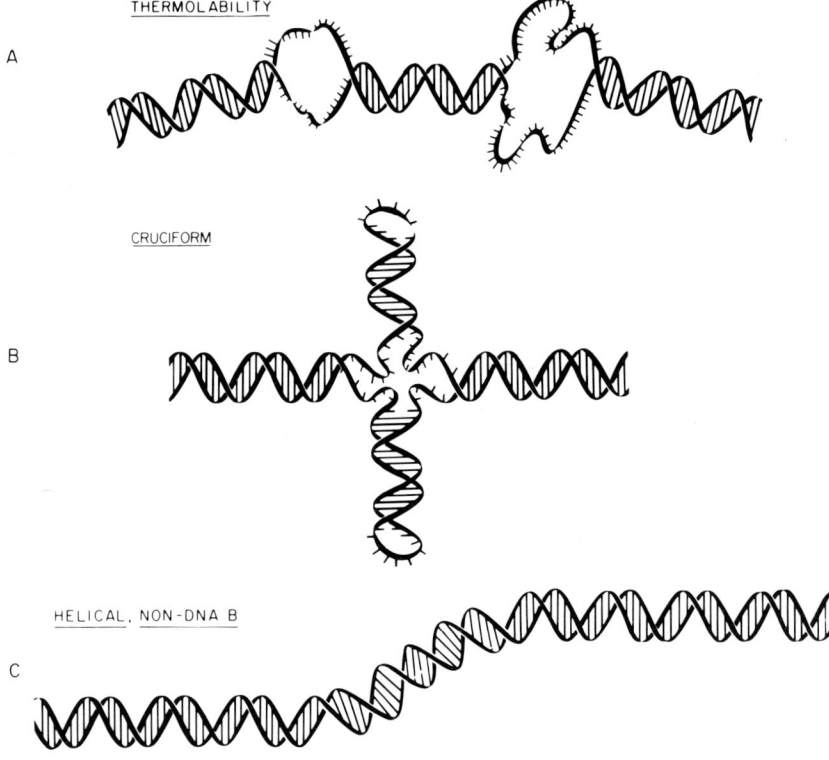

FIG. 1. Models for structural irregularities in duplex DNA that might facilitate specific protein–DNA interactions. Redrawn from Wells *et al.* (*1*). The lower figure shows "bent" DNA, also shown in Fig. 3.

stabilized by DNA binding proteins (reviewed in Section VI, C) and thus may be more generally accessible for the binding of other proteins, for example, the *dna* G protein for the initiation of replication. These sites could be generated by a region of low G+C content or by the presence of modified nucleotides that destabilize the helix.

  b. Helical Non-DNA-B Conformations. It is possible that certain segments of chromosomes contain non-DNA-B conformations, as schematically shown in Fig. 1C. These unique regions would be helical but would contain different geometric conformations from most of the DNA chromosome. For example, the tilt of the base-pairs, the sugar conformation, or the helical dimensions would be somewhat at variance with most of the DNA, which is in a DNA-B structure. Such helical variances have been recognized in a variety of DNA model systems (described in Section II, A), and the "microheterogeneity" of DNA conformations has recently been recognized by single-crystal X-ray studies on small oligonucleotides.

  c. Cruciforms. The cruciform structure (Fig. 1B) has been a favorite hypothetical conformation for serving as a DNA regulatory site. The observation that a number of genetic recognition sites contain varying lengths of inverted repeat sequences encouraged this speculation. Although these structures may exist transiently, there is no direct evidence, to the best of our knowledge, to support the existence of this type of configuration in a natural DNA for any substantial period of time. If a cruciform exists transiently, it could be stabilized by a suitable protein. Since there are nonpaired nucleotides in both the loop and branch-point regions, these would provide ample sites for protein recognition. Negative supercoils in a covalently closed circular DNA will tend to stabilize cruciform structures.

  d. Conformational Distortion. The binding of a number of proteins to DNA causes a deformation of the DNA structure (considered in Section VI). The capacity of a region of DNA to be distorted may be an important component in the recognition mechanism. Moreover, the binding of a second protein may be influenced by the prior binding of another protein; this influence may be transmitted through the DNA molecule, which thus could act as a conduit of changes in static structure or of alterations in the dynamics.

2. SINGLE-STRANDED GENOMES

  Many DNA viral genomes (ϕX174, M13, fd, f1, St-1, etc.) that are considered primarily to be "single-stranded" (that is, do not exist as double-stranded or replicative form within the virus particle) contain a substantial amount of potential double-helical hairpin-loop regions. The observations supporting this contention are considered in Sections VII and VIII.

  These double-helical regions in the single-stranded viral genomes may serve as important recognition sites for such biological processes as replication and transcription. Additional specificity for protein recognition would be provided at the ends of the duplex regions, as the nonpaired polynucleotide chains at the end of the duplex regions would contain a number of binding sites.

II. DNA Structure

A. Duplex DNA Conformations

1. X-Ray Studies on Fibers

Our knowledge of the static structure of DNA is derived mainly from X-ray studies on DNA fibers of synthetic polymers as well as natural se-

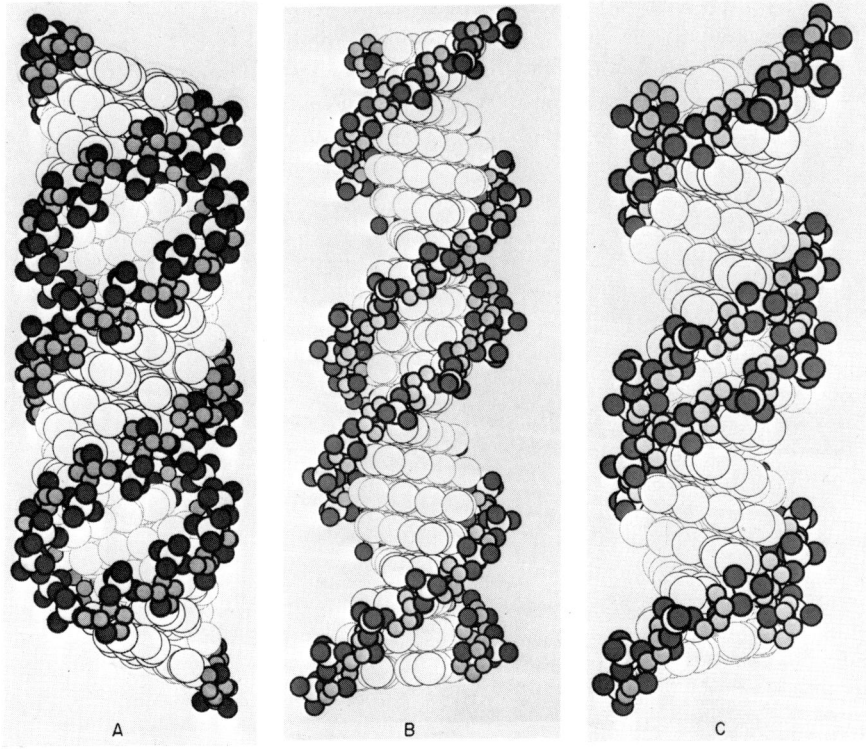

FIG. 2. Helical projections for (A) A-DNA, (B) B-DNA, and (C) D-DNA. Two helical turns are shown for each duplex conformation. All atoms of the bases are indicated as large circles, whereas atoms of the sugar–phosphate backbones are shown as smaller circles with carbon atoms smallest, oxygen atoms intermediate in size and shaded, and phosphorus atoms largest and unshaded. Published with permission from Wells et al. (1).

quences. These results have been reviewed extensively (1, 5, 6) and are therefore not covered in detail in this article. Three of the characterized conformations are shown in Fig. 2, and some of their conformational characteristics are listed in Table I. Owing to the substantial variation in the architecture of these duplex conformations, it is not difficult to envisage how regulatory proteins might recognize different regions of a chromosome if the various DNA structures are present at different loci.

It has been widely agreed that DNA in aqueous solutions of low ionic strength adopts the B-conformation (7); however, other conformations can also exist, depending on the environment. As the ionic strength of the solution is increased, the circular dichroism (CD) spectrum of the DNA changes in the long wavelength range. This has been interpreted as a continuous, noncooperative transition of the B-conformation to the C-conformation (9, 10), with possibly a small portion in the A-form (11). By increasing the ethanol concentration in a DNA solution of very low ionic strength, the double-helix dehydrates and undergoes a cooperative transition to the A-conformation (12–14). This transition is supported by the presence of polyamines like spermine, spermidine, etc., as it occurs at lower ethanol concentrations when the solution contains these polycations (15).

TABLE I
STRUCTURAL PARAMETERS OF DUPLEX CONFORMATIONS OF POLYNUCLEOTIDES[a]

Helix conformation	Helix symmetry	Rise per residue (Å)	Base-pair parameters			
			Tilt (°)	Twist (°)	γ (°)	R (Å)
A-DNA	11_1	2.56	19.3	−3.2	20.2	4.5
B-DNA	10_1	3.38	−5.9	−2.1	6.3	−0.2
C-DNA[b]	28_3	3.31	−8.0	1.0	8.0	−0.9
D-DNA	8_1	3.03	−16.0	5.6	16.4	−1.8
B'-DNA	10_1	3.29	−7.9	−1.0	8.0	−0.1
A-RNA	11_1	2.82	16.0	−6.9	17.4	4.4
A'-RNA	12_1	3.00	10.0	−7.6	12.5	4.5

[a] Selected conformational features of DNA and RNA double-stranded helices are listed. Helix symmetry is given as N_m (i.e., N residues per m turns of helix). Marvin et al. (505) describe tilt and twist angles in greater detail. Tilt is essentially the angle that base-pair planes make to the helix axis, and twist is the angle between base planes in a base-pair (bases are not exactly coplanar in a base-pair, but are related like the blades of a propeller). Tilt and twist angles can be easily visualized in Fig. 4 in ref. 1. The parameter γ is a composite of tilt and twist and is described in ref. 1. R is the x coordinate of purine C-8 in the helix [as described in (506)] and represents the distance of base-pairs from the helix axis. Positive and negative values of R denote opposite directions from the helix axis, as can easily be seen in Fig. 4 of ref. 1.

[b] Values are for a representative C-DNA structure (1).

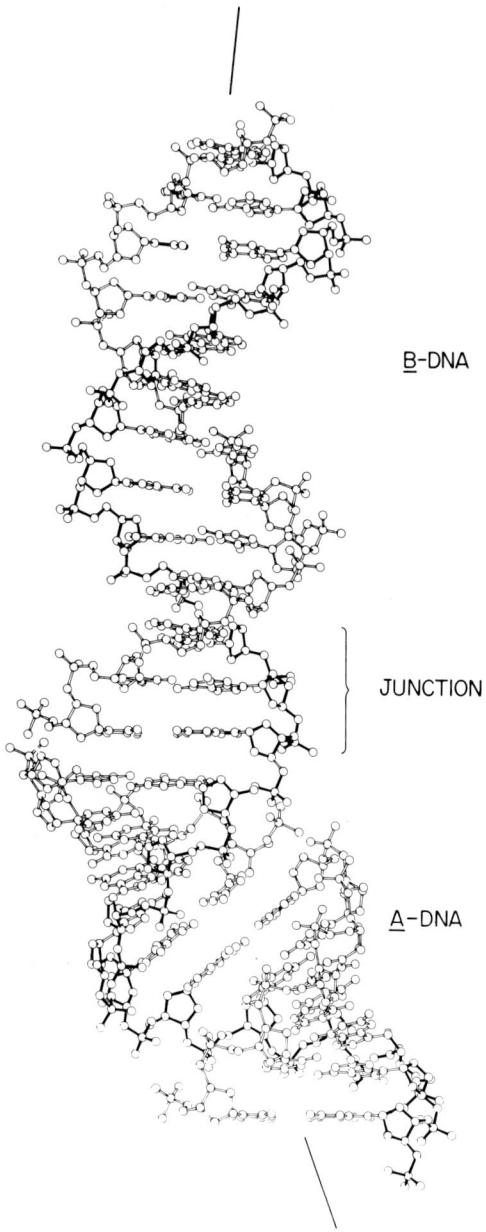

Fig. 3. Bent DNA: the result of neighboring A-DNA and B-DNA helical segments. One helical turn of both A- and B-DNA is depicted, and the axes of the joined helices are shown. The base-pairs encompassing the junction region are indicated. The bonds of one sugar–phosphate strand of this duplex are depicted as solid for clarity. Reprinted, with permission, from Selsing et al. (19).

Both high local charge density and a hydrophobic environment can be provided by DNA binding proteins, which therefore may influence the DNA conformation.

In contrast to DNA, RNA cannot adopt a B-conformation, owing to steric interaction of the 2′-hydroxyl group and the bases (16). A DNA·RNA hybrid, which may be an intermediate of transcription, can therefore exist only in the A-conformation. The synthesis and characterization of $dG_n \cdot (rC_{11}dC_{16})$ showed indeed that the hybrid (dG·rC) part of this complex adopts the A-conformation, whereas the dG·dC part maintains the B form (17–19). These molecules were the first examples of A–B junctions; a visualization is shown in Fig. 3. The junction region comprises only one base-pair and the two neighboring internucleotide linkages, and exhibits full hydrogen-bonded base-pairing, full base-stacking, and unexceptional stereochemistry. In addition, the junction has a mixed sugar-ring pucker with the junction base-pair adopting C(2′)-endo and C(3′)-endo furanose rings in the complementary strands. Since the junction is fully base-stacked, the differences in base tilt between A-DNA and B-DNA result in a bend of 26° in the duplex at the junction. This study demonstrates that structural diversity within one molecule can be accomplished, at least in the case of an A–B junction, without the energetically unfavorable loss of hydrogen bonds or base-stacking interactions. It can be derived from Fig. 3 that a repeating structure of alternating A-, B-conformations results in a highly bent molecule, which is similar to other recently developed models to be discussed below.

2. Non-DNA B-Conformations in Solution?

Studies by Wells *et al.* (1, 20) on duplex DNA polymers with repeating nucleotide sequences first indicated that the B-conformation was not the only structure in solution. Investigations were performed on homopolymers, on repeating sequences with self-complementary structures, and on repeating dinucleotide, trinucleotide, and tetranucleotide polymers (reviewed in *1* and *20*). The physical and biochemical studies performed included absorbance–temperature transitions, absorbance–pH, analytical buoyant density in CsCl and Cs_2SO_4 gradients, binding of actinomycin D, binding of netropsin, the specific formation of three-stranded complexes between double-stranded DNA and single-stranded RNA and their selective inhibition of transcription, viscosity, X-ray diffraction, circular dichroism, ultraviolet spectra, high-resolution nuclear magnetic resonance (NMR) studies, *in vitro* replication, *in vitro* transcription, *lac* repressor binding and interferon induction. *All of these determinations clearly demonstrated that the sequence of nucleotides dictated the properties and, in some cases, the conformations of the DNA* (*1, 20*). Similar types of studies are presently underway on DNA restriction fragments of defined sequence.

Other considerations also question the existence of a "pure" B-DNA conformation in solution. The necessity of folding DNA into highly condensed structures in nucleosomes has led to the proposal of kinks interrupting the base-stacked double helix in a regular manner, with preservation of the hydrogen bonds (21, 22). However, comparison of the NMR signals assigned to the hydrogen-bonded protons obtained from spectra of nucleosome core particles and protein-free 140-base-pair fragments of DNA revealed no differences that could support the proposal of a kinked structure of either type (23). The conclusion, therefore, is that DNAs free in solution and in nucleosome core particles exhibit similar secondary structures that maintain continuous base-stacking. The kink hypothesis was also challenged by propositions that short DNA fragments could bend smoothly with a much shorter radius (24) than was previously concluded from the frequency of forming circular ligation products from fragments of various lengths (25). The conclusion on the smooth bend was independently derived by building models (26), by calculating energy minima (27), and by allowing only small individual changes for the otherwise "rigid" (28-30) nucleic acid backbone bonds (31, 32). It has also been suggested that neutralization of the negative phosphate charges on only one side of the double helix by a basic protein leads to bending as a result of the asymmetrical loss of charge repulsion (33).

These concepts revived interest in the structural behavior of DNA in solution, which led to the application of a variety of new techniques to this problem. Most of the recent results described below indicate that DNA is more varied conformationally than was recognized some years ago. The *Nature* editorial (34) quoted at the beginning of this essay emphasizes this notion.

a. Helical Repeat. The length of the helical repeat of DNA has been studied recently by several different approaches. The contour lengths of ϕX174 duplex DNA and RNA·DNA hybrid molecules were measured by electron microscopy, which revealed that the DNA double-helix in dilute solution had a rise per residue of 2.9 Å and a helical repeat unit of 10.5 base-pairs (35). These values differ considerably from the 3.4 Å rise per residue and the 10.0 base-pairs per helical repeat for the B-conformation.

Nearly the same result was obtained by analyzing the distribution of superhelical forms generated by "topoisomerase" in plasmids containing inserts with an increasing number of base-pairs. This method measures the helical repeat of the insert, which in this case, after the necessary size corrections, turns out to be 10.4 ± 0.1 base-pairs per turn (36, 37).

Moreover, nuclease digestion of the DNA in histone core particles revealed that approximately every tenth phosphodiester bond is more sensitive to the cleaving activity than the others (38). A more detailed analysis of this cleavage pattern also leads to a value of 10.33–10.40 base-pairs as the helical repeat unit in this complex (39).

However, a recent attempt to detect the alteration of the helical repeat number for diluted DNA by X-ray studies of DNA fibers containing increasing amounts of water indicated no difference in the conformation of the double-helix under these experimental conditions (40).

b. Internal Mobility. In agreement with the theoretical considerations of conformational freedom of the phosphodiester and deoxyribose bonds mentioned above as one basis of conformational flexibility in the DNA double-helix, recent NMR studies show that the phosphodiester backbone possesses a much greater internal mobility than was previously believed. The NMR relaxation measurements of the ^{31}P signal in free DNA and DNA complexes with histones showed in both cases that internal motion occurred with a correlation time of about 4×10^{-10} second (41). A similar result was obtained from an analysis of the ^{1}H NMR signals of DNA more than 200 base-pairs in length. The correlation times were less than 3×10^{-7} second for the reorientation of base-pairs and less than 5×10^{-8} second for the motion of sugar protons relative to the aromatic hydrogens (42).

c. Alternating Conformations. Alternating conformations of the phosphodiester bonds and deoxyribose puckers were recently observed for the crystal structure of $(dA-dT)_2$ (43). Figure 4 shows the tetranucleotide in this "alternating B-form." A model built on the basis of the crystal structure of $(dA-dT)_2$ extends this special feature to poly $(dA-dT)$ in solution (44). The possibility of an alternating B-conformation was discussed by Klug *et al.* (44) in light of the increased binding of the *lac* repressor protein to derivatives of $(dA-dT)_n$ in which the methyl group of the thymidine residues was replaced by bromine. Also the specificity of pancreatic DNase I to cleave between A and pT rather than between T and pA, which results in a distribution of even-numbered oligonucleotides, can be explained more easily by an alternating conformation of the polynucleotide. Although it may be somewhat dangerous to extrapolate helical parameters from a tetranucleotide structure having a staggered arrangement of the single strands (43) to a model for DNA in solution, this concept gains some support from ^{31}P NMR measurements on oligo- and polynucleotides with alternating purine–pyrimidine sequences (discussed in Section II, A, 2).

Determination of the crystal structure of $(dC-dG)_3$ revealed a very unusual conformation of the nucleotides in this self-complementary hexamer. Also, the double-helix formed in the crystal showed novel features (45). This oligomer crystalizes in a helix with antiparallel strands forming Watson–Crick hydrogen bonds. The helical symmetry, however, is left-handed and the distribution of nucleotides along the helix axis is not as even as in

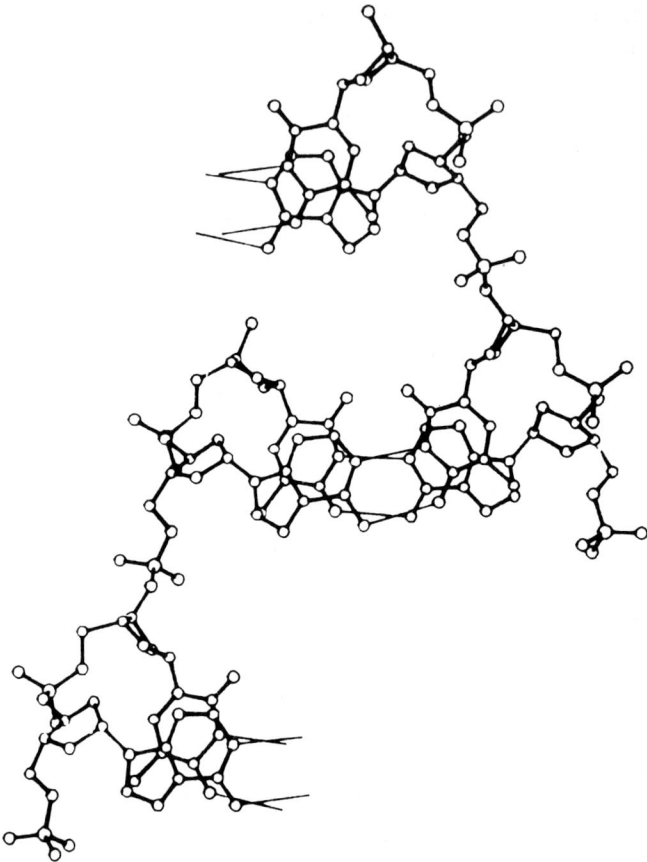

FIG. 4. View of the minihelix perpendicular to the bases found in the crystal structure of p(dA-dT)$_2$. The sugar–phosphate backbone exhibits the deoxyribose conformations alternating between C(2′)-endo and C(3′)-endo. The phosphodiester linkages alternate between the regular *gauche–gauche* conformation for the dA-dT part and the *trans–gauche* conformation for the dT-dA part of the tetranucleotide. Thin lines indicate hydrogen bonds. Reprinted, with permission, from Viswamitra (*43*).

B-DNA. Whereas all common DNA structures described so far require the nucleotides in the *anti* conformation for the base with respect to the deoxyribose, all guanosine residues in this structure exhibit the *syn* conformation. The oligomer also has an unusual deoxyribose pucker; it is C(3′)-endo for the internal two guanosines of the hexamer and C(2′)-endo for the guanosine at the end. The torsion angle around the C(4′)-C(5′) bond is *gauche–trans* for the dG residues instead of the usually observed *gauche–gauche* conformation of the rC residues.

Despite these variations from "normal" structures, (dC-dG)$_3$ forms a regular double-helix in the crystal with only minor deviations at the ends of the hexanucleotides, which is possibly due to the lack of phosphates. It is obvious, from the unusual conformation of the nucleotides, that only a different helix conformation could incorporate all these features. The result is a left-handed helix, which the authors call the Z-form, with several unusual parameters. Figure 5A compares space-filling models of B- and Z-helices. Two of the most striking features of the Z-DNA are the lack of a major groove and the zigzag distribution of the phosphates around the helix. The authors conclude that this is the high-salt form of (dG-dC)$_n$ (see below) and that the high ionic strength is necessary to compensate for the increased charge-repulsion of the phosphates.

Arnott et al. (46) have interpreted fiber X-ray diffraction patterns on several DNA polymers containing repeating purine–pyrimidine sequences including (dG-dC)$_n$ (47–49, 1, 5) and (dT-dG)$_n$ · (dC-dA)$_n$ (1, 5, 50, 47). Although these polymers usually assume the common A- or B-conformations, they also seem to adopt a Z-DNA structure (the left-handed, double-helical conformation) under some conditions. The proposed conformation for the G residues in these structures are *syn* rather than *anti*, the rotation at C(4')-C(5') is *trans* as opposed to *gauche*$^+$, and the conformation along the P-O bonds are *gauche*$^+$ rather than *trans* or *gauche*$^-$. This left-handed helix would also be free from nonbonding contacts.

In 1970, Wells et al. (51) showed that (dI-dC)$_n$ adopts an unusual conformation, which was interpreted as a left-handed double-helix on the basis of CD and X-ray studies on oriented fibers. It was proposed (51) that these types of structures are recognition sites for regulatory proteins. Moreover, the conformational feasibility of left-handed helices has been documented (52, 53). Hence, the notion of left-handed double helices is not new. It is likely that those regions in natural DNA that adopt the Z-conformation are composed of alternating purine–pyrimidine sequences. However, there is disagreement on the nature of the junction between B-DNA and Z-DNA. Arnott et al. (personal communication) propose a smooth junction, whereas Rich et al. (45) mention the possibility of a kinked structure at the junction.

^{31}P NMR measurements performed on the double-stranded (dG-dC)$_8$ in high concentrations of salt showed two distinct peaks of roughly the same intensity for the phosphate resonance (54). Although not yet proved, the most likely interpretation is that alternating phosphodiester bonds adopt different conformations. The same split of chemical shifts is also apparent for the resolved sugar protons, indicating that the ring puckering is also different for the alternating nucleotides (54). For (dG-dC)$_n$, these conclusions provide a possible explanation for the Pohl–Jovin transition, which was previously observed in the CD spectra of poly (dG-dC) at high concentration of

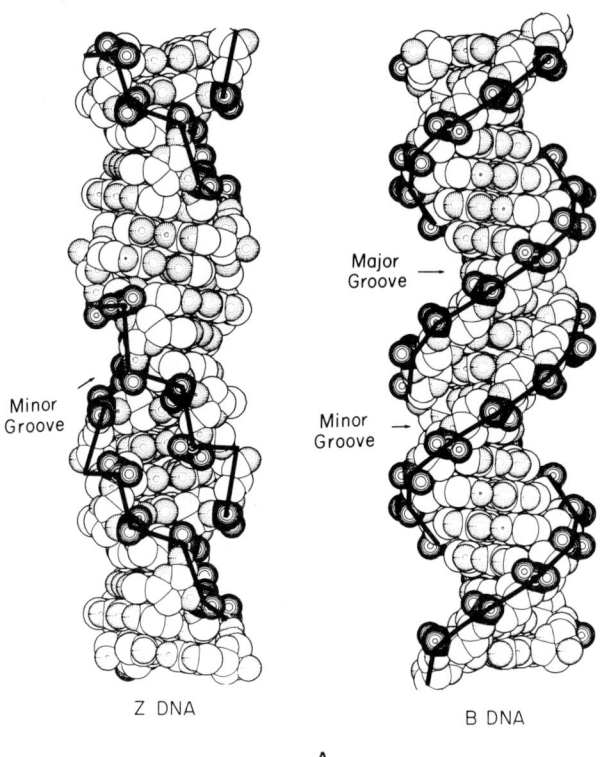

FIG. 5. (A) van der Waals side views of Z-DNA and B-DNA (45). Two views of Z-DNA that are 30° apart in orientation about the helix axis are shown. The irregularity of the Z-DNA backbone is illustrated by the heavy lines that go from phosphate to phosphate along the chain. This includes positions where the phosphates are missing in the crystal structure but would be occupied in a continuous double-helix. The minor groove in Z-DNA is quite deep, extending to the axis of the double-helix. In contrast, B-DNA has a smooth line connecting the phosphate groups and two grooves, neither one of which extends into the helix axis of the molecule.

salt (55, 56), and also for crystals of $(dC-dG)_2$, which adopt different space groups when grown from low- and high-salt solutions (57), and they are in agreement with the crystal structure of $(dC-dG)_3$ described above.

A similar splitting of the phosphate signals has also been observed for $(dA-dT)_n$ samples when the chain length was short enough to yield a sufficiently narrow line width to make the split signal observable (58). This finding was also interpreted as the result of an "alternating B-conformation" (44). Hence, there is good agreement in general between the NMR and the X-ray results.

These studies also strongly suggest "microscopic conformational

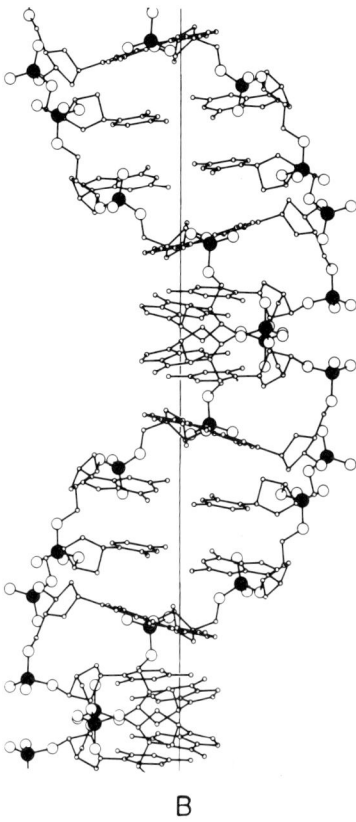

(B) A two-pitch length segment of Z-DNA viewed perpendicular to the helix axis and showing the disposition of the phosphates, whose atoms are depicted with larger circles. Reprinted, with permission, from Arnott et al. (46).

heterogeneity" along DNA double-helices due to unique local nucleotide sequences. Future studies may reveal a substantial number of conformations unlike the traditional DNA-B form of the double helix.

The relationship of these determinations on small oligonucleotides to high-molecular-weight DNA is always uncertain. However, tools are now available for incorporationg such sequences within natural molecules (59, and described in Section II, D). Thus, we can expect these questions to be addressed in a rigorous manner in the not-too-distant future. Moreover, the finding that a small amount of internal flexibility of single backbone bonds, which should not be sequence-dependent, that adds up to an overall conformational freedom of a large DNA molecule may also change our view of structure–function relationships of biologically important sites.

B. Temperature-Dependent Conformational Changes

1. PREMELTING CHANGES

The subject of premelting changes in DNA conformations has been thoroughly reviewed (60). Subsequent studies (11) indicate that the amount of DNA in the C-form decreases in favor of the B- and possibly A-form as the temperature increases. This conclusion was derived from CD studies and was made possible because the influence of temperature and salt on the number of superhelical turns that originate from differences in the winding angle had been characterized (61).

2. MELTING CHANGES

a. High-Resolution Thermal Denaturation Studies. The helix-coil transition in DNA has been the subject of numerous reviews (2, 62, 63). Since virtually all important biological processes (such as replication, transcription, and recombination) on DNA involve disruption of the duplex strands, a thorough understanding of the mechanism of strand separation is of fundamental importance. Improved instrumentation and data interpretation (63) justify optimism for our understanding in the near future of the segmental melting of DNA with resolution of at least one turn of helix (approximately 10 base-pairs) and perhaps even at the level of individual base-pairs. This degree of resolution is 3- to 10-fold higher than achievable by electron microscopy.

Since RNA polymerase causes a local strand separation (see Section IV), the thermodynamic properties of promoter regions may be quite important in the efficiency of transcription. Several investigators have noted that promoters are frequently located in or immediately adjacent to regions of DNA that contain unusually high percentages of dA and dT (64-67). However, this is not the case of the class II promoters of bacteriophage T7 (68). This richness in dA and dT may allow the transient opening of the helical structure, either under conditions of torsional strain induced by superhelicity (67) or in conjunction with some protein-DNA interaction that alters the local environment of DNA (69). It has been proposed that this aids in promoter recognition by allowing RNA polymerase to bind preferentially to transiently single-stranded regions of the DNA duplex, and subsequently to migrate along the DNA to the transcription start site where an initiation complex could then be formed (reviewed in 70). Though high (dA+dT)-content is not in itself sufficient to permit specific polymerase binding and the initiation of transcription, the degree of (dA+dT)-richness may affect promoter efficiency and facilitate RNA-polymerase-mediated separation of the DNA strands (64, 66). Supporting this model is the demonstration that *Escherichia coli* RNA polymerase binding unwinds the DNA template (71-73) in the region of the promoter (74). For these reasons, the

DNA STRUCTURE AND GENE REGULATION 183

relative structural stability of regions along the DNA helix may be of considerable biological importance, at least in the process of transcription.

Differences may exist between the type of helix disruption involved in biological processes and that observed in DNA melting studies. Hydrogen-bonding and base-stacking are the two main factors contributing to the stability of the double-helix. During thermal denaturation of DNA, both interactions are disrupted. In the resulting single strands, hydrogen-bond donor and acceptor sites are immediately occupied by water molecules. The energy difference between these states contributes to the stability of the double-stranded conformation. Therefore, stacking interactions are the main stabilizing force. On the other hand, in a biological process such as transcription, disruption of hydrogen bonds and exposure of functional groups is possible without greatly affecting stacking interactions.

Some of the questions yet to be answered are the following.

What is the influence of the arrangement of DNA bases on melting properties (position and size of transitions)?

What is the influence of mutations (deletions and even point mutations) on thermal properties?

What is the range and magnitude of long-range thermodynamic effects (telestability effects)?

What is the influence of ligand binding on thermodynamics of transitions in DNA? These ligands could be specific proteins (repressors, RNA polymerase, positive transcription regulatory factors, proteins involved in DNA replication) or drugs (actinomycin, adriamycin, netropsin), carcinogens (such as polycyclic hydrocarbons) that bind to DNA, or cross-linking agents (such as mitomycin C or nitrous acid).

What is the possible correlation between cooperative transitions and genetic function?

What is the influence of degree of supercoiling on thermodynamic properties of a given region and, therefore, on genetic function of this region?

The highly cooperative denaturation of DNA has been widely studied in the past by ultraviolet spectroscopy as well as by other techniques. A number of investigators have reported that the melting of DNA fragments and viral DNAs consisting of at least 1000 base-pairs does not occur in a single transition but is split into a number of subtransitions (63). In some cases, the relative locations of the subtransitions are sensitive to the ionic strength of the medium (75, 76). (For reviews of salt effects on nucleic acids, see 77 and 78.) With the advent of more sensitive instrumentation and the application of computers to the problems of data gathering and interpretation, the quality of high-resolution thermal denaturation studies has improved dramatically. The most recent instrumentation, techniques, and theory involved in these studies has already been reviewed in considerable detail (63). The present review deals with high-resolution thermal denaturation as a probe for DNA stability and indicates its potential for fine-structural stability mapping.

Michel and co-workers employed high-resolution DNA melting and reassociation experiments to analyze yeast mitochondrial DNAs from 13 *petite* mutants (79). By comparing the melting profiles of mitochondrial DNAs having known combinations of genetic marker deletions, they were able to assign the DNA region conferring erythromycin resistance to two peaks melting at 72°C and 75°C. This is considered to be caused by a selective enrichment in the various mutants for gene-specific sequences. Michel has also used the hysteresis and the partial irreversibility of DNA denaturation to construct a model describing the distribution of (A+T)- and (G+C)-rich blocks in the above-mentioned mutants (80). This experimental approach has considerable power in mapping the physical locations of cooperatively melting regions in large unsequenced DNAs. The theory and technique of this approach, as well as preliminary data analyzing the leftmost 40% of λ *plac*5 DNA, has been reviewed (81).

A number of studies have been published on the complex pattern of thermal stability exhibited by DNA from bacteriophage lambda (82–87). From the standpoint of relating the knowledge gained through high-resolution thermal denaturation studies on lambda to gene structure, the experiments of Gotoh *et al.* (83) are the most informative. Figure 6 shows the results of one of their experiments in which the melting of λ DNA was compared to that of a deletion mutant. The deletion mutant was found to be lacking several of the cooperatively melting regions found in the undeleted DNA. Owing to the size, complexity of melting, and lack of a complete nucleotide sequence for lambda, a more extensive interpretation of the rich detail shown in these melts is not yet possible.

Studies such as these are important from the standpoint of providing a benchmark for characterizing the sensitivity of the instrumentation system and data-handling procedures. Aside from this, however, without additional characterization such as by the technique of Michel or analysis of a more extensive collection of deletion mutants, the melting data per se add little to our knowledge of how DNA structural stability may play a role in gene regulation.

Tachibana *et al.* (88) have employed high-resolution thermal denaturation to map the thermal stability of the whole genome of fd phage by examining the melting of the linear replicative form and a number of overlapping fragments produced by restriction-enzyme digestion of the phage DNA. The object of this study was to devise a map that physically located cooperatively melting regions of the DNA, and to compare the locations of these regions

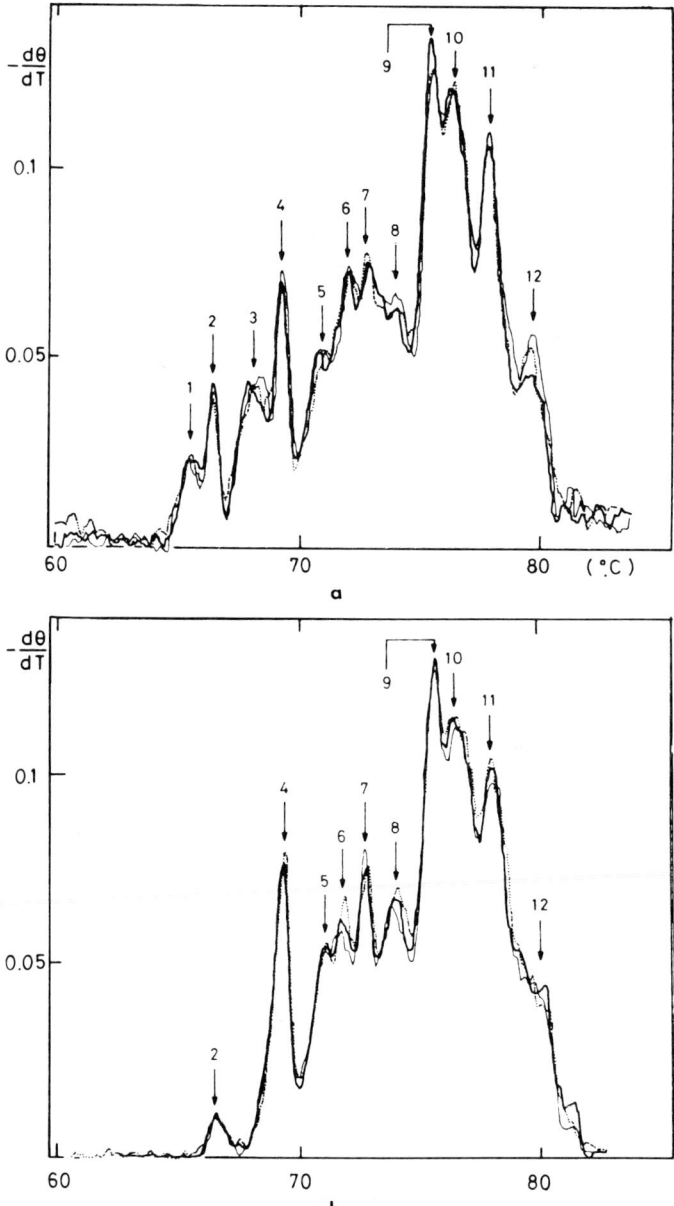

FIG. 6. (a) Differential melting profile of λcI857S7 DNA in 15 mM NaCl, 1.5 mM trisodium citrate. (b) Differential melting profile of deletion mutant λcIb2b5 DNA in 15 mM NaCl, 1.5 mM trisodium citrate. The three curves in each represent independent determinations. Reprinted, with permission, from Gotoh et al. (83).

with the genetic map. Although this study was accomplished before the entire nucleotide sequence of fd phage was known, and subsequent analysis revealed a number of errors in assignment (89), the assignment of regions of thermal stability was largely correct. The results from this line of experimentation show in some cases meaningful correlations between the boundaries of genes and cooperatively melting regions (89). Additional studies on a larger number of DNAs will be necessary before it is known how significant this correlation is, but the strong implication (81, 89, 90) is that the stepwise melting of DNA may reflect genetic structures.

Detailed melting studies have also been performed on the replicative form of ϕX174 DNA (91, 92), and attempts have been made to compare the theoretical and experimental melting profiles (90, 93, 94). A detailed discussion of the calculation of theoretical melting curves from DNA sequences has been presented elsewhere (63) and will not be presented here. It is likely that this will be a lively area of nucleic acid research in the near future as a large number of nucleic acid sequences are presently being published and as it may be possible to improve present theoretical calculations.

The thermal stability of small DNA restriction fragments of known sequence carrying genetic elements from the lactose operon have been examined (95). The rationale for using small fragments is threefold. First, by obtaining a higher concentration of a specific sequence, small melting regions in that sequence constitute a larger percentage of the total observed hyperchromicity. This allows them to be more easily distinguished from the background "noise" inherent in spectrophotometer measurements. Second, because a smaller piece of DNA containing fewer cooperatively melting units is being analyzed, the data are easier to interpret; thus, a more definitive assignment of cooperatively melting regions to physical locations on the DNA is possible. Third, the sequences of all fragments under study are known. An additional strategy employed in these studies (95) has been the use of overlapping fragments. This allows further confidence in distinguishing cooperatively melting transitions and assigning them to their physical locations on the DNA molecules.

Most high-resolution denaturation studies require multiple determinations, each using relatively large amounts of highly purified DNA. Though an exception to this can be found in the novel microcell designed by Grachev and Perelroyzen (96), a lack of material has limited many studies to types of DNA that could be readily obtained and purified in large quantities. In view of this, our studies have been made possible only because of the techniques of gene cloning and the ability to separate and purify large amounts of DNA restriction fragments by sucrose gradient sedimentation or RPC-5 column chromatography (97–101).

Figure 7 shows some melting data for DNA restriction fragements carrying the lactose operon control region. These results show that an 81-basepair region containing the lactose promoter is unexpectedly sensitive to

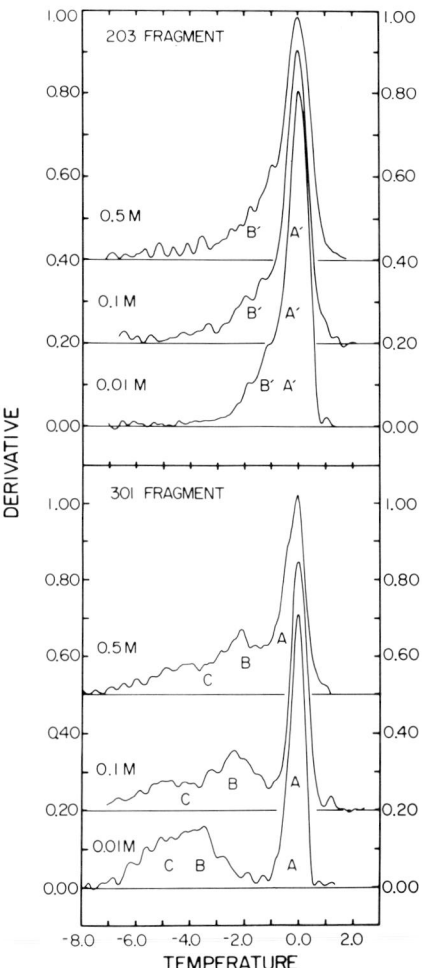

FIG. 7. High-resolution melting profiles on DNA restriction fragments carrying control regions of the lactose operon. *Upper three panels:* Differential melting curves of the 301-base-pair fragment. The vertical axis represents $d\theta/dT$, where θ is the hyperchromicity at 260 nm normalized to a total of 1. The horizontal axis represents degrees centigrade displacement from the position of the maximum derivative value. Concentrations of Na^+ are as indicated. Potassium phosphate buffer, 1 mM, was used in all determinations. The results of duplicate experiments are shown for 0.5 and 0.1 M Na^+. The 0.5 and 0.1 M curves are offset by 0.5 and 0.2 unit, respectively, on the vertical axis. The t_m's of the major transitions were 71.4°, 86.2°, and 94.5°C for 0.01, 0.1, and 0.5 M Na^+, respectively.

Lower three panels: Differential melting curves of the 203-base-pair fragment. The curves at 0.5 and 0.1 M Na^+ are offset by 0.4 and 0.2 units, respectively, on the vertical axis. The t_m's of the major transitions were 67.6°, 83.1°, and 91.8°C at 0.01, 0.1, and 0.5 M Na^+, respectively.

The melting data shown are interpreted in the case of the 301-pair fragment as showing three peaks, A, B, and C; in the case of the 203-pair, as showing two partially overlapping peaks labeled A' and B'. These letters relate the cooperatively melting regions to the physical and genetic map shown in Fig. 8. The behavior of the 81-pair region containing the promoter (B) is unusual because lowering the salt concentration from 0.1 M to 0.01 M causes an approximately 1.5°C shift in the melting temperature of this region relative to the major transition.

changes in salt concentration. This was determined by comparing the melting of the region at various salt concentrations in a fragment 301 base-pairs long. The melting of a 203-base-pair subfragment was consistent with the behavior found in the large fragments. Figure 8 shows the arrangement of these fragments with respect to each other and the genetic map and summarizes our conclusions showing how thermal stability is distributed in this genetic control region. Theoretical calculations of melting transitions of these fragments have succeeded in reproducing the stepwise melting as well as the melting temperatures. Hence, the interpretation given in Fig. 8 was confirmed by these studies (R. M. Wartell, personal communication). Apparently the cooperative forces in the region of the *lac* promoter are balanced so that its stability is particularly sensitive to change in salt conditions. Inspection of the nucleotide sequence in this region shows that the 81-base-pair region consists of adjacent (A·T)- and (G·C)-rich blocks. The unusual thermal stability and sequence arrangment may be related to the biological

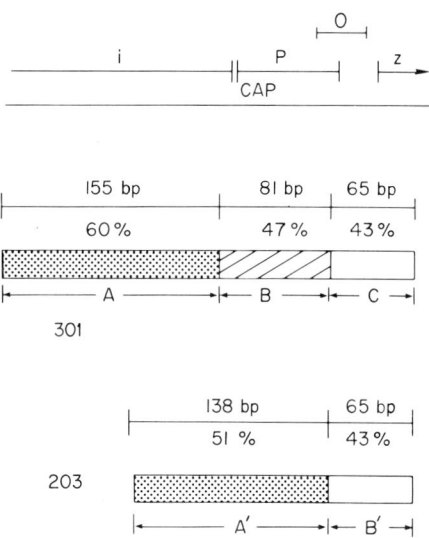

FIG. 8. Map of cooperatively melting regions in the 203- and 301-base-pair fragments. The genetic loci are shown at the top. Symbols are: i, region coding for C-terminal of *lac* repressor; P, promoter; CAP, CAP binding site (also abbreviated CRP); O, operator, A, region coding for N terminal of β-galactosidase. Stippled boxes represent the higher-melting cooperative regions, open boxes indicate the lower melting cooperative regions, and the hatched box in the 301-pair fragment represents the intermediate-melting cooperative region observed in this fragment. The letters beneath these boxes relate these regions to the derivative melting data shown in Fig. 7. The sizes of the regions (in base-pairs) and their composition (in percentage of dG+dC) are also shown. Reprinted, with permission, from Hardies et al. (95).

functions of this region [promoter, catabolite–gene-activator protein (CRP) binding site, operator, i-gene terminator, z-gene start site].

A more detailed discussion of promoter function and the role of protein–DNA interactions in gene regulation is presented elsewhere in this review. However, it is appropriate to mention here that the stability of the promoter could be sensitive to the binding of CRP as has been suggested (102, 103). The CRP binding could stimulate transcription by destabilizing the DNA that RNA polymerase must denature during the initiation of transcription. These results show that the CRP binding site is at the boundary between two melting regions where it could influence the degree to which the adjacent (G·C)-rich region can stabilize the promoter.

b. Conformational Changes on Strand Separation. Upon melting of the double-helix, the single strands adopt a random-coil structure. The rather limited degrees of conformational freedom of the glycosyl torsion angles in the double-helix are spread over a wide range in the single strands. Nuclear magnetic resonance has been a powerful tool for evaluating the secondary and tertiary structures of nucleic acids (104,16) and for the study of the conformational changes in the backbone accompanying denaturation. The results, however, seem to depend to some extent on the source of the DNA. For example, while calf thymus and salmon sperm DNA seem to have the *gauche–gauche* conformation over the phosphodiester bonds in both cases (105), poly(dA-dT) seems to shift from the *gauche–gauche* to the *gauche–trans* conformation upon melting (106). Unfortunately, it is not possible, even in short double-helices, to assign the ^{31}P signals to individual phosphodiester bonds (107).

The comparison of ^1H chemical shifts and line widths in the melted and double-stranded conformations allows conclusions about the geometry of base overlap and the dissociation rate constant for strand separation (106, 107). The melting of the (associated) block polymers $d(C_{15}\text{-}A_{15}) \cdot d(T_{15}\text{-}G_{15})$ proved that the dA·dT base-pairs melted slightly before the dG·dC pairs (108), although there was no noticeable difference in the transitions for the two ends of the molecule when monitored by UV hyperchromicity (109,110). Aside from dinucleotides (111), it is as yet impossible to resolve and assign all signals of even short DNA fragments, which would allow a complete conformational description of the molecule (112).

3. PROBES FOR NONPAIRED REGIONS

In order to detect small regions of nonpaired nucleotides in regions of dynamic "breathing" or in large chromosomes, it is necessary to have an exquisitely sensitive probe for nonpaired nucleotides. As indicated immediately below, it has been possible to make some progress in this area. However, an ideal probe has not yet been characterized. Moreover, when an investigator places great reliance on the results from a very sensitive measurement, he must also know a great deal about the mechanism of the probe itself. For most of the agents described below, this information is fragmentary at best.

a. Endonucleases. DNA endonucleases specific for nonhelical DNA have been used to indicate that regions of duplex DNA undergo transient "breathing". Some years ago (*113,114*), it was demonstrated that micrococcal nuclease attacked helical DNA at (dA·dT)-rich regions, whereas little preference for specific sites was found when a denatured DNA was used as substrate. This observation suggests that the enzyme sought out transiently open regions to the substrate. These results have been greatly extended in recent years, but nucleases were used that exhibit a very high preference (several thousandfold) for nonhelical DNA, such as S1 (*115*) and mung bean (*116*) nucleases.

The potential of using these enzymes as probes for dynamic structure of DNA was proposed (*116*) and has been exploited by determining the sensitivity to them of specific regions of lambda *plac* DNA (*117,118*). Lambda *plac*5 DNA was treated with various concentrations of nuclease, and the modified DNA was analyzed for its *lac* repressor binding activity as well as the number of "nicks." The repressor binding capacity of the λ *plac* DNA was maximally reduced by as few as 2 to 5 cuts per molecule by either S1 or mung bean nuclease. However, approximately 300 cuts per molecule by any of three nonspecific agents (micrococcal nuclease, pancreatic DNase, or sonication) were necessary to give the same effect. Mapping experiments (*118*) showed that a specific nick was present approximately 100 nucleotides from the end of the 789-base-pair fragment containing the *lac* operator–promoter. The amount of nicked fragment was commensurate with the extent of loss of repressor binding. These results clearly demonstrate that a DNA region near the position of binding of the repressor is uniquely sensitive to single-strand-specific nucleases. The reason for this sensitivity is unknown at present but is under investigation with smaller and better-defined fragments (*97–101*). An extension of these studies on λ *plac*5 DNA showed a direct correlation between the nuclease susceptibility of a fragment and its dA·dT content as revealed by electron microscopic denaturation mapping (*119*).

Interestingly, mung bean nuclease produces gene-size segments of DNA on digestion of linear duplex DNA (*120*). The reason for this behavior is uncertain but may be related to the distribution of base-pairs within the genomes.

Venom phosphodiesterase produces specific cleavages in superhelical φX174 DNA and superhelical PM2 DNA (*121, 122*), generally in (dA·dT)-rich regions. However, not all such regions of the genome are cleaved by the enzyme.

A fascinating new enzyme has recently been described from *Alteromonas espejiana* (previously thought to be *Pseudomonas* BAL31). It is a sensitive endonuclease (122a,b), one that can be used to detect lesions, such as those produced by ultraviolet light or by treatment with carcinogens, in duplex DNA. Because of their high specificity, enzymes of this sort will probably find wide use in future studies.

A useful modification of single-strand-specific nuclease reactions on duplex DNAs has recently been described (123). In order to enhance the specificity of cleavage of disrupted regions of DNA, duplex molecules were treated with chloroacetaldehyde (which reacts at nonpaired sites) under conditions where both types of base-pairs could react and where the melting of the DNA was not base-pair-specific. The nuclease then attacked the regions "fixed" in single-stranded form, becoming a probe for defects in native DNA structures. Under these conditions, T7 DNA denatures principally from the ends, and a heteroduplex molecule showed denaturation from the site of the deletion loop (123).

When single-strand-specific nucleases are used as sensitive probes to detect structural anomalies in DNA, it is necessary to understand fully the mechanism of recognition by the enzymes in order to derive rigorous interpretations. As one approach to this, the number of nonpaired nucleotides necessary for cleavage by S1 or mung bean nuclease were determined (1,124,125). Model DNA polymers containing heteroduplex regions of known sequence and size were synthesized using polynucleotide phosphorylase and calf thymus terminal transferase (124). Heteroduplexes were of the form $(dG)_n \cdot d(C_{12}-A_m-C_{\bar{x}})$, where m = 1 to 6, and $(dG)_n \cdot d(C_{10}-G_m-C_{\bar{x}})$, where m = 1, 3, 4, or 5. Thermal melting studies of the model DNAs indicated that the heteroduplex regions did not disrupt the cooperative interaction between the flanking regions of $dG \cdot dC$ base-pairs. Thus it is possible that the heteroduplex nucleotides are accommodated in a stacked helical structure. The sensitivity of the model DNAs containing $dA \cdot dG$ and $dG \cdot dG$ (i.e., dA's and dG's opposite dG's in) heteroduplex regions of defined length to S1 and mung bean nucleases was tested (125). Single-base mismatched heteroduplexes were extremely resistant to these nucleases, although low levels of cleavage of the heteroduplex nucleotide were observed at very high nuclease concentrations. The nuclease sensitivity of the $dA \cdot dG$ region increased gradually as its length increased from 1 to 6 nucleotides. The sensitivities of $dG \cdot dG$ heteroduplexes 3 to 5 nucleotides long were considerably greater than that of the single $dG \cdot dG$ mismatch. Moreover, three heteroduplexes were formed between wild-type 789-base-pair *lac* promoter fragments and promoter point mutations with established sequences (1). No specific cleavage by the single-strand-specific nucleases was observed with these natural systems. A similar conclusion has been reached by R. Schleif (personal communication) with fragments of the arabinose operon.

S1 nuclease is active in aqueous solutions containing >60% formamide, 50% dimethyl sulfoxide, 30% dimethylformamide, and 2% formaldehyde (126), which makes it possible to use this single-strand-specific nuclease as a probe for DNA structure in denaturing solvents. It was also shown that the enzyme recognizes the same regions observed by optical melting profiles. Moreover, mung bean nuclease is a heat-stable enzyme (127) and can be used as a probe at elevated temperatures.

b. *Electron Microscopy.* Electron microscopy of DNA after partial denaturation and fixation with formaldehyde has provided the most graphic evidence for locally unstable regions in DNA. This procedure has been quite useful for identifying interesting features of DNA and correlations with genetic function. However, the resolution of this technique is not nearly as great as that described above for single-strand-specific nucleases. Whereas it is possible to detect several nonpaired nucleotides with such nucleases, it is necessary to have a "bubble" 40 to 100 base-pairs long for rigorous detection by electron microscopy. Studies have been performed on a variety of DNAs including T7, PM2, λ, P22, and others. A recent paper (128) shows a comparison of partial denaturation maps with known base sequences of SV40 and φX174 replicative-form DNA. The denaturation maps show a good correlation with experimental maps based on percentage of dA·dT pairs. Thus the regions of DNA to denature first are, in fact, those regions with the highest dA·dT content (128, 101). Denaturation mapping on PM2 DNA (6×10^6 molecular weight) indicates that it contains one to three denaturation sensitive regions (129). This study is particularly interesting because the early-melting regions were also detected with chemicals, proteins, and an endonuclease specific for unpaired DNA.

c. *Chemical Agents.* A variety of chemical agents and techniques have been used to probe DNA for local disruptions in structure, including the following: iodination, methylation, water-soluble carbodiimide, tritium exchange, formaldehyde, glyoxal, and chloroacetaldehyde. All of these show a specificity (to variable extents) for nonhelical nucleic acids.

The thallium-catalyzed iodination of random-coil DNA proceeds 100 to 200 times faster than that of native DNA (130). This reaction was proved to be specific for nonpaired nucleotides by studies on a heteroduplex molecule containing a nonpaired region 73 nucleotides in length. The product of the iodination reaction is 5-iodocytidylic acid. Although this reaction shows high specificity, it is much less than that exhibited for single-strand-specific nucleases, which have a 10^3- to 10^4-fold preference for single-stranded compared to duplex DNA. However, an advantage of the iodination procedure is that radioactive iodine can be used to increase the sensitivity of detection of disrupted secondary structures.

Another promising chemical agent is dimethyl sulfate, which methylates the N-7 position of guanine and the N-3 of adenine in double-stranded DNA.[1] However, in single-stranded DNA, the N-1 of adenine and the N-3 of cytosine can also be methylated (74): in duplex DNA these positions are not accessible because they are involved in hydrogen bonding. A recent note (74) describes the use of this reaction to determine the number of base-pairs unwound by *E. coli* RNA polymerase on early T7 phage promoters. Methylation of N-1 of adenine or N-3 of cytosine prevents re-formation of the duplex structure. S1 nuclease cleaves at regions of the DNA methylated at the N-1 position of adenine. DNA sequence analysis then reveals the position of methylation. Ethylnitrosourea may also be a useful chemical agent in the future.[1]

Another chemical agent that shows great promise for future studies is a water-soluble carbodiimide that reacts preferentially with single-stranded regions of DNA. A very small amount of reaction by this agent on supercoiled PM2 DNA (131) or SV40 DNA (132, 133) gives a large inhibition of transcription with *E. coli* RNA polymerase. Hence this agent has been successfully used as a probe for regions of disrupted helix caused by supercoiling. An interesting feature of this agent is that the large bulky group reacts at base-pairing positions on nucleotides, thus further destabilizing the helix. However, the lack of a readily available radioactive probe detracts from its general usefulness.

Local fluctuations ("breathing") in DNA have been studied by hydrogen-tritium exchange in helical DNA and RNA (134 and references therein). The transient opening of base-pairs and the exposure to solvent is measured. Englander *et al.* proposed that the open structure consists of a large number of non-hydrogen-bonded but stacked bases and a disoriented polynucleotide backbone, which leads to hindrance of opening and closing. They have also proposed the existence of relatively long-lived and abundant (0.5%) "traveling loops."

Formaldehyde has also been thoroughly studied as a probe for single-stranded regions of DNA (2, 135–137). A substantial specificity for single-stranded polynucleotides is shown by this agent, since it reacts with the amino functions of the bases. However, its utility is limited by the reversibility of the reaction.

The reaction of two other agents, glyoxal (138, 139) and chloroacetaldehyde (123), also appear to be promising. Both of these agents react with single-stranded polynucleotides at a substantially greater rate than duplex molecules (J. Wetmur, personal communications).

d. Physical Probes. Only a few physical probes have been used to attempt to detect nonpaired regions in DNA. By far the most promising tech-

[1] See articles by Singer and Kröger in Vol. 23 and Singer in Vol. 15 in this series [Ed.].

nique involves the electrochemical reduction of adenine and cytosine (60). In native DNA, the double bonds of adenine and cytosine, representing the primary sites of electroreduction, are necessary for hydrogen bonding and are not accessible; native DNA thus shows no polarographic reduction step at room temperature. However, nonpaired nucleotides can be reduced. This is a sensitive method of monitoring helix–coil transitions as well as premelting changes in DNA (60). A sensitive variant of polarography is derivative (differential) pulse polarography in which the polarographic signal of double-stranded DNA can be observed even below room temperature. Further developments with this method on polynucleotides with defined structural variations may give useful information.

Other techniques explored involve t_m curves, mixing curves with synthetic RNAs, and fluorescence (140, 124, 125).

C. Long-Range Interactions in DNA

"Telestability" describes cooperative thermodynamic effects in DNA. The term encompasses all contributions from other parts of a DNA molecule that extend nearest-neighbor interactions. The result may be either the stabilization or destabilization of a given DNA sequence. Some of these additional contributions may be enthalpic, like next-nearest-neighbor interactions, whereas others may be entropic. Examples of entropic contributions could be the loop entropy, influences from already denatured ends at one end of a cooperatively melting unit, and possibly other factors that must be introduced in theoretical calculations of the helix–coil transition in order to improve the agreement between theory and experiment. The fundamental concepts were derived from studies on synthetic oligonucleotides but have recently been extended to natural recombinant DNAs (see Section IV). The effect is thermodynamic in nature and may or may not be accompanied by conformational changes in DNA; the telestabilization observed for the DNA block polymer (described below) is not accompanied by a conformational change. The range and magnitude of telestability is still uncertain but clearly extends over at least 35 base-pairs, and recent data (see following paragraphs) suggest that it may extend more than 100 base-pairs. The possible role of long-range thermodynamic transitions in gene regulation have been reviewed previously (1).

Telestability was first studied using molecules of $(dI-dC)_n$ of varying chain lengths. Because the high-molecular-weight polymer has an "inverted" CD spectrum, it is possible to monitor small conformational changes. The studies showed that approximately four turns of helix were necessary for an oligomer to have a spectrum comparable to that of the polymer; this suggested (49) that "previously unrecognized long-range structural forces may be important for DNA."

Quantitative notions on telestability were an outgrowth of biochemical and physical studies on several duplex block polymers synthesized with 10 or 15 A·T base-pairs attached to either 15 or 20 G·C base-pairs (reviewed in *1*). The thermal denaturation of these polymers showed that the melting of these polymers was cooperative from end to end. For example, $d(C_{20}\text{-}A_{15}) \cdot d(T_{15}\text{-}G_{20})$ melted in a single monophasic transition with a t_m of 51°. The dG·dC portion of the duplex block polymer stabilized the dA·dT portion to 34° above the calculated t_m of isolated $d(A)_{15} \cdot d(T)_{15}$. The observed t_m was even 3° above that found for high-molecular-weight $d(A)_n \cdot d(T)_n$. Thus, the dA·dT pairs in the polymer were given a thermodynamic character quite different from a typical isolated dA·dT pair, even when it is maximally stabilized in a polymer of dA's and dT's. In addition, the 20 dG·dC basepairs were destabilized by the dA·dT portion of the helix. This duplex block polymer melted 8° below the calculated t_m of isolated $(dG)_{20} \cdot (dC)_{20}$. In a natural DNA with many types of sequences, this behavior must also take place. However, in this particular polymer, with its highly restricted sequence, it was experimentally possible to show that regional cooperativity extended over at least 35 base-pairs.

Telestability may be important in DNA–protein interactions. The studies described above indicate that the ability of a portion of a natural DNA helix to melt can be influenced by the nucleotide sequence of adjacent regions. Hence, alterations of the nucleotide sequence (via deletions or base-pair changes) or of regions adjacent to an actual DNA–protein interaction site could influence the recognition of that site, particularly if the stability is important for protein binding (see Section VI).

Additional information on telestability was obtained using base-pair-specific drugs that bind to only one end of the block polymers. Actinomycin binds only to the dG·dC portion and netropsin binds only to the dA·dT portion of the synthetic polymers (*1, 110, 102*). The increased thermostability caused by drug binding extended to the nonbinding portions of these DNAs. For example, thermal denaturation of a block polymer in the presence of actinomycin increased the t_m by 12° but still yielded a single monophasic transition, indicating that both ends of the molecule melted simultaneously. Degradation of the dA·dT portion of $d(C_{15}\text{-}A_{15}) \cdot d(T_{15}\text{-}G_{15})$ by *E. coli* exonuclease I (which recognizes single-stranded character in DNAs) was inhibited when actinomycin was bound to the dG·dC end. This finding was consistent with the thermal denaturation studies—i.e., ligand binding to one region of a DNA can influence the dynamic properties (e.g., "breathing" frequency) of adjacent regions of the DNA.

Calculations showed (*141*) that nearest-neighbor base-pair interactions could not explain the observed extent to which one DNA region influenced the thermostability of adjacent regions.

Proton magnetic resonance analyses at 300 MHz of $d(C_{15}\text{-}A_{15}) \cdot d(T_{15}\text{-}G_{15})$ (108) showed that the long-range thermodynamic effect was not due to a large conformational change of one end of the molecule on the other. The dA·dT portion of the polymer existed in two conformations, the major portion (11 to 12 base-pairs) having the same conformation as $d(A)_n \cdot d(T)_{25}$ while 3 to 4 base-pairs at the junction of the block were induced into an altered conformation by the adjacent dG·dC block. The dG·dC base-pairs existed in only one conformation; however, the conformation of the dG·dC portion varied at different salt concentrations, whereas that of the dA·dT portion was not influenced by variation in the salt concentration. Actinomycin binding to the dG·dC portion of the polymer did not alter the conformation of the majority of the dA·dT pairs.

Similar studies have recently been performed on $dG_n \cdot (rC_{11}\text{-}dC_{16})$ (17–19), with conclusions in complete agreement with those described above. It is possible that telestability in some cases may be accompanied by conformational changes; however, this is not the case for the synthetic molecules examined to date.

These studies have served as the foundation of several models relating long-range thermal effects in DNA to gene regulation. Wartell (69) has calculated from theoretical considerations that melting and stabilizing proteins can alter the t_m of base pairs as much as 20 to 100 base-pairs away. The magnitude and range of this effect is strongly influenced by the base-pair composition and sequence of the protein binding site and the immediately adjacent DNA regions.

Reiss et al. have also attempted to calculate the effect of certain mutations on the melting properties of segments of DNA molecules (81, 142, 143). Although it is clear from these calculations that long-range interactions are to be expected, it has not yet been possible to verify experimentally the thermal denaturation properties of small restriction fragments (less than 300 base-pairs) at the base-pair level. The first high-resolution studies of the thermal denaturation of fragments ranging in size from 100 to 300 base-pairs has just been reported (95).

Recent electron microscope studies (B. Funnell, R. B. Inman, and R. D. Wells, unpublished) on HindII-linearized pRZ2 DNA have greatly influenced our notions of the range of telestability. These determinations show that the large dA·dT block at one end of this DNA (101) has a substantial destabilizing effect on the neighboring 425-base-pair fragment. When the denaturation mapping studies are performed on the isolated fragment (101) as well as on the intact pRZ2 DNA that embodies this fragment, there is a destabilizing effect over at least 100 base-pairs caused by the neighboring (dA·dT)-rich region. This is the longest and most marked teledestabilizing effect known to date.

Further studies are necessary to determine what influences will disrupt telestability. We believe that a nick (a single-strand break) in the DNA double helix will interfere with long-range interactions. Moreover, any element of single-strandedness, such as a cruciform or other single-stranded region, may also influence long-range thermodynamic effects. Moreover, if the binding of a ligand (such as a carcinogen) causes a disruption in a DNA, it may thus influence these long-range interactions. Thus there are a number of structural changes in DNA that may influence gene regulation via telestability.

Long-range thermodynamic or conformational effects have been invoked to explain a number of genetic and biochemical observations in natural regulatory systems, such as the following: arabinose operon, T7 promoters, the *lac* promoter and catabolite-gene-activating-protein effect, the high efficiency promoter for the 11p gene (the lipoprotein promoter), high efficiency promoters for ribosomal RNAs. These effects are described below under promoter effects and have been (in part) reviewed previously (*1*, *3*). However, convincing physical or biochemical evidence for long-range thermodynamic or conformational effects in these systems is lacking.

Drugs as Probes

Several recent drug-binding studies have indicated long-range conformational or thermodynamic effects in DNA. No attempt is made herein to review the voluminous literature on drug binding to DNA; only those papers relevant to the discussion of long-range interactions are discussed.

The phenomenon of allosterism in DNA has been considered for several years from a variety of determinations (*55*, *56*, *145*, *146*). Studies on synthetic DNAs (described above) were pivotal in these considerations. In addition, it was found (*56*) with $(dG-dC)_n$ (*47*) that ethidium binding affects the balance between the double-helical forms of the DNA. Drug binding at high salt concentrations was markedly cooperative; the observation was considered to be a long-range effect over hundreds of base-pairs.

The binding of distamycin[2] to calf thymus DNA induces a cooperative transition to a new form with higher affinity for the drug and with altered structural properties, and substantially reduces the affinity of the DNA for ethidium. These observations were interpreted as an allosteric conversion of the DNA. The effect takes place over hundreds of base-pairs.

A similar conclusion was reached (*147*) from the interaction of netropsin[2] with $(dA-dT)_n$. High resolution proton NMR measurements show that netropsin induces structural changes at adjacent antibiotic-free base-pair regions when the antibiotic complex is formed at a low concentration of the

---

[2]See Zimmer in Vol. 15 of this series [Ed.].

drug. The structural perturbation is primarily at the glycosidic torsion angles. These and other NMR studies on drug–nucleic acid interactions have been reviewed recently (*148*).[3]

van de Sande and his associates (unpublished) have recently studied the capacity of a variety of drugs to inhibit the cleavage at some specific sites by DNA restriction enzymes. The drugs that showed most promise are olivomycin and Hoechst 33258. At an appropriate drug/DNA ratio the cleavage of one *Eco*RI site on lambda DNA is completely inhibited whereas cleavages at other sites are uninhibited. Studies have also been performed on the *Hpa*I reaction of PM2 DNA and on the *Hha*I reaction on M13 replicative-form DNA. Owing to the relatively low concentration of drug necessary to cause these specific inhibitions, these workers suggest that the inhibitions must be by a neighboring effect and are interpreted in terms of long-range interactions in DNA. Prior studies (*149, 150*) show that 6,4′-diamidino-2-phenyl-1*H*-indole, distamycin, and actinomycin can cause preferential inhibition at some restriction sites with several different restriction enzymes.

The gel electrophoretic mobility of certain DNA restriction fragments is influenced disproportionately by the binding of netropsin, distamycin, actinomycin D, Hoechst 33258, and olivomycin (*151*). Although the binding of these drugs to a variety of restriction fragments altered the electrophoretic mobility in native acrylamide gels, some fragments were influenced to a much greater extent than others. There seems to be little correlation between the known binding affinities of the drugs and the base composition of the fragments most affected (G. Staffeld, G. T. Horn, and R. D. Wells, unpublished observations). The reason for the differences in the effects of some drugs on the electrophoretic behavior of the fragments is unclear, but these differences may be a result of a perturbation in the structure of bihelical DNA or be due to the amount of drug bound.

The binding of either daunorubicin or doxorubicin (adriamycin) to $(dA-dT)_n$ facilitates the binding of actinomycin D (*152*). In the absence of either, actinomycin D does not bind to $(dA-dT)_n$. However, CD measurements indicate that cooperative binding between the two intercalating drugs is probably due to distortions, induced in the double-helix and transmitted over several base-pairs (*152*). The equilibrium constant for the binding of actinomycin to $(dA-dT)_n$ in the presence of daunorubicin is approximately 4×10^5 M^{-1}. Further studies will be required to determine the mechanism of this synergistic effect. Neither ethidium bromide nor acridine orange facilitates the binding of actinomycin to $(dA-dT)_n$ (*153*).

Conformational transitions in duplex DNA can be damped out over a

---

[3] See also Suhadolnik in Vol. 22 and Singer and Kröger in Vol. 23 of this series [Ed.].

relatively short distance (3 to 5 base-pairs) in some cases (154, 155). It is possible for conformational differences as large as that between the B- and A-conformations to coexist side by side in a bihelical DNA (18, 19). However, in other cases, it is possible that the conformational effects described above may occur over long distances.

Actinomycin stimulates the ability of bleomycin[4] to degrade SV40 DNA; ethidium bromide under the same conditions inhibits the bleomycin cleavage reaction. However, both ethidium bromide and actinomycin enhance the bleomycin-induced breakage of native DNA. The mechanism of these synergistic effects is uncertain at present.

It is likely that these types of antibiotics will receive increasing attention in future years, as the studies described above indicate that the mechanisms of binding of these compounds to DNA is poorly understood. Moreover, biological data (157, 158) indicate that at least actinomycin and netropsin may bind to specific segments of DNA with far greater specificity than has been previously recognized. Future studies may identify the molecular basis of these highly specific inhibitions.

D. Cruciforms

The consideration of nonpaired regions in cruciforms (159) in biologically important control regions has been greatly stimulated by the frequent occurrence of inverted repeat sequences (1, 3, 160, and Section IV). These may form cruciform structures that could serve as protein recognition sites and as initiation points for DNA and RNA synthesis. In general, it is considered unlikely that cruciform structures will be found in isolated DNA fragments, since the formation of cruciforms is accompanied by a loss of base-pair interactions. However, there is no general agreement on this point (161), and it is also possible that the binding of certain proteins to cruciform structures may cause substantial stabilization.

Alternatively, the strain generated by the superhelix in superhelical DNA may be relaxed by "looping out" inverted repeat sequences into "rabbit-ear" type structures.

To the best of our knowledge, only two reports describe characterized palindromic sequences of substantial magnitude, other than inverted repeated sequences that may be involved in recombination. Selsing and Wells (59) described the synthesis and characterization of $dG_i\text{-}dA_j\text{-}dC_k \cdot dG_k\text{-}dT_j\text{-}dC_i$ by gene cloning techniques. The recombinant DNA molecules were stable and could be isolated from amplified cell cultures (59). In addition, Gellert et al. (162) described the formation of a head-to-head recombinant molecule from pBR322 plasmid DNA. This molecule is a "perfect inverted

---

[4] See Müller and Zahn in Vol. 20 of this series [Ed.].

repeat." Further physical and biochemical studies will be needed to determine the biological role of these types of sequences (see Note Added in Proof).

III. Preparation of Large Amounts of DNA Restriction Fragments

As mentioned in the Introduction, there is a transition at this point in this review from studies on DNA properties to studies of a biochemical and genetic nature on regulatory sites such as promoters and origins of replication. This transition is necessitated by the lack of studies on small DNA restriction fragments (20 to 300 base-pairs in length) of established nucleotide sequence with characterized biological functions.

Two recent developments are rapidly increasing availability of small fragments. First, gene cloning techniques have made such enormous progress that virtually every issue of most distinguished biochemical and genetic journals are filled with interesting innovations. It is anticipated that this progress will continue; therefore, no attempt is made here to review this subject. However, gene cloning alone will not provide a large quantity of a purified DNA restriction fragment since techniques must be devised for freeing the insert from the vector.

The second important procedure is chromatography on RPC-5 columns. This procedure was applied to the separation of DNA fragments in 1976 (*164*), and several further improvements have been reported (*98–101* and reviewed in *165*). To the best of our knowledge, this is the only procedure presently available that can fractionate multimilligram quantities of DNA restriction fragments; as much as 500,000 times more DNA can be fractionated on this type of column as by gel electrophoresis (*166-177*). The order of fractionation is not always the same as produced by gel electrophoresis (see the following paragraphs); therefore, the sequential use of these two methods is more powerful than either method alone. In some cases, RPC-5 column chromatography can completely resolve duplex restriction fragments of the same size.

Figure 9 shows the profile of a preparative RPC-5 column fractionation of a *Hae*III digest of pRZ2 DNA using sodium acetate as the eluting salt; 17 fragments are produced by *Hae*III ranging in size from 43 to 850 base-pairs. Also, polyacrylamide gel electrophoretic analyses on a number of column fractions are illustrated in the upper panel. Each peak in the profile (Fig. 9) corresponds to the elution of a different fragment, and 15 of the fragments were separated, including K (102 pairs) and L (98 pairs). Thus, a 4% difference in size is sufficient to provide at least partial separation of these two fragments. In other studies (*1, 99*), fragments of identical size are in fact separated. Figure 9 demonstrates that fractionation of the digests on RPC-5

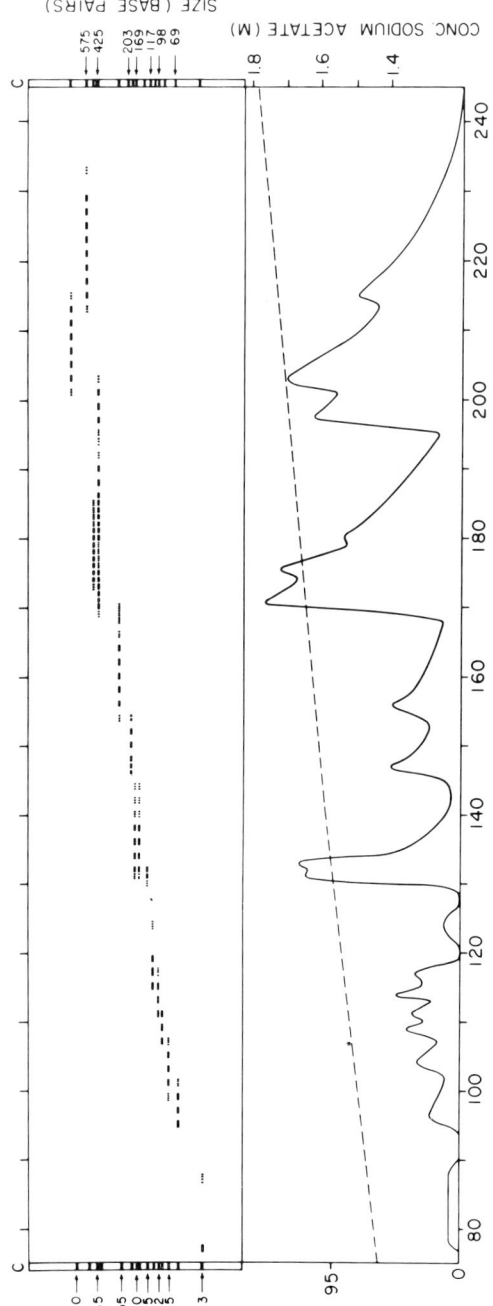

FIG. 9. Fractionation of a *Hae*III digest of pRZ2 DNA by reversed-phase (RPC-5) column chromatography. The upper panel is a drawing of the polyacrylamide gels run on certain fractions across the gradient. A continuous line represents a strong band; a dotted line represents a weak one. On either end of the upper panel, the control digests (C) are shown and the bands are labeled according to size (in base-pairs). Electrophoresis was from top to bottom. Details of the chromatography methodology have been reported (100). Reprinted, with permission, from Patient et al. (100).

was due primarily to fragment size, but in addition another parameter was involved since the three D fragments (425 pairs) were completely resolved from each other. Also, fragment B (575 pairs) eluted after fragment A (850 pairs). This behavior is due to the base composition of the fragments (*100, 101*).

Figure 10 further illustrates the resolution capability of this technique

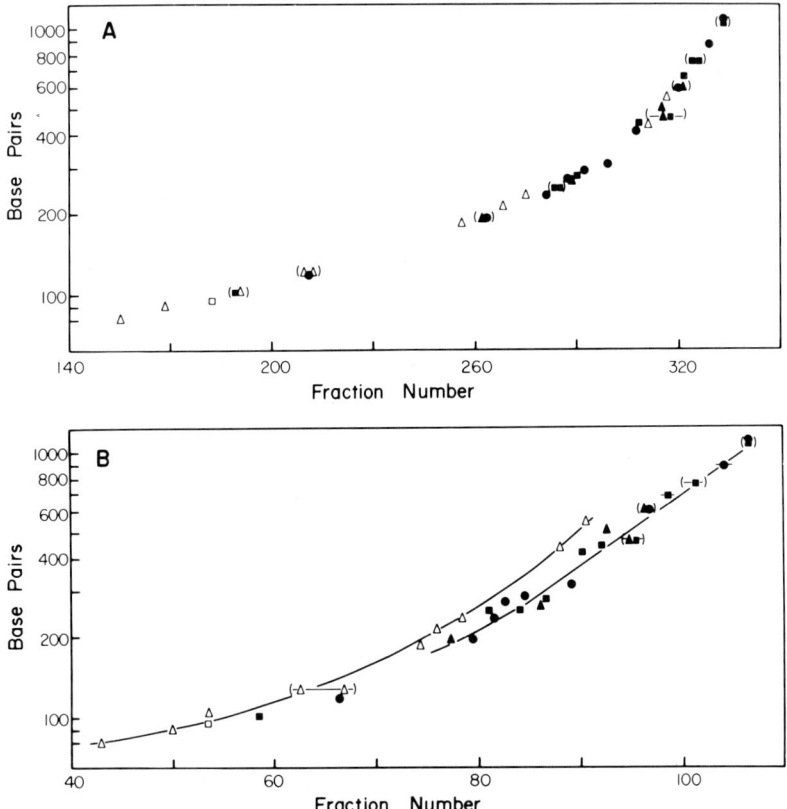

FIG. 10. Reversed-phase (RPC-5) elution behavior of *Hae*III fragments of φX174 RF, G4 RF, and pBR322 DNAs. An equimolar mixture of *Hae*III digests of φX174 DNA (○, ●), G4 RF DNA (□, ■), and pBR322 (△, ▲) was chromatographed on RPC-5. Open symbols represent restriction fragments with a (dG+dC)-content of >55%; filled symbols, a content of <55%. Parentheses indicate that two fragments were too close in size and elution position to be distinguished from each other. Fragments eluting in more than several fractions are indicated by a line. Panel A: Chromatography was performed on a 0.15 by 100 cm column at 43°C with a 100 ml linear gradient of 0.55 to 0.85 M KCl containing 10 mM TrisCl (pH 6.8), 2 mM sodium thiosulfate, 0.1 mM EDTA. Panel B: Chromatography was performed on a 0.15 by 80 cm column at 25°C with a 100 ml linear gradient of 1.3 to 1.8 M sodium acetate containing 10 mM Tris acetate (pH 8.2), 2 mM sodium thiosulfate, 0.1 mM EDTA.

and reinforces the influence of the eluting salt on the properties of the fractionation. Figure 10A shows the composite behavior of HaeIII fragments from three different viral or plasmid DNAs on RPC-5 when KCl was used as the eluting salt. A relatively smooth curve can be drawn through the relationship between the size of the fragment and its position of elution, irrespective of base composition. However, as shown in Fig. 10B and as indicated previously (99–101, 165), when sodium acetate is the eluting salt, fragments rich in dA and dT bind more tightly to the column than expected on the basis of size. Moreover, the distribution of dA·dT pairs within a fragment can have a substantial influence (100, 101).

In summary, fragments elute from RPC-5 on the basis of their size, base composition, the presence or absence of and type of sticky ends, as well as at least one other factor (99, 165).

RPC-5 column chromatography has also been an invaluable technique for separating insert DNAs from vectors in large quantity (98). In addition, sucrose gradient centrifugation can be used (98, 163) if the size difference between the vector and the insert is sufficiently large. However, this technique is quite laborious and must be performed in a repetitive fashion in order to obtain milligram quantities. Preparative polyacrylamide gel electrophoresis has also been explored for the large-scale fractionation of DNAs (109, 178). However, the capactiy, resolution, and purity of the ensuing fragments are far inferior to those from RPC-5.

IV. Transcription Recognition Sites

A. Bacterial Promoters

A brief review of the properties of promoters is presented in order to establish the background for later discussion of DNA structural influences on the process of transcription. The majority of studies described in this section were conducted with the *E. coli* DNA-dependent RNA polymerase; however, some discussion of other bacterial and phage RNA polymerases is also included.

1. GENERAL PROPERTIES OF PROMOTERS

Promoters are specific regions of DNA that are sites for the initiation of mRNA transcription following binding by DNA-dependent RNA polymerase holoenzyme (core plus sigma). These regions have historically been defined by *cis*-acting mutants that affect the production of a single species of mRNA (179). Thus a promoter can contain not only the site for RNA polymerase recognition and binding, but also binding sites for effector molecules involved in transcriptional regulation. These effector molecules can bind DNA sequences that are continuous with or even overlap the RNA polymerase binding site (*103, 180*).

Promoters are of substantial interest, as it is believed that a significant amount of gene regulation, at least in prokaryotes, is controlled at the transcriptional level. Numerous regulatory proteins bind to certain promoters, even though the promoters are relatively short sequences of DNA. This concentration of interesting molecules that interact with DNA affords a unique opportunity to both the geneticist and biochemist to study gene regulation at the molecular level. An example is the lactose operon control region where key regulatory proteins (cAMP-CRP, *lac* repressor, and RNA polymerase) bind within a 150 base-pair sequence (*103*).

An extensive literature dealing with promoter structure and function attests to the importance and interest in this field. DNA binding sites for the regulation of specific promoters including *lac*, *gal*, and *ara* have been reviewed elsewhere (*181 ,3*). In addition, the prokaryotic transcription literature has been reviewed from the following perspectives: RNA polymerase interactions and kinetics (*182, 174, 183*); promoter sequence and function (*3, 184, 185*); specific analysis of various promoter mutants, especially *lac* (*103, 186*); and semiquantitative analysis of various DNA–protein interactions, with emphasis, in part, on promoter function (*2 ,4*). Despite this sizable body of literature, our understanding of promoter function at the molecular level is incomplete (*4*).

2. Model for Transcription

One general scheme for transcription is illustrated in Fig. 11. Phase I, the initial interaction of *E. coli* RNA polymerase with promoter DNA, consists of three steps: recognition, "melting-in," and the initiation of polymerization of the mRNA (*70, 187*). The recognition site for RNA polymerase is believed to be from -28 to -35 base-pairs from the mRNA start site. (The start site is designated $+1$; the upstream sequence is negative; the downstream sequence, i.e., the direction of transcription, is positive.) RNA polymerase binding to the recognition site does not involve detectable unwinding of the double-helix, thus this interaction is referred to as the "closed complex" (*187*). The closed complex is sensitive to polyanions and is easily disrupted by heparin and single-stranded DNA (*187*). RNA polymerase is believed to enter the DNA between positions -6 and -12 and unwind the double-helix forming the "open complex" (*187*), which is not sensitive to polyanions. Transcription ensues upon binding of the first ribonucleoside triphosphate followed by mRNA chain elongation (phase II). Most transcripts are terminated (phase III) at specific termination sequences with release of the mRNA and disengagement of RNA polymerase from the DNA. Some polymerase molecules are able to proceed through the termination

DNA STRUCTURE AND GENE REGULATION

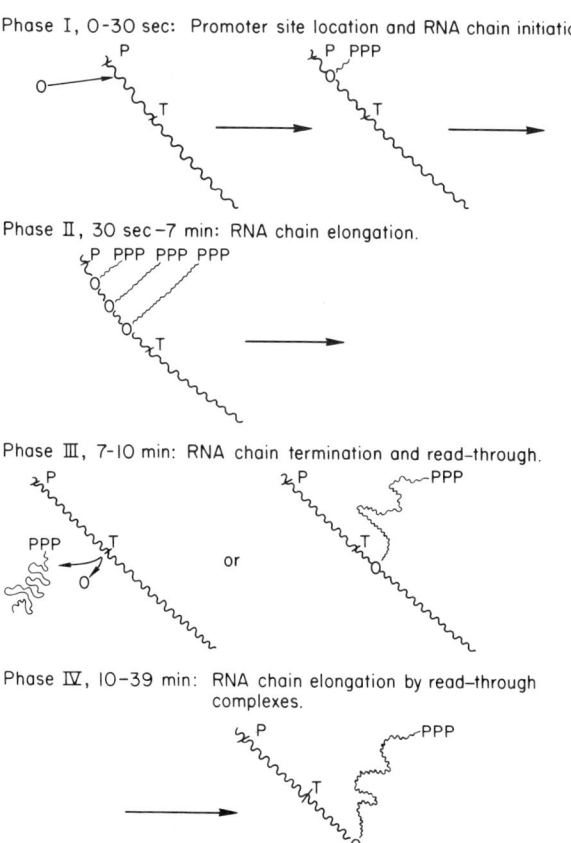

FIG. 11. Outline of transcriptional program for *Escherichia coli* RNA polymerase acting with T7 DNA at 30°C. Reprinted, with permission, from Chamberlin *et al.* (*188*).

sequence continuing mRNA chain elongation (phase IV), a process called "read-through." An assay has been devised to quantitate these phases with a variety of prokaryotic RNA polymerases on T7 DNA (*188*). The process of transcription is complex, especially when additional *in vivo* factors are considered (*189, 191*).

3. SEQUENCE HOMOLOGIES

It was initially thought that only a single species of RNA polymerase exists in the bacterial cell and that the DNA sequences of various promoters would be homologous. However, variations in the strength of promoters, both *in vivo* and *in vitro*, suggested instead that the sequence of promoters is heterogeneous, or that there may be modified species of RNA polymerase (*182*). In fact, a comparison of the sequences of promoters (*3, 184, 186*) has indeed shown marked heterogeneity, whereas the existence of various RNA polymerases is still in doubt (*189*).

Three relatively common regions have been suggested for a majority of promoters. These are the recognition region (−28 to −35), the "melting-in" site (−6 to −12), and the transcription start site (+1). Identification has been based on chemical probes of RNA polymerase contact sites (*3, 186, 192, 193*) and the sequencing and biochemical analysis of *cis*-acting mutations (*186, 194*) in conunction with a correlation of these data with sequence homologies of numerous promoters.

a. −6 to −12. In the 46 promoters tabulated by Rosenberg and Court, most seem to have a common 7-base-pair sequence (TATAATG) in the −6 to −12 region. Homology in this region was originally pointed out by Schaller *et al.* (*195*) and Pribnow (*196, 197*) and is often called the "Pribnow Box." The frequency of occurrence of each base in this sequence, using the 46 sequenced promoters, is: T 0.87, A 0.89, T 0.54, A 0.63, A 0.65, T 1.0, and G 0.37. The third T (sixth nucleotide) in the sequence is invariant and may represent an RNA polymerase contact point common to all promoters. Mutations in this region have a marked effect on promoter efficiency (*186, 194*). Studies with chemical probes strongly suggest that a number of contact points are made by polymerase with the DNA in this region (*193, 192*).

b. −28 to 45. The importance of the −28 to −35 region has been derived from sequence homology, a number of promoter mutations, and two contact points with RNA polymerase, as demonstrated by protection of the DNA from reaction with dimethyl sulfate (*195, 198, 184, 3*). Using a computer to analyze 17 sequenced promoters, Arnott *et al.* (*199*) detected a partial sequence homology of GTTGACATTT near the −30 region. Rosenberg and Court (*3*), using the invariant T at the start of the Pribnow Box as a reference, were able to line up 46 sequences and obtain homology in the same region. They found a conserved trinucleotide TTG followed by a less conserved ACA. The frequency of occurrence of each base of this 6-base-pair sequence is T 0.87, T 0.80, G 0.76, A 0.59, C 0.63, and A 0.57.

c. +1. The third region of homology is at the transcription initiation site. Arnott *et al.* found that a trinucleotide CAT occurred frequently. Indeed, Rosenberg and Court's compilation shows that the entire CAT triplet appears in 11 out of 46 promoters, each base appearing with a frequency of C 0.48, A 0.48, and T 0.50.

In addition to these three regions of homology, Arnott *et al.* (*199*) pointed out that both ends of the 17 promoters studied were generally flanked by (dA+dT)-rich regions. This observation will be discussed in subsection 6 below.

4. INITIATOR NUCLEOTIDES

The intial triphosphate for mRNA transcription is generally a purine (A or G) regardless of the distance from the Pribnow Box (3). However, several promoters—rrnE, rrnA (200), rrnD and rrnX (201), and c17 in phage lambda (202)—give transcripts that begin with CTP or UTP. Young and Steitz (201) have suggested that some of the pyrimidine starts may be *in vitro* artifacts. It should be noted that these pyrimidine start sites are preceded by several purines in positions that are used as start sites in other promoters; the reason for this is unclear. In addition, there are several promoters that have nearly identical sequences around their start sites, but that begin transcription at different nucleotides. Examples of this phenomenon are the *gal* P$_1$ and the *ara* BAD promoters, whose start sites are two base-pairs apart even though their sequences are nearly identical, and each gives a transcript beginning with ATP. (The invariant T in the Pribnow Box is marked by an asterisk and the start site is indicated by an arrow.)

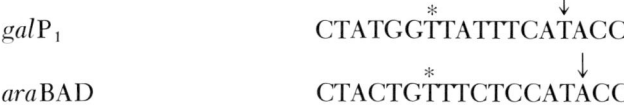

It is likely that small differences in sequence place a significant constraint on the start point of polymerase (3). The large number of promoters with markedly different sequences suggests that RNA polymerase has the ability to accommodate sequence variations by binding to subsets of a large number of contact points within the 40-base-pair promoter region (4).

5. MUTANT STUDIES

Although the literature on various promoter mutants has been reviewed (103, 184, 186, 185, 3), a brief discussion of the effects of altering the basic promoter sequence on promoter efficiency is appropriate. It should be pointed out that "promoter efficiency" is a composite term that can be a manifestation of a number of molecular events. Some of the assays that have been employed are RNA polymerase binding studies (203), *in vitro* transcription assays (204), and determination of the expression of a given gene product (205). These assays may or may not analyze the specific molecular events that are responsible for the effect on promoter efficiency. Differing levels of transcription can be due to the alteration of a contact point that may cause premature termination or abortive initiation [specific assays to study these phenomena are being developed (206, 207)]. A deletion may remove a specific contact point or alter the separation between noncontiguous regions of the DNA–polymerase interaction. Some properties of the DNA may be altered, such as the (dA+dT)-content, which may prevent or slow down the "melting-in" of polymerase. Any of these mutations may change the rate-limiting step of transcription initiation for specific promoters (208).

It is interesting that mutations that increase promoter efficiency more closely approximate the Arnott model promoter, whereas those that have a negative effect deviate from it (199). Changes in (dA+dT)-content of various promoter mutations may be responsible for their activity. *lac* p^{r}1a is a mutant with a G/C to A/T change in the Pribnow box. This change may affect the ability of RNA polymerase to melt into the double-helix (186, 194). Substances that destabilize the double-helix enhance mRNA initiation (209, 210, 182). RNA polymerase is a helix-destabilizing protein capable of unwinding the double-helix some 140 degrees, opening approximately 7 bases. This work (71) has recently been extended (74) by localizing the region of unwinding for an early promoter on T7 DNA, which is transcribed by the host RNA polymerase, from −9 to +2, a total of 11 base-pairs. The unwound region begins near the middle of the Pribnow Box.

By melting a series of small fragments (97–101) containing overlapping regions of the promoter, it has been possible to localize an 81-base-pair region of unusual thermodynamic properties near the Pribnow Box (described in Section II,B,2,a). This region was sensitive to salt, more so than predicted from the base composition. This sensitivity may be related to the biological function of this region. It is possible that alterations of the (dA+dT)-content of this region may increase the efficiency of transcription by providing better "melting contacts" (position of entry of the enzyme into the double helix) with RNA polymerase. Several mutations that alter the (dA+dT)-content of the Pribnow Box [p^{r}1a, p^s, and uv5 (194)] do show increased promoter efficiency.

Some effects caused by other mutations are not clear. Maquat and Reznikoff (194) found a mutant that would indicate that sequences beyond the initiation site are important. P^{r}111 is a C/G to A/T transversion at +10. The mutant occurs in a block of (dG+dC)-rich DNA and was found to increase the level of *in vivo* transcription. However, *in vitro*, the production of mRNA is less than wild type, although RNA polymerase binds the promoter more tightly.

6. Effect of Neighboring Regions

Some, but not all, promoters are likely to be found in (dA+dT)-rich regions of the genome (67, 64, 66). In addition, theoretical calculations based on sequenced DNAs show that regions between genes are thermolabile (211). Highly efficient promoters such as those for the *rrn* operons (ribosomal RNA) and *11p* gene (lipoprotein) are preceded by (dA+dT)-rich stretches of DNA. In the case of the *11p* promoter, the (dA+dT)-content is 70% to the −261 position (212). Several authors have suggested that these regions facilitate promoter function by altering the thermal lability when under torsional strain of supercoiling (see Section IV,B). This melting-out of the DNA either increases the binding constant for RNA polymerase or allows numerous polymerase molecules to queue upstream from a given promoter (213). In fact, as many as 30 RNA polymerase molecules have been found to bind in the region immediately upstream from the rRNA promoter initiation sites (213).

Recent evidence (200, 201, 214) suggests an explanation for the high level of efficiency of the rRNA promoters of *E. coli*. *In vitro* assays for promoter activity correlate well with the *in vivo* observation that their RNA reaches 50% of the cell's total stable RNA, although these genes comprise only 0.4% of the bacterial genome. Recently, four of the rRNA operon gene promoters were sequenced: rrnA and rrnE (200, 214) and rrnD and rrnX (201). Each of these operons has two *in vitro* start sites. Sequence analysis shows that there are two promoters in each of them, separated by about 110 base-pairs. The first promoter start site (P_1) in these operons is preceded by an (dA+dT)-rich sequence (in fact 17 dA's and dT's between -40 and -58), the P_1 promoters all have significant homology in the 19 base-pairs preceding the initiation site, and all start transcription with GTP. The second promoter (P_2) located 110 base-pairs downstream from P_1 has little homology and starts the transcript with CTP.

Aside from the fact that these promoters reside in a (dA+dT)-rich region of the genome, another mechanism may account for their efficiency. Since RNA polymerase is a helix-destabilizing protein, it may be that the binding of one molecule to one or both contiguous promoters alters the conformation or stability of the adjacent DNA sufficiently to facilitate the binding of additional polymerase molecules. Given the dA and dT richness (and ease of melting) of the region upstream from these promoters, the cooperativity effects of binding multiple RNA polymerase molecules would force the "queueing up" of numerous polymerases, accounting for the high efficiency of these promoters (201). Thus, the actual DNA content of a region surrounding a regulatory site could have a pronounced effect on its level of expression.

The availability of large amounts of promoter DNA (see Section III) for spectroscopic, crystallographic, and kinetic studies will allow quantitative studies on promoters to test some of these hypotheses.

B. Effect of Supercoiling on Transcription

A subject of increasing interest is the dependence of promoter activity on degrees of superhelicity of the DNA template. The relevance of superhelicity *in vivo* has been reconsidered since the discovery of DNA gyrase. Superhelical DNA, previously considered to be an artifact of isolation (215), may actually exist *in vivo* and be involved in pleiotropic regulatory mechanisms *(216–218)*.

A general discussion on superhelical DNA can be found in several excellent reviews *(219–221)*. Reviews have also been published on topoisomerases *(222, 208)*, enzymes that passively decrease superhelical density, and DNA gyrase *(223)*, which uses ATP hydrolysis to increase superhelical density. Hence, only a brief outline of the properties of gyrase is given.

1. DNA GYRASE

DNA gyrase was originally discovered as a necessary cofactor in the *in vitro* assay for lambda integrative recombination *(224)*. It was soon found to generate negative superhelicity in covalently closed circular DNA, using ATP hydrolysis as an energy source. A tetrameric enzyme, gyrase has two copies each of two separate polypeptides produced from unlinked genes *(225)*. One subunit, Pnal, is produced by the nalA gene and is the target of action for the antibiotics nalidixic acid *(226, 227)* and oxolinic acid. This subunit has a "nicking-closing" activity and is thought to relax an intermediate in the gyrase reaction that is positively supercoiled *(223, 228–230)*.

The other subunit, Pcou, is produced by the cou gene *(231, 232)* and is the target for the antibiotics coumermycin and novobiocin. This protein is thought to catalyze the ATP-dependent unwinding of part of the DNA molecule, thereby segregating positively and negatively supercoiled domains *(223, 228–230)*. Mutations in either of these genes can produce subunits resistant to their respective antibiotics. In addition, a temperature-sensitive mutant in nalA was recently reported *(233)*.

Coumermycin competes for the ATP binding site on Pcou *(223)*, while nalidixic acid seems to trap the gyrase–DNA complex in a covalently attached intermediate *(234)*. Deproteinization of this trapped intermediate leads to site-specific cleavage of the DNA.

Since all of the above antibiotics efficiently block DNA synthesis *(235–237)*, gyrase is implicated as a necessary component of the replicative process. It is still unclear whether it only unwinds the positive supercoils in front of the replication fork, or goes beyond this and negatively coils the replicating DNA (which would then facilitate the extension of the fork). A curious finding was that inhibition of gyrase by nalidixic acid blocked replication of phage T7 *(238, 239)*, which is not thought to replicate via a circular intermediate *(240)*. However, it is argued that this inhibition is caused by a side reaction of the drug-inhibited gyrase, since the temperature-sensitive gyrase does not block T7 growth *(233)*.

DNA gyrase and its specific inhibitors can be used to probe and alter the superhelical state *in vivo*. Although superhelicity has not yet been directly measured *in vivo*, changes in superhelical density have been measured on isolated DNA after inhibiting the gyrase *in vivo*. For example, lambda DNA isolated from cells treated with coumermycin has only about 15% of the superhelical density found in DNA taken from untreated cells (*231*). Similarly, the bacterial chromosome itself appears to have only about 25% of the normal superhelicity after treatment with coumermycin (*241*). Both of these results suggest that superhelicity exists *in vivo* and is reduced upon inhibition of gyrase, rather than just being an artifact of isolation and deproteinization of the DNA (*215*).

2. Gyrase, Superhelicity, and Transcription

The relationship between gyrase, superhelicity, and transcription has only recently been investigated. Early studies found small reductions of transcription at levels of drug higher than those needed to block replication. The effect on transcription was therefore thought to be nonspecific (*242*). Soon, however, selected cases of more pronounced effects were seen. Transcription of phage S13 was quite sensitive to inhibition by nalidixic acid (*243*), as was transcription from the virion N4 (*244*) and certain *E. coli* genes (*245*). Negative superhelicity of the promoter enhances RNA polymerase binding, as this enzyme is thought to unwind a segment of the helix as part of its binding mechanism. Negative superhelicity, which energetically favors this unwinding, would then contribute to the binding energy of the polymerase (*246, 247*). Thus, the superhelical form of phage ϕX174 RF was shown to be a better template for transcription than the relaxed form (*248*); similar results were found for the DNAs from PM2 (*249*), fd (*250*), and lambda (*216, 67*). These last papers pointed out a further complexity, namely, that supercoiling can stimulate different promoters to different extents.

It was suggested in 1973 (*216*) that some promoters show a preferential stimulation by negative superhelicity. In studies on the overall RNA production *in vitro* from phage lambda, it was found that RNA produced from relaxed DNA had a higher precentage of early RNA than transcripts from supercoiled DNA. Increasing amounts of superhelicity, generated *in vitro*, steadily increased the overall levels of transcription and reduced the proportion of early RNA, suggesting that the early promoters were less sensitive to superhelicity than were other promoters on the phage. A potential role for superhelicity in promoter regulation was also discussed.

In 1976, Botchan reported a more detailed study (*67*) on the same effect. Using electron microscopy to localize nascent transcripts, he showed that all of the promoters were stimulated by negative superhelicity with the P_R promoter being stimulated about 1.6-fold and P_L being stimulated 5-fold by a superhelical density of 0.072. Promoters on the superhelical template, some of which were not active on the linear DNA, were generally associated with (dA+dT)-rich regions as measured by partial denaturation mapping. This association was also seen by other workers (64-66). Botchan postulated that such regions adjacent to promoters might act in conjunction with negative superhelicity to aid the unwinding of the DNA helix by RNA polymerase.

The selective effects of superhelicity on some promoters were also found in studies on the rates of synthesis for six enzymes in the absence or the presence of nalidixic acid at 50 µg/ml (245). Two enzymes, the lambda receptor and maltodextrin phosphorylase, showed fivefold and threefold reductions in transcription with the antibiotic, respectively. However, reduction of the drug to 10 µg/ml, did not affect transcription (but still blocked replication). A third enzyme, β-galactosidase, showed a twofold reduction of synthesis at the higher drug concentration, while three other enzymes [tryptophanase (see the following paragraphs), glucose-6-phosphate dehydrogenase, and homoserine dehydrogenase] were not affected by the drug. Since the existence of DNA gyrase was not known at this time, a mechanism of inhibition involving direct binding of the drug to promoter sites was proposed (also see 243).

In studies on the inhibition of transcription at various concentrations of nalidixic acid, the tryptophanase gene, previously thought to be unaffected by nalidixic acid (see above), showed a tenfold reduction in synthesis, the *lac* promoter and three promoters for the *mal* operons showed sevenfold reductions, while the *gal* promoter(s) showed only a twofold reduction (217). The *lac* and *gal* effects occurred at higher drug concentrations than the inhibitions of the other promoters. The *gal* promoter effect may be questioned because two promoters are now thought to transcribe the same genes (251). These results suggested a correlation between the sensitivity of a promoter to CRP stimulation and its sensitivity to nalidixic acid. Conversely, two CRP-independent anabolic operon promoters for the threonine and tryptophan synthetic enzymes were unaffected by the drug. Finally, two mutant promoters were actually stimulated by nalidixic acid, the *lac* I^Q promoter (showing fourfold stimulation) and the *lac* UV5 promoter (discussed in following paragraphs).

Further evidence for this promoter-specific effect of superhelicity comes from the use of both the phage lambda P_L promoter and the *trp* promoter to transcribe the anthranilate synthetase gene, under conditions to selectively repress either promoter, which showed that the *trp* promoter is relatively insensitive to either nalidixic acid, coumermycin, or novobiocin, whereas the phage promoter is inhibited by a factor of 10 under identical conditions (252). Although this seems to argue for a direct involvement of gyrase in transcription, it may be that gyrase acts indirectly on the promoters by altering the *in vivo* superhelicity of the phage DNA.

Another report (253) shows that the phage lambda promoter P_R is relatively insensitive to gyrase inhibitors, whereas P_L and P_{RE} are more sensitive. This supports the observation (67) that P_L is stimulated by negative superhelicity three times more than is the P_R promoter. The *trp* promoter is more sensitive to oxolinic acid or coumermycin than to nalidixic acid (253).

This drug-specific effect on promoter activity is difficult to explain if the drugs are only inhibiting the supercoiling activity of gyrase. Side reactions of the inhibited gyrase (233) might explain these anomalous results. The use of the temperature-sensitive gyrase or *in vitro* transcription assays could avoid complications produced by side reactions.

Novobiocin inhibits the transcription of the colicin E1 promoter by a factor of three, whereas another ColE1 promoter is inhibited by a factor of two (218). Other promoters in ColE1 do not seem to be affected by the drug. Again, the *trp* promoter is inhibited by only a small amount, whereas the *lac* promoter is reduced sevenfold. Coumermycin and oxolinic acid inhibit the *lac* promoter, but to different extents. Furthermore, the ribosomal promoter for the rrnB cistron shows a sixfold inhibition by novobiocin, as did the promoter for the tetracycline resistance genes on the plasmid pBR322. Another promoter on the same plasmid, P_{AMP}, shows no inhibition by novobiocin.

A third ColE1 promoter is also activated by superhelicity (254). This promoter is for the 104-base transcript (255) thought to be involved with the initiation of ColE1 replication (256). It is yet unclear whether the effects of superhelicity on transcription can regulate replication as well as gene expression.

3. Summary of Effect of Supercoiling on Transcription

The main point of this section is that *in vitro* characterization of promoters on relaxed templates may not truly mimic the situation in the cell because the promoters *in vivo* are most likely supercoiled and are either enhanced, unaffected, or inhibited by superhelicity. A specific case that emphasizes this point appears in Fig. 12 which shows the effect of nalidixic acid on the wild-type and UV5 mutant *lac* promoters (217). The activity of the wild-type promoter drops with increasing drug concentration, whereas the mutant promoter is activated. An assumption that *in vitro* transcription assays, on relaxed DNA mimic the high-drug portion of this plot, would lead to an erroneous conclusion concerning relative promoter strengths. The *in vitro* assay would indicate that UV5 was seven times stronger, whereas it would actually be two times weaker *in vivo*. [Note that these data may not be in accordance with Reznikoff (203).] Similar discrepancies can occur in the comparison of other promoters if superhelicity effects are not taken into consideration.

As described in Section (IV,A) much effort has gone into studying se-

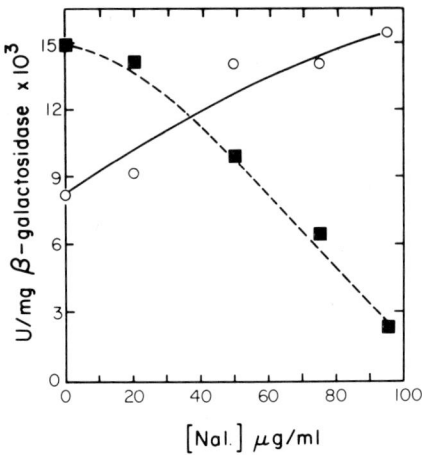

FIG. 12. Comparative effect of nalidixic acid on the synthesis of β-galactosidase in strains 3000 (with a wt *lac* promoter) and UV 5. Cultures were induced with 1 mM isopropyl-β-D-thiogalactoside. The differential rates of β-galactosidase synthesis are plotted as a function of nalidixic acid concentration [Nal.]. Open symbols, strain UV5; filled symbols, strain 3000. Reprinted, with permission, from Sanzey (217).

quence similarities that might define the unique sites and strengths of RNA polymerase binding. However, no consideration has been given to the fact that some of the promoters compared are much stronger than others, and, as demonstrated above, the measurement of relative promoter strengths must also necessarily include a consideration of superhelicity. Hence, additional work in well-defined systems will be required in order to deduce the preferred binding sites for RNA polymerase.

C. T7 Late Promoters

In contrast to the numerous uncertainties described above for other cellular and viral promoters, the late promoters of phage T7 are remarkably simple and understandable. Sequence studies (see below) on eight class II and class III promoters indicate that a 23-base-pair sequence contains all the necessary information for both recognition and initiation by the T7 RNA polymerase.

When the virulent bacteriophage T7 infects *E. coli*, the bacterial RNA polymerase transcribes only the class I (early) genes located at the leftmost 20% of the 40,000-base-pair DNA. One of the early genes (gene 1) codes for the T7 RNA polymerase, which transcribes the remaining 80% of the phage DNA. The T7 polymerase is a single polypeptide chain of about 107,000 MW and exhibits almost absolute specificity for the transcription of the late genes of T7, both *in vivo* and *in vitro*. The late genes are divided into two groups.

Class II genes are located between 15% and 45% of T7 DNA and are expressed between 6 and 15 minutes after infection, whereas class III genes are located between 45% and 100% and are expressed from 6 to 8 minutes until lysis. The transcription of T7 DNA has been carefully reviewed (257–260).

The determination of the first complete restriction map of T7 DNA in 1978 (261) led to rapid progress in our knowledge of the structure and function of the late promoters. On the basis of the lengths of transcripts synthesized with restriction digests or purified fragments of T7 DNA as templates, six class II late promoters were discovered and localized at 14.6, 14.8, 15.9, 18.9, 22.6, and 27.9% of the genome (262–264). The exact locations of four class III promoters were also identified at 46.5, 55, 57, and 87% (265), and the approximate position of a fifth one was placed at 67 to 73% (266). These positions correspond to the origins of the major class III transcripts previously observed *in vivo* and *in vitro* (267–271). It is expected that additional class II promoters will be identified in the 30 to 45% region.

Determination of the DNA sequence around three of the class II promoters (262, 68) and five of the class III promoters (265, 266) revealed a striking homology (Fig. 13). Four of the class III promoters share an identical 23-base-pair sequence around the transcription initiation point (265), and

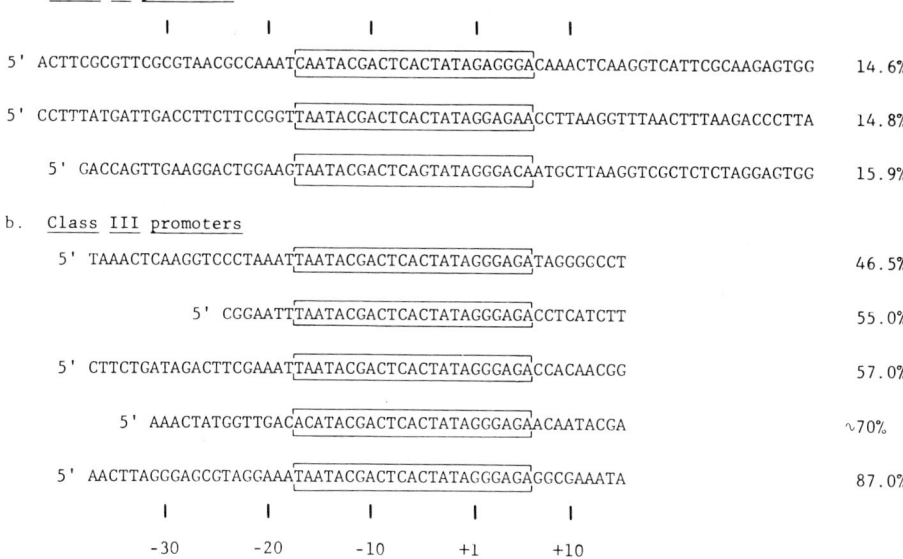

FIG. 13. DNA sequences around the T7 late promoters. The highly preserved sequence that consitutes the recognition-initiation site is indicated by the solid lines. Positions are marked from the initiation point at +1.

the sequence of the fifth promoter at ~70% differs only by 2 base-pairs at the left end of the common sequence (266). The same sequence is also shared by the class II promoters, with some variation at the right-hand side between positions +2 and +6 (68).

Considering all eight promoter sequences, the following 23-base-pair sequence can be proposed for the T7 late promoters:

$$(5') \text{ t-a-A-T-A-C-G-A-C-T-C-A-G-T-A-T-A-G-R-R-R-n-A } (3')$$
$$\phantom{(5') \text{ t-a-}}-17\phantom{\text{-A-T-A-C}}-10\phantom{\text{-A-C-T-}}-5\phantom{\text{-C-A-G-T}}+1\phantom{\text{-T-A-G-R}}+6$$

where small letters indicate nucleotides that are somewhat variable and R indicates "a purine nucleoside." Although this sequence is highly preserved among the eight different promoters, variation at the indicated nucleotides is apparently well tolerated. For example, the sequences of the promoters at 14.6 and 14.8% differ in their nucleotides at positions −17, +2, +3, and +5. Yet, in experiments in which the two promoters were assayed in parallel, comparable amounts of discrete transcripts were synthesized (264). In contrast, the sequence of the promoter at 15.9% has unique nucleotides relative to the other promoters, at positions −5 (in the middle of the common sequence) and at +5 (where variation did not appear to have a detectable effect). The 15.9% promoter was found to be approximately fivefold weaker than either of the promoters at 14.6 and 14.8% (68, 264), and this decrease in activity may be the result of the G → C transversion at position −5.

The complete lack of homology in the DNA regions outside the common sequence, as opposed to the highly preserved homology inside, suggests that position −16 to +6 constitutes the promoter site. Direct evidence in support of this notion came from experiments in which the DNA upstream and/or downstream from the common sequence was either replaced by vector DNA (in cloned T7 fragments) or removed by restriction cleavage (68). The results of these experiments indicated that the DNA sequence outside the common region is not essential for promoter activity. By these criteria, the 23-base-pair contiguous sequence constitutes a fully defined recognition-initiation site for T7 RNA polymerase.

The sequence of the recognition site contains very little twofold symmetry. Only the 6-base-pair sequence between positions −5 and +1 is symmetrical, and additional symmetry around the same axis depends on the nature of the nucleotides in the -R-R-R-m-part of the sequence. In this regard, the 14.6% promoter contains much more symmetry than the 14.8% promoter, but this feature does not appear to affect the relative expression of these two promoters (68). The lack of extensive twofold symmetry is also observed with bacterial promoters (199).

In promoters of *E. coli* RNA polymerase, a 7-base-pair site of partially preserved sequence (the "Pribnow Box") is located 7-base-pairs upstream from the transcription initiation point, and a second region of partially preserved sequence also exists approximately 30 base-pairs farther upstream (see preceding section). In contrast, T7 promoters consist of one contiguous recognition site that includes the initiation point, and no other sequence further upstream appears to be required for complete function. In this respect, T7 promoters are quite distinct from their bacterial counterparts. The physical separation of the two recognition sites and the initiation site in the *E. coli* promoters, as opposed to the integration of these elements into one contiguous site in T7 promoters, may reflect the size difference between the two enzymes (the *E. coli* polymerase is approximately five times the size of the T7 RNA polymerase).

The well-defined sequence of the recognition-initiation site, combined with the high specificity and single subunit composition of the RNA polymerase, render the T7 promoter system an attractive addition to the bacterial system for the elucidation of the mechanisms underlying promoter selection and the effect of DNA structure on promoter function.

D. Terminators

The regulation of mRNA synthesis is dependent on both the initiation and termination of transcription. Efficient termination of the transcript is required to allow independent expression of adjacent operons. Recent studies have expanded the role of terminators in genetic regulation, in that a terminator can serve as fine control (i.e., attenuator) of gene expression when it is interposed between a promoter and structural genes.

A detailed description of termination is beyond the scope of this review, and only the structural aspects will be mentioned. Several excellent reviews on general aspects of termination are available (272–275, 3).

As with promoters, a comparison of the sequences of known points of termination (3, 276) has not demonstrated extensive homologies. Many of the stronger terminators, however, have a (dG+dC)-rich block followed by an (dA+dT)-rich region in which the RNA is actually terminated. Symmetries within the (dG+dC)-rich block allow the postulation of hairpin structures. Although these could exist either in either DNA or RNA, RNA is favored from studies of T1 RNase resistance (273, 277), base analog effects on termination frequency (278), and the common proposed mechanisms for attenuation in the *trp* (272), *thr* (279), *leu* (280), *his* (281), and *phe* (282) operons. However, DNA structural influences may also have an effect on termination, as termination mutants have been isolated 60 base-pairs downstream from the termination point (276) in an untranscribed region.

V. Recognition at the Origins of RNA-Primed DNA Replication

In most systems, DNA replication originates at a unique site of the genome. Even for the simplest plasmids and phages, this process involves a number of proteins whose identities and functions are just beginning to emerge. The enormous accumulation of information regarding the initiation, elongation, and termination of DNA synthesis *in vitro* and *in vivo* have been the subject of numerous recent reviews (*283-293, 222*). The scope of this section is to review the latest information on the mechanism by which a DNA site is recognized as the origin of replication.

Initiation of DNA replication requires a DNA site that will serve as the origin; a "priming" or "nicking" enzyme DNA polymerase and several auxiliary proteins. These components are briefly described below.

A. Location of Origins

The origin is located either within an intercistronic (untranslated) region and/or in close proximity to a gene coding for one of the initiation proteins. The origins of (−) strand synthesis [the conversion of the (+) viral strand to the double-stranded replicative form (RF)] of the isometric phages G4, St-1, ϕK, and α3 are located in intercistronic regions (*294*). In contrast, (+) strand synthesis from the RF form of ϕX174 originates at a nick (a single-strand cleavage) located inside the phage gene A; the nick is introduced by the gene A protein itself (*295*). Similarly, replication of the lambdoid phages λ, 434, ϕ21, and ϕ80 begins inside the phage gene O (*296-298*). In plasmid R6K, one of the two origins is adjacent to gene II (*299*). The protein products of these genes are directly involved in the initiation of DNA replication, and their functions are discussed below. The origin of replication of bacteriophage T7 was placed, on the basis of electron microscopic evidence, at approximately 17% of the genome (*300*). Subsequent studies on cloned restriction fragments place the origin to the left of 16%, between 12 and 16% (*301*). T7 genes 1 (the phage RNA polymerase), 1.1 (unknown function), and 1.3 (the phage ligase) are located in this region (*257*).

The DNA sequences of several origin regions have been determined. The common structural features of these sequences is discussed in connection with the proteins that may recognize them.

B. Primase

The priming protein "primase" synthesizes a short RNA transcript that is subsequently extended by the DNA polymerase complex. The *E. coli dnaG* protein (*302*), a rifampicin-resistant enzyme, appears to serve this function for the (−) strand synthesis of the isometric phages G4 (*303*) and ϕX174 (*304*,

305), as well as the lambdoid phages (306). The *E. coli* RNA polymerase appears to prime DNA synthesis for the (−) strand synthesis of the filamentous phages (307, 308) and for the plasmids colicin E1 (292) and R6K (299).

C. Auxiliary Proteins

Several auxiliary proteins are required for the initiation of DNA replication. These proteins appear to select the site that will serve as the origin, ensure that the origin is available for recognition by the primase, direct the primase to the correct site by providing appropriate contacts, and, finally, regulate (turn on or maintain off) the initiation event. The regulation of the initiation of DNA replication was the subject of the "replicon" model proposed several years ago (309). This model defined the unit of replication (the replicon) as a DNA segment composed of a specific site (the replicator) and a structural gene whose product (the initiator) acts upon the replicator. More recently, the phenomenon of autoregulation was proposed [suggesting that synthesis of the initiator is self-regulated (310, 311)]. Examples of autoregulation may be found in the replication of the lambdoid phages and plasmid R6K. The topid of regulation has been extensively reviewed very recently (293).

The specific functions of most auxiliary proteins are not known. However, some information on their apparent involvement in recognition at the origin is available. The best-studied systems are the filamentous phages, the isometric phages, and lambdoid phages.

1. fd

Recognition of the origin of (−) strand synthesis of the filamentous phage fd requires (at least *in vitro*) the priming protein (RNA polymerase) and only one auxiliary factor, the "DNA binding protein I" (DBP) (312). The following events have been postulated to occur during recognition at the origin (313, 314). DBP binds on the viral (+) strand and covers it completely, except for a short segment ($\sim$ 120 bases). This segment forms a stable hairpin secondary structure that cannot be melted by DBP. RNA polymerase now binds and initiates transcription of the primer, 6 bases upstream from the stem of the hairpin loop. Transcription through the stem disrupts its secondary structure and engages the transcribed strand into a DNA–RNA hybrid. In this manner, the segment forming the opposite side of the stem becomes single-stranded and binds DBP. The polymerase, which is progressing, eventually encounters the newly bound DBP and terminates 25 to 35 bases downstream. The DNA polymerase III holoenzyme then extends the primer.

This model (313, 314) suggests a triple function for the auxiliary protein. DBP directs the priming enzyme to the origin (several promoters elsewhere on the DNA are blocked by DBP), causes termination of the primer, and facilitates the action of the DNA polymerase III holoenzyme (*315*). According to this model, a DNA origin sequence is *not* selected directly by the priming protein. Instead, the origin is specified by the formation of the stable hairpin and the binding of DBP on other possible promoter sites, so that the RNA polymerase initiates at the origin by necessity. However, the possibility that DBP bound on the DNA interacts and directs the RNA polymerase to the correct initiation site has not been excluded.

Several lines of evidence are consistent with this model. The facts that the DNA sequences of the origins of f1 (*316*) and M13 (*317*) share extensive homology with the sequence of fd, and that the potential for hairpin formation is maintained, suggest that the same mechanism may apply to all filamentous phages.

2. G4, St-1, α3, φK

The isometric phages G4, St-1, α3, and φK contain a single-stranded circular DNA that replicates to the RF much like the filamentous phages. Initiation of this reaction requires only DBP and the priming enzyme, which, in this case, is the *dnaG* protein rather than RNA polymerase (*302*). Initiation occurs at a unique site both *in vivo* (*318*) and *in vitro* (*319*). Determination of the exact initiation sites and their DNA sequences revealed (*294*) that, for all four phages, the (−) strand initiation site occurs within an intercistronic region of approximately 135 bases. Extensive sequence homology exists in this region, but not in the adjacent coding regions. Highly conserved DNA sequences occur in two regions 42 and 45 bases long, separated by 13 bases of divergent sequence. The conserved sequence regions have the potential to form hairpin structures, and the initiation point for primer synthesis is located 5 bases upstream from the stem of the hairpin loop. In this respect, recognition at the origin of these isometric phages appears to follow the same mechanism with the filamentous phages, despite the substitution of RNA polymerase with *dnaG* protein as the priming enzyme. Again, the priming enzyme appears to be directed to a DNA site protected from the DBP by the formation of a stable secondary structure. Nevertheless, the extensive sequence homology observed at the origin might serve one of two possible functions: to preserve a recognition site for the priming protein, and/or to preserve the capacity of the single-stranded DNA to form a stable secondary structure.

In an attempt to distinguish between these two alternatives, the sequences of the origins of the filamentous phages were compared with the

---

FIG. 14. The DNA sequence at the origins of replication of *Escherichia coli* and bacteriophages St-1 (φK, α3), G4, λ (φ21, 434), and φ80 drawn into similar secondary structures. Reprinted, with permission, from Sims *et al*. (*294*).

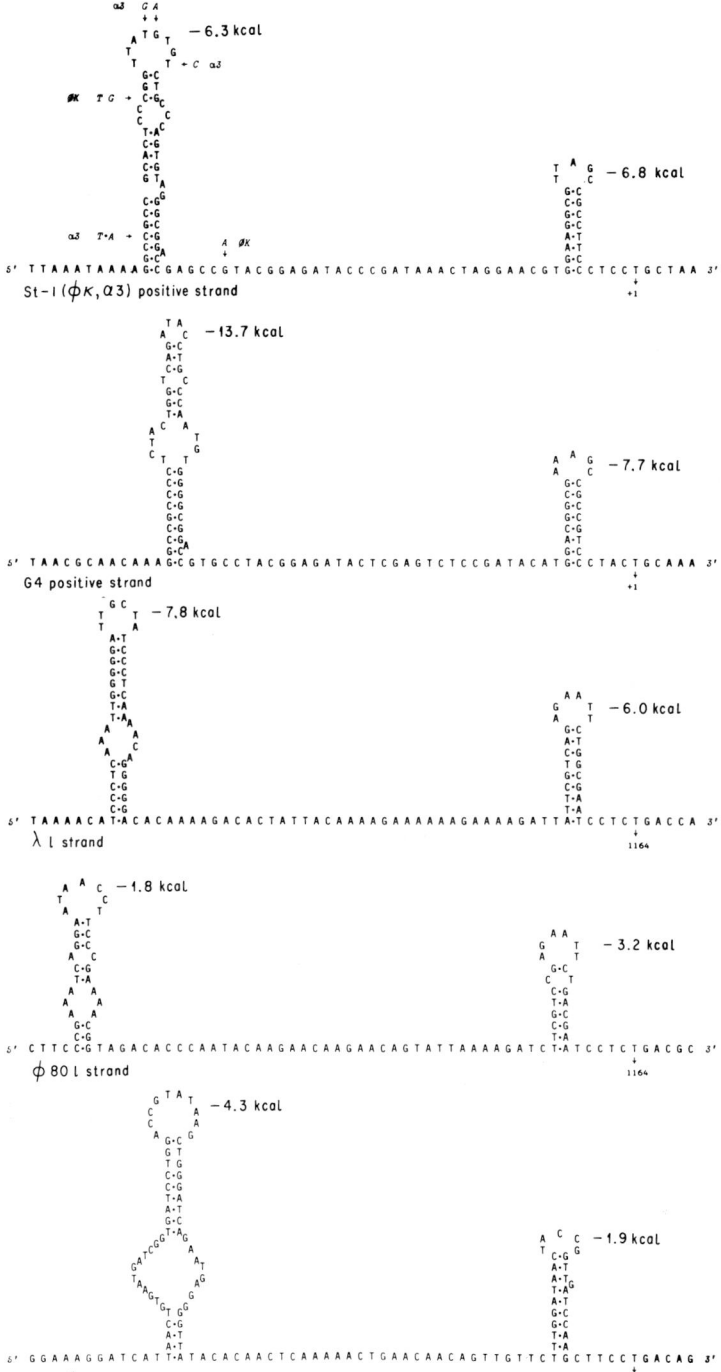

sequences of the origins of the lambdoid phages (296, 298, 320) and E. coli (321, 322), as the latter systems also require dnaG protein for their replication (although dnaG protein may be required only for elongation). This comparison (Fig. 14) revealed only limited homology among the sequences of the nine different origins, but a striking capacity to form similar hairpin sequences was clearly maintained. In every case, two long hairpins could be formed, separated by a single-stranded segment. On the basis of this evidence, it was postulated that an origin for dnaG-dependent strand initiation may consist of (a) one large and one small hairpin separated by 30–40 bases; (b) a short pyrimidine-rich sequence ending in dT-dG. This sequence is located immediately upstream from the smaller loop, and codes for the primer (294).

3. LAMBDOID PHAGES

The notion the dnaG is not directly involved in recognition at the origin but, rather, is directed to the correct site by the auxiliary proteins appears to apply to the priming of double-stranded DNA replication as well.

Replication of the double-stranded DNA of the lambdoid phages initiates at a unique site, proceeds bidirectionally and is carried out by several host and phage proteins. The bacterial proteins DNA polymerase III, RNA polymerase, dnaB, dnaG, dnaJ, dnaK, and the phage proteins O and P are required for replication in vivo (for recent reviews see 316, 292, 293). Some evidence suggests that initiation is primed by the dnaG protein (323). The exact point of initiation by the primer is not known but is believed to be inside gene O around the position shown in Fig. 15. In addition to RNA polymerase and the dnaG protein, gene O, gene P and dnaB proteins participate in initiation. Indirect evidence suggests that gene O, gene P, and dnaB proteins may act as a complex (292). It appears likely that protein O binds directly to the DNA with its N-terminal region and to protein P with its C-terminal domain (324).

Four different processes seem to regulate λ DNA replication (Fig. 15). First, the interaction of protein O with the origin region inside gene O appears to be an example of positive autoregulation (324). Second, transcription of genes O and P, which lie downstream from the main rightward promoter (P_R) is controlled by the cI and cro repressors (325). Third, rightward transcription may have the additional function of inducing strand sep-

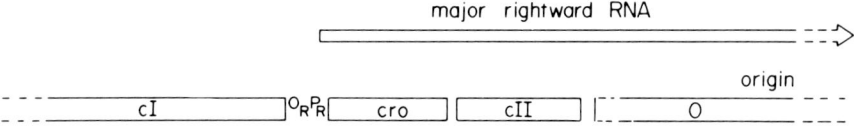

FIG. 15. Genetic map of the origin region of phage λ (298).

aration at the origin [transcriptional activation (326)] in a situation analogous to that encountered with the filamentous and isometric phages discussed above. Although these three processes may regulate primer initiation, a fourth level of control may be exerted at the termination of leftward transcription of the primer at a site that also signals the DNA–RNA junction for leftward leading-strand synthesis. This site was termed the "inceptor" (323, 327).

Determination of the DNA sequence around the origins of the lambdoid phages 434 and $\phi 21$ revealed extensive sequence homology with λ. In contrast, the sequence of $\phi 80$ differed from the other lambdoid phages for most of its length, except for several specific stretches within the functional limits of the ϕ origin (320). The major stretch of 31 homologous base-pairs contained two internal inverted repeats and could, therefore, be drawn in the form of a large hairpin loop. Furthermore, the DNA sequence between about 40 to 130 base-pairs to the left of this hairpin could be drawn in the form of elaborate base-paired cloverleaf-like structures in both λ and $\phi 80$, despite the fact that little sequence homology exists in this region. On the basis of this observation, it was postulated that during initiation, RNA polymerase transcribes from P_R through the origin region and, in the process, denatures the DNA and engages the codogenic strand into a DNA·RNA complex, in a situation analogous to the single-stranded phages. Then, the complementary single-stranded stretch of the DNA can assume the elaborate cloverleaf structure, which is presumably recognized by gene O protein, in connection with gene P and, perhaps, dnaB proteins. Hence, the primase (dnaG protein) can recognize some site around the smaller loop and begin leftward transcription (320). Finally, termination of primer transcription occurs approximately 600 base-pairs to the left, at a site reminiscent of a transcriptional termination signal. Extension of the primer by DNA polymerase III begins at this "inceptor" site (323, 327).

The main weakness of this rather detailed model is that it involves, by necessity, a number of assumptions. However, it can account for most of the available information on the mechanisms of initiation, and, in addition, it assigns a specific role to the extensive dyad symmetry elements observed at the origin sequences. A major implication (and test) of this model is whether double-stranded DNA can indeed assume elaborate cloverleaf structures that are stabilized by protein binding. Calculations of the necessary stabilization energy yielded "not inaccessible" values (320) although the exact figures were not reported. Although it is topologically feasible for double-stranded DNA to be converted from a linear duplex conformation to a cloverleaf-like structure without nicking, the presence of topoisomerases might facilitate this process. DNA gyrase (328) and a novel nicking-closing enzyme (329) have recently been implicated in DNA replication.

In addition to the extensive inverted repeats that may be involved in hairpin formation and a generally high (dA+dT)-content that is believed to facilitate strand separation, most DNA origin regions display several direct repeats of a specific sequence. In the lambdoid phages λ, φ80, and φ82, an average sequence of 18 pairs is repeated four or five times (*298*) in the origin region. In plasmid R6K, seven 22-pair direct repeats joined in tandem have been postulated to function as a recognition site for the gene II protein (*299, 293*). The latter is required for initiation and, also, inhibits transcription of its own gene in an autoregulatory operator–repressor type system. In the origin region of *E. coli*, a tetranucleotide sequence is repeated 14 times (*321, 322*), and this sequence is preserved at the origin region of *Salmonella typhimurium* (*330*). The direct repeats found at the origin are believed to represent multiple binding sites for one of the proteins involved in initiation, but no direct evidence has yet been reported.

VI. Effect of Protein Binding on DNA Conformations

A. General Considerations

Utilization of the genetic information encoded in a DNA molecule requires the interaction of various proteins with that DNA. The cellular processes (such as replication and transcription) related to and regulated by protein–DNA interactions are multifaceted and highly varied.

To obtain a stable interaction between a protein and DNA molecule, the free energy of the protein–DNA complex must be lower than for the separated components. The decrease in entropy must be overcome.

In addition to binding nonspecifically to DNA, some proteins recognize and bind to specific DNA base sequences by having a higher binding affinity at that segment of DNA with the appropriate base sequence.

The attainment of favorable free energies of binding between proteins and nucleic acids requires the involvement of a number of potential intermolecular forces including (*a*) hydrogen bonding; (*b*) charge–charge (electrostatic) interactions; (*c*) hydrophobicity; (*d*) aromatic ring stacking interactions; and (*e*) van der Waals forces. Of these forces, charge–charge and stacking probably provide the greatest contribution to the free energy of binding.

For most DNA binding proteins, there is a strong salt dependence; binding affinities decrease with increasing ionic strengths. This is due to a competition between cations in solution and the protein for the negatively charged phosphates of the DNA backbone (*331*). It also demonstrates the importance of charge–charge interactions in stabilizing protein–DNA binding. At physiological pHs, positively charged basic amino acids, such as lysine and arginine, have the potential to interact with the DNA phosphate backbone provided they can be aligned in close proximity to one another (reviewed in 2).

In addition to electrostatic interactions, aromatic ring stacking is thought to provide the other major contribution to binding free energy. The planar ring amino acids tryptophan, tyrosine, and phenylalaine all present the possibility of intercalating between nucleic acid bases. Studies of the interaction of these amino-acid residues with DNA polymers suggest that intercalation plays an important role in protein–DNA binding and that it may account for a portion of the sequence-specificity in binding (332–338).

Hydrogen bonding may perform a discriminating function rather than adding to the favorable free energy of binding. Since the free energy of hydrogen bonding between protein and DNA is approximately the same as that between the macromolecules and solvent, the free energy difference between complexed and dissociated components is thought to be negligible. However, if the protein–DNA complex cannot form the same number of hydrogen bonds between each other as present with solvent in the dissociated form, then hydrogen bonding represents a destabilizing effect. Thus hydrogen bonding is important in binding in the sense that failure to form proper hydrogen bonds makes an unfavorable contribution to the free energy of binding.

The same intermolecular forces that allow formation of a stable DNA–protein complex are responsible for sequence specific binding. The presumption is that a protein exhibiting specific binding properties has those amino-acid residues involved in DNA binding in a conformation such that the most favorable intermolecular interactions are obtained when these residues are aligned with a specific nucleotide sequence. Thus, binding to the proper nucleotide sequence leads to maximal DNA–protein affinity. To obtain this specific interaction, there must be structural features in the DNA, which depend on the base sequence, that are required for protein interactions. The parts of the bases exposed in the DNA grooves may provide specific interaction sites. The possibility that the protein may recognize conformationally altered segments of DNA has been previously discussed (for a more complete review, see 4).

Since both proteins and DNA are flexible molecules to some extent, another aspect of their interaction entails the conformational changes which may be induced as a result of their binding. Such conformational changes probably stabilize the protein–DNA interaction by better aligning the functional groups involved in binding, lowering the free energy of the complex. An interaction that alters DNA conformation, such as the opening of the double-helix presumed necessary in DNA replication or transcription, might serve a mechanistic function.

This section deals primarily with some of the work on proteins that alter DNA conformation, especially at the nucleotide level. Attention is focused on (a) the nature of the DNA conformational change; (b) how the protein might induce this change; (c) whether the conformation change appears relevant to the cellular function of the protein; and, when possible (d) what the conformational change indicates about the role of DNA sequence and conformation as related to protein–DNA interactions in general.

B. Amino Acids and Polypeptides

Amino acids and oligo- and polypeptides are not biologically important DNA binding proteins, but, because of their simplicity, they provide models useful in establishing the effects of different amino-acid residues that bind to DNA.

Polylysine distorts the DNA helix (for a review, see 2), decreasing the maximum and increasing the minimum amplitudes of the DNA CD spectra (339). This has been attributed to dehydration and charge neutralization of the DNA leading to a DNA conformational change from B to a combination of B and C (339). The apparent importance of lysine in protein–DNA electrostatic interactions may make it a common contributer to protein–induced helix deformation.

Intercalation of aromatic planar amino acids alters the conformation of single- and double-stranded DNA polymers (332, 333, 335, 336, 340, 341). One might suspect that intercalation of amino-acid residues like that of dyes, would increase the distance between nucleotides and alter the base stacking. Tyramine intercalation in calf-thymus DNA decreases the amplitude of the maxima and minima of the CD spectra of the calf-thymus B conformation which has been interpreted as a decrease in base-stacking interactions (340). Intercalation of tryptophan and tyrosine also appears to result in unstacking of the nucleotide bases (333, 335, 336, 341).

In addition to altering DNA conformation, tryptophan and tyrosine appear to exhibit a strong preference for single-stranded DNA (334, 338). A degree of nucleotide preference has also been observed. Intercalation of tyrosine in double-stranded DNA is more favorable for (dA+dT)- than for (dG+dC)-rich regions (332). Intercalation of tryptophan seems to be favored at adenosines in single-stranded DNA polymers (341). Also, increasing tyrosine content in lysine–tyrosine copolymers increases the specificity of the polypeptides for dA·dT base-pairs (337). Whether the base specificity is related to the ease of distorting the helix or just the electronic interaction in stacking is unknown.

These studies involving amino acids and model oligo- and polypeptides demonstrate that conformational changes in DNA as a result of protein binding may be due in part to the fundamental interactions of amino-acid residues with functional groups of the DNA.

C. Helix-Destabilizing Proteins

Helix-destabilizing proteins evoke obvious DNA conformational changes. They aid in the transition of DNA from helix to coil by binding to single-stranded DNA with higher affinity than to double-stranded, thus shifting the conformational equilibria to the single-strand form. In all such proteins studied thus far, the coil-protein conformation does not appear to be random. The different helix-destabilizing proteins allow comparisons of differences and similarities of the binding mechanism that alters the DNA conformation. The conformation and properties of random-coil DNAs are discussed in Sections II,B and VII.

1. T4 Gene 32 Product

Gene product 32 (gp 32) of bacteriophage T4 is a 35,000 MW protein that binds preferentially and cooperatively to single-stranded DNA, facilitating melting and denaturation of $(dA-dT)_n$ (*342, 343*). GP 32 does not facilitate the melting-out of native DNAs, which has been attributed to a kinetic block in the formation of complexes of protein and single-strand DNA (*344*). Gp 32 has been implicated as a functional element in DNA replication (*345, 346*), genetic recombination (*347, 348*), protection of DNA from nuclease digestion (*349*), and repair of radiation damage (*350*).

The general binding characteristics of gp 32 have been determined in a number of studies. Gp 32 undergoes a conformational change in binding to single-stranded polynucleotides (*351*). The size of the gp 32 binding site has been determined by a number of methods to be 5 (*352*), 6.7 to 7.5 (*353*), 10 (*343*), and 11 nucleotides per protein monomer (*354*). The binding of gp 32 to both native DNA and single-stranded polynucleotides is dependent upon ionic strength, stressing the importance of electrostatic interactions in binding (*353*). On the basis of oligonucleotide binding, it was concluded that the principal binding interaction involves a dinucleoside monophosphate moiety (*352*). A tryptophan residue has been implicated in the DNA binding site of the protein (*355*). The binding of gp 32 shows no preference for deoxyribo- as opposed to ribooligonucleotides (*352*). Electron microscopy of denaturation patterns of native λ in the presence of gp 32 suggests that gp 32 preferentially invades (dA+dT)-rich regions (*356*). The cooperativity of binding probably involves a protein–protein interaction (*357–359*) and may involve collaboration of a multimeric complex to distort the DNA helix (*352*).

Many studies have demonstrated that gp 32 significantly distorts the conformation of single-stranded DNAs on binding. It has been postulated that gp 32 might bind to phosphate groups to leave nucleotide bases in the complex exposed (343, 346, 352). The exact type of conformational change has not been uniformly resolved in light of conflicting data and anomalous interpretations of CD data. Single stranded fd phage DNA complexed with gp 32 sediments as though it were held in a rigid extended conformation (343). Electron microscopy of the same complex reveals a flexible extended rodlike conformation with 4.6 Å per nucleotide. The DNA backbone is extended, but not to its fullest extent (356).

Since CD is particularly sensitive to changes in base positioning, it would seem well suited for studying DNA conformational variations. While CD does give a clear indication that a conformational change in the DNA has occurred, rigorous conformational interpretations are sometimes difficult; a given spectrum can be open to different interpretations. The CD spectral change that occurs when gp 32 binds to single-stranded fd phage DNA is shown in Fig. 16. Spectral changes in binding gp 32 to the same nucleotides are generally very similar between different research groups, although the interpretations vary. Jensen *et al.* (353) interpreted a UV hyperchromic absorption shift and a decrease in the CD maxima for gp 32 binding with single-stranded polynucleotides and poly[d(A-T)] to be the result of total base unstacking and extension of the phosphate backbone. In contrast, when

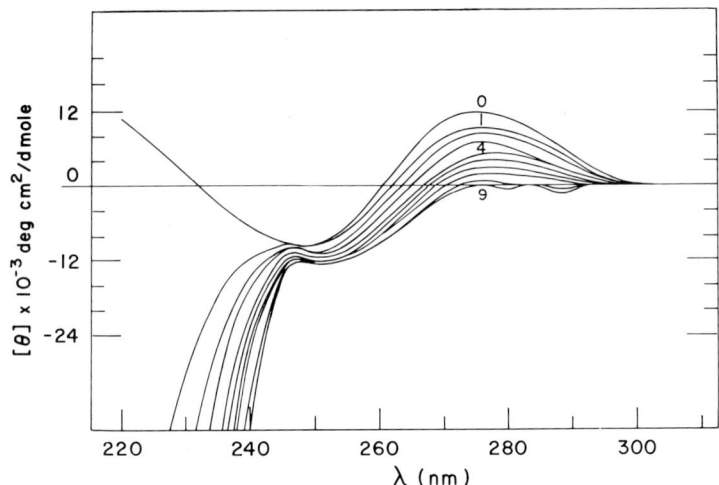

FIG. 16. Circular dichroism (CD) spectra of fd DNA showing changes induced by the addition of DNA binding proteins. fd DNA (25 μM) was intially present in 2.4 ml of 10 mM TrisCl (pH 8.0) at 25°C. Changes in CD on addition of 10- and 20-μl aliquots of gene 32 protein (84 μM) in the same buffer to yield final protein concentration of 4.1 μM are indicated by curves 1-9. For curve 9, the molar ratio of protein to DNA-base is 1:5.8. Reprinted, with permission, from Anderson and Coleman (354).

binding gp 32 to various polynucleotides, Greve *et al.* *(360)* interpreted the decrease in the CD maximum, but retention of its shape, as indicating that the base planes remained stacked and in fixed position to one another. The gp 32 complexed single-strand conformation was thought to be the same as for one of the helical complementary strands in poly(dA-dT) at a high LiCl concentration.

A decrease in positive ellipticity in the fd DNA spectra upon gp 32 binding may be a result of dehydration of the DNA helix and electrostatic bonding of phosphates with positively charged groups of the protein *(354)*.

Thus CD, UV absorbance, and electron microscopy have yielded conflicting results, which are difficult to resolve. One certainty is that a conformational change in a single strand of DNA does occur upon gp 32 binding.

The molecular bases for conformational changes are difficult to establish, given the contradictory nature of some of the experimental evidence. One can assume that the protein binds single-stranded DNA preferentially because the conformation of single-stranded DNA is better suited to stable binding, or its conformation can be altered more easily to fit the binding site of the protein. Amino-acid intercalation may have a role in the single-strand preference of gp 32. Tyrosyl residues in gp 32 have been implicated in DNA binding. Nitration of five of the nine tyrosyl residues in gp 32 destroyed its DNA binding, but prior binding of gp 32 to DNA blocked tyrosyl nitrification *(354)*. The implication that tyrosine is important in gp 32 binding may fit with other observations. It has already been suggested *(354)* that intercalation of tyrosyl residues between bases may explain some of the helix-lengthening observed by Delius *et al.* *(356)*. The facts that tyrosine intercalation is preferred in (dA+dT)-rich polymers *(332)*, and that binding shows a single-strand preference *(334, 338)* suggest that such intercalation could be responsible for some of the (dA+dT) and single-strand DNA specificity of gp 32. Thus gp 32 may provide information about the role of amino-acid intercalation in site specificity and DNA conformational alteration.

2. fd GENE 5 PROTEIN

The fd gene 5 protein (gp 5) is a helix-destabilizing protein that binds preferentially and cooperatively to single-stranded DNA *(361, 362)*. Unlike T4 gp 32, gp 5 will destabilize native DNAs *(362)*. It is necessary for single-strand synthesis from the fd phage RF template *(363-365)*. One of its functions may be to keep newly synthesized DNA single-stranded until it can be packaged *(366)*.

Gp 5 is a protein of molecular weight 9700 with 87 amino-acid residues *(367, 368)*. In contrast to T4 gp 32, which forms large multimeric aggregates *(343)*, gp 5 aggregates only to a dimer, producing two DNA binding sites on opposite sides of the dimer *(369)*. Each monomer of gp 5 covers four nucleotides on binding (*362, 370, 371*). The observation that gp 5 binds p(dA)$_4$ but not trinucleotides suggests that four nucleotides are involved in the active binding site (*371*).

Chemical modifications of the protein have demonstrated that lysyl, tyrosyl, and cysteinyl residues are located near the polynucleotide binding site and may interact with the DNA (*371*). Lysines in the protein are involved in electrostatic interactions with DNA (*372*). Three of the five tyrosyl residues have been implicated in the interaction with DNA by Anderson *et al.* (*371*). In addition, the three surface tyrosines of gp 5 interact with tetra- and octanucleotides, and NMR results indicate that these interactions involve intercalation (*372*). Alteration of the CD of the protein on binding to DNA also implicates the tyrosines in DNA binding (*370*).

DNA conformational changes as a result of gp 5 binding were first demonstrated by electron microscopy. Binding gp 5 to single-stranded fd DNA coalesces the circular DNA into rod-shaped or branched structures (*362, 373*). Cavalieri *et al.* (*369*) proposed a model in which the dimeric gp 5 protein binds the single-strand DNA at an oppostie side of the dimer to form hairpin-like structures, which would form a complex between double-stranded DNA and protein when extended through the whole strand.

The DNA conformational change at the nucleotide level has been investigated using UV and CD spectroscopy. Ultraviolet hyperchromic shifts in the DNA indicate that the bases in single-stranded DNA-protein complexes are unstacked (*361, 362, 370*).

The CD changes observed when gp 5 binds to phage fd single-strand DNA and other polynucleotides are very similar to those for T4 gp 32 (*370, 371*). As with gp 32, CD spectral changes are open to different interpretations. Day (*370*) interpreted his results as reduced stacking. Anderson *et al.* (*371*) note that the change in spectra is similar to dehydration and charge neutralization of a double-stranded DNA helix. Such changes in environment have given similar CD changes in double-stranded calf thymus DNA, which was interpreted as a B-to-C conformation change (*9, 10*).

With the exception of the gross DNA conformation seen by electron microscopy [which may be a consequence of the dimeric form of the protein (*369*)], the results of the binding studies with gp 5 are similar to T4 gp 32. Perhaps this indicates a similarity in the basic binding mechanism. The same arguments presented to explain gp 32 sequence specificity and induced DNA conformational changes could be applied to fd gp 5. Coleman *et al.* (*372*) proposed a model based on predicted protein structure and amino-acid residue modification studies involving an intercalation of a phenylalanine and possibly three tyrosines at the DNA binding site. The model also proposes a twist in the DNA by rotating the bases, which would be consistent with the electron microscope data.

Recent studies have resulted in a more refined model for the interaction of gp 5 with DNA. X-Ray diffraction investigations of gp 5 with a number of different oligodeoxynucleotides indicates a fundamental unit of six gp 5 dimers, which form a ring around which the DNA may be spooled in a double-helical fashion (374).

X-Ray diffraction of the protein itself has revealed a cleft that appears to be the most probable region for the DNA binding site (375). The interior of the cleft has a large number of basic amino acids whereas the external edges contain a number of aromatic amino-acid side groups that may be in a position to stack on the DNA bases. A binding mechanism involving electrostatic interactions and intercalation of aromatic amino-acid residues that is consistent with the structure and previously observed physicochemical properties has been discussed (375).

3. *Escherichia coli* UNWINDING PROTEIN

Escherichia coli unwinding protein is another helix-destabilizing protein that binds cooperatively to single-stranded DNA and represses the melting temperature of double-stranded DNA (376). It functions in ϕX174 replication (377), where it is necessary for conversion of the (+)-strand into RF (378, 379). It has also been implicated in phage fd RF formation (380). (The role of this protein in replication is discussed in Section V.) The *E. coli* unwinding protein stimulates DNA synthesis by DNA polymerase II (376, 380). This stimulation may involve unwinding of the double-stranded helix and also formation of a protein–protein complex with DNA polymerase II to facilitate DNA binding (381, 382). The molecule has a molecular weight of 18,500 (383) and appears to bind to single-strand DNA as a tetramer (383, 384). It has been suggested that the cooperativity of DNA binding may be a function of the tetramer (385). Eight nucleotides are covered per monomer (376) and other results agree with that figure for tetramer binding (383). Circular dichroism results give a value of 14 nucleotides covered per protein monomer.

Modification studies demonstrate that lysine residues are essential to DNA binding (386). Unlike T4 gp 32 and fd gp 5, *E. coli* unwinding protein contains no nitratable tyrosyl residues, suggesting that no tyrosines are involved in DNA binding (354; also 386); likewise, arginine and cysteine are not involved. Fluorescence studies indicate that tryptophans are located at or near the DNA binding site (386).

It was proposed that unwinding protein bound principally to the DNA sugar–phosphate backbone based on equivalent binding constants with different DNA homopolymers. Weiner *et al.* (383) found a polynucleotide specificity, poly(dT) and poly(dC) being preferred over poly(dA). In the same study, a comparison of deoxyribopolynucleotides with ribopolynucleotides led the authors to conclude that it was the secondary structure of the polynucleotides, not the bases, that led to specificity of unwinding protein binding. *Escherichia coli* unwinding protein has a preference for melting (dA-dT)-rich regions of bacteriophage λ DNA (*376*). Thus, the nature of binding specificity is unclear at this time.

Besides its helix-destabilizing function, *E. coli* unwinding protein appears to maintain single-strand DNA in a conformation different from random coil. By electron microscopy, fd DNA saturated with unwinding protein remains a circle but is 35% shorter than uncomplexed fd DNA, which has a spacing of 1.8 Å per nucleotide (*376*). A regularly folded DNA conformation with a broad helix was postulated to explain this observation. The protein–fd DNA complex has local flexibility (*386*). The change in the spectra of fd DNA upon titration with protein is very similar to that obtained for other helix-destabilizing proteins, and similar conformational changes were postulated (*354*).

4. OTHER HELIX-DESTABILIZING PROTEINS

Other helix-destabilizing proteins from both prokaryotic and eukaryotic organisms are known (*222*). The nature of any conformational changes induced, other than helix destabilization, has not been studied except for bovine pancreatic ribonuclease. (It is doubtful that this protein has a biologically relevant function as a helix-destabilizing protein.) This noncooperative destabilizer brings about a hyperchromic shift upon binding to poly(dA), but no shift is observed when it binds to double-stranded DNAs. This suggests an unstacking of 20% of the bases at saturation levels of protein.

5. COMPARISON OF HELIX-DESTABILIZING PROTEINS

The three helix-destabilizing proteins studied extensively, T4 gp 32, fd gp 5, and *E. coli* unwinding protein, provide an opportunity to compare the binding mechanism of three proteins that bind to DNA and serve a similar function. All but T4 gp 32 are capable of destabilizing native DNAs. Since there appears to be somewhat regular DNA conformation in the complex of protein and single-stranded DNA, the failure of gp 32 to destabilize native DNA may be caused by difficulty in obtaining a single-stranded loop large enough for the action of several protein monomers to distort a single strand into its final conformation (*353*). Thus the failure of T4 gp 32 to destabilize native DNA may be a function of its mechanism of cooperativity and/or the transition state energy of the conformational change involved in binding. This suggests that the conformational change of gp 32 is different from that of gp 5 and *E. coli* unwinding protein.

Further evidence of differences in the DNA conformations in the complexes comes from electron microscope (EM) data (*356, 376, 362*). Gp 32 shows an extended DNA–protein complex whereas gp 5 and *E. coli* unwinding protein show a contraction of single-strand length, possibly the result of a supercoiling- or packaging-like function. These differences may be a result of DNA binding to the multimeric forms of different proteins, as each has been shown to yield a different native multimeric form. The CD spectral data are in opposition to the EM data. The spectral changes resulting from the formation of complexes between helix-destabilizing protein and DNA are similar for all three proteins.

The discrepancy between the CD spectra and EM results could be resolved by several considerations: (*a*) the EM results are artifactual; (*b*) CD is insensitive to the actual DNA conformation changes; or (*c*) the EM results may represent not only the conformation at the protein binding site, but also a packaging structure influenced by the multimeric protein–protein interactions. At this time, differentiation between these possibilities is not possible. The most consistent interpretation of the CD spectra is that the binding of all three destabilizing proteins induces a conformational change that results in some base-unstacking and may maintain the bases in a single strand in a conformation similar to those of bases in a double-stranded helix in a C-like conformation. Circular dichroism spectra are, however, open to alternative interpretations.

A common feature of all three helix-destabilizing proteins is their preference for (dA·dT)-rich regions (*376, 362, 356*). This specificity may be a function of base specificity and/or the frequency of transient single-stranded regions in a "breathing" double-helix. Such a preference has ramifications with regard to the base composition around origins of replications; one might suppose that a (dA+dT)-rich origin of replication should facilitate the initial opening of the helix by providing a preferred site for the binding of helix-destabilizing protein. In fact, most origins appear to be rich in dA·dT, as described in Section V.

D. *E. coli* Lactose Repressor

The *E. coli* lactose repressor binds specifically to the lactose operator and thereby inhibits transcription of the lactose operon genes by RNA polymerase. The *lac* repressor may also bind DNA nonspecifically. The *lac* repressor has been the subject of a substantial amount of research (for a recent review, see *387*).

The DNA binding site of *lac* repressor appears to be in the amino terminal of the protein and probably involves the first 60 amino acids (*388–390*). The binding of repressor to DNA involves a tetramer of identical subunits, but the exact number of subunits making contact with the DNA is not firmly resolved (*387*). The best indication is that there are two subunits in contact with the operator or nonspecific binding site (*391–394*).

Electrostatic interactions are involved in specific and nonspecific binding. Eleven ion-pair bonds appear to be formed between DNA phosphates and basic amino acids per tetramer bound to nonoperator DNA (395), while approximately 8 ion-pairs are formed in repressor-operator binding (396). These observations indicate that specific and nonspecific repressor–DNA binding interactions may be substantially different. Modification of lysines and tyrosines results in loss of operator binding activity (397). The analysis of certain mutant repressor molecules indicates that substitution of certain amino acids in the amino-terminal end results in i⁻ phenotypes (390). Threonine 5, tyrosines 7, 17, and 47, glutamines 18 and 54, and leucine 45 were implicated in the operator binding site in such experiments. It is suggested that tyrosines 7 and 17 were probably directly involved in DNA binding (390). Histidine residues have been implicated in nonspecific repressor binding but not in specific binding, again illustrating possible differences of mechanism in the two types of interactions (393, 395, 398).

The affinity of *lac* repressor for single-stranded polynucleotides is substantially less than for double-stranded DNAs (399). Repressor stabilizes nonoperator DNA to thermal denaturation (399). Nonspecific binding also is not cooperative (395). The number of base-pairs covered per tetramer is 12 if repressor binds to only one side of the DNA, or 24 if it binds to both sides (391).

Another fundamental difference between specific and nonspecific binding is in the contacts between the protein and the DNA. Nonspecific binding may make contact with only the minor groove of double-stranded DNA (400–402), whereas in the specific repressor–operator interaction there appears to be contact with both minor and major grooves (403). Most of the specific contacts in repressor–operator binding are thought to be contained in a 17 to 21 base-pair segment (404, 405). Some symmetry elements are present in the minimal recognition sequence, but whether subunits of the repressor tetramer use the sequence symmetry to bind in a symmetrical fashion is unclear at this point (394, 389, 392). However, it is clear that flanking segments of DNA play an important role in repressor binding, as the smaller fragments containing the operator do not bind as well as larger fragments (118).

Wang *et al.* (406) first demonstrated the induction of a conformational change when *lac* repressor binds at the operator site. Superhelical turns in the DNA helix increase the rate of association slightly, up to a given superhelical density, and also stabilize the repressor–operator complex. The conclusion was that binding of repressor to operator unwinds the *lac* operator 40° to 90°. These results along with studies demonstrating that the inverted repeat sequences are unnecessary for repressor binding (404, 405) exclude the possibility that repressor induces or traps a cruciform-like structure in the operator. In contrast to these results (406) is the report that repressor bound a chemically synthesized 21-base-pair operator fragment with higher affinity when the fragment was part of a linear, as opposed to a supercoiled, plasmid. These results may indicate that base-pairs outside the 21-base-pair core fragment are involved in the unwinding effect induced by repressor.

DNA denaturing agents increase repressor affinity for operator and nonoperator DNAs (408, 395). The increased affinity for operator DNA is a result of a decrease in the dissociation rate, suggesting that local DNA destabilization might facilitate specific repressor–DNA interactions by exposing more features in the DNA grooves. This result is consistent with the supercoiling effect of Wang et al. (406).

Possible DNA conformational changes resulting from nonspecific DNA–repressor interaction have been examined by CD. For bacteriophage λ DNA, calf thymus DNA, and poly[d(A-T)], a similar change in spectra is observed in which the 270 nm maximum peak increases dramatically (409, 391). The CD change may be due to a base tilting, possibly toward a more A-like conformation or a superhelical twisting of the helix that improves the coupling between bases (391). Binding of repressor to poly(A) decreases the maximum of the CD spectra, and it is suggested that the result is similar to that seen on heating poly(A) or on quenching due to aromatic amino acids stacking with bases (410). However, the nonspecific and specific repressor–DNA interaction are different in many physical characteristics previously discussed, and these CD changes may not be relevant to operator binding by repressor. One can state that there is a definite conformation change resulting from repressor binding to both specific and nonspecific DNAs. It would be interesting to know if the conformational change for specific and nonspecific binding are similar. The availability of milligram amounts of small restriction fragments containing these genetic loci will permit this study to be undertaken (97–101 and Section III).

E. *E. coli* RNA Polymerase

RNA polymerase has been known for some time to open the double-stranded DNA helix to provide a single-strand template for transcription. RNA polymerase function is described in Section IV. One investigation concludes that RNA polymerase interacts weakly along the DNA grooves, making preferential contact with the minor groove in nonspecific interactions (411). Another reports specific promoter binding interactions to be predominantly associated with the major groove (192). The different results may be a reflection of differences in procedures or differences in nonspecific and specific binding interactions.

RNA polymerase binds single-stranded DNAs preferentially, especially those with the least structure (reviewed in 70). Richardson (412) did not observe it to be a melting protein (in the sense of the term as used for helix-destabilizing proteins). Some results suggest that the holoenzyme nonspecific binding is essentially all electrostatic with the formation of approximately 11 ion-pairs (413). Lysines may represent some of the binding amino-acid residues, as some lysines in RNA polymerase are not modified when polymerase is bound to DNA (414). Cross-linking studies revealed that the σ and β subunits are in contact with the DNA in specific polymerase–promoter interactions, while β' and α are not (415). The σ subunit has been shown to be cross-linked specifically to the antisense strand of a promoter (416).

It has now been demonstrated in numerous studies that RNA polymerase does indeed unwind the DNA helix. RNA polymerase unwinds the DNA helix by 240° at 37°C, and the angle decreases with increasing salt or decreasing temperature. Nonspecific unwinding has also been demonstrated with λ DNA (417). In another nonspecific binding experiment, binding to T7 DNA and poly(dA-dT) opened 15 base-pairs (411). Specific opening of the helix at promoter sites was demonstrated by observing hyperchromic shifts induced by RNA polymerase binding to DNA fragments containing early T7 promoters (73). Under the conditions used, DNA fragments without promoters did not show the same hyperchromic shift. Hseih and Wang (73) concluded that the hyperchromic shift they obtained suggested a helix disruption of about 10 base-pairs. Siebenlist (74) using DNA modification techniques appears to have localized the opened helix region in the T7 A3 promoter to an 11-base-pair segment located in the Pribnow Box transcription initiation region. Little alteration is seen in the CD spectra when RNA polymerase is bound to DNA (73, 418).

There is evidence that the opening of the helix may be the rate-limiting step in initiation of transcription from promoters (186). Thus a (dA+dT)-rich promoter would be presumed to be desirable for an efficient promoter. Evidence that such richness and ease of denaturation is important for the function of some promoters is presented in Section IV,A.

Just how RNA polymerase opens the helix is unknown. It is possible that different subunits bind separate strands to trap a single-stranded region in a "breathing" DNA, or induce the unwinding in an active manner.

F. Other Proteins That Alter DNA Conformation

Remarkably little work has been reported for other proteins that bind to DNA and may, therefore, alter its conformation (further discussed in Section IX). The *E. coli* catabolite-gene-activating protein, or cyclic-AMP-receptor protein (CRP), alters the CD spectra of DNA (nonspecific binding mode) in a way similar to that seen for helix-destabilizing proteins (419). This does not necessarily imply that CRP is a melting or helix unwinding protein, although such a function has been postulated (*110, 186, 1, 103*). Since it is now possible to isolate large amounts of small fragments containing the CRP specific binding site (*97–101*), further work should be facilitated. Also, the *Micrococcus luteus* DNA polymerase (*420*) has been reported to decrease the magnitude of the CD spectra on binding to poly(dA-dT) (*421*).

Proteins such as DNA gyrase and "swivelases" alter the topological features of closed circular DNAs. Their properties are discussed in Section IV,B.

G. Conclusions: Protein Binding and DNA Conformations

It is clear that DNA conformational alterations connected with protein binding are not a rare event. Some conformational changes are obviously functionally related, as in the case of the helix-destabilizing proteins and RNA polymerase. Other more subtle conformation changes at the DNA binding site of proteins may be related to forming stable binding complexes. It is possible that most protein-DNA binding interactions require a conformational change in the DNA and/or the protein to obtain maximal fit and interaction between actively binding structural units. This conformational changes induced by the protein binding may have the potential to alter the DNA conformation at sites distant from the actual binding site. The conformational change transmitted through the DNA could alter binding affinities of other proteins at those distant sites.

The bases of these conformational changes at the amino acid and nucleotide level are not clear. Circular dichroism, which would seem to have the greatest potential, appears to be capable only of detecting changes, not of delineating the precise nature of the conformational change. The theory for interpreting CD changes with DNA structure has not been well worked out. Electrostatic interactions appear to be important in all the proteins studied.

The dehydration and charge neutralization of the DNA helix does not appear to be a dominant cause of conformational alteration since different CD changes are observed in different protein binding interactions, all of which include electrostatic interactions. Intercalation of planar aromatic amino-acid residues may be a means of giving proteins a single-strand preference, as they are implicated in all the helix destabilizing proteins and show single-strand preference. Intercalation has another functional aspect in that it might be expected to unwind and expand the helix somewhat. The slight unwinding of the DNA helix by the *lac* repressor may involve its tyrosine residues, which are believed to participate in DNA binding.

Specificity in binding may, in part, be determined by the ability of the protein–DNA complex to undergo a conformational change to stabilize binding that is not permitted for nonspecific binding sites. This may be a function of obtaining base-specific protein–DNA interactions that induce a conformational change and/or an inherent lower stability in a specific DNA sequence itself. The *lac* repressor may be in this category since it has different specific vs nonspecific binding characteristics.

It is anticipated that the availability of large amounts of cloned small fragments containing the operator (see Section III) will allow the spectral analysis of the specific binding interaction to determine whether the conformational change differs from that found for nonspecific binding.

VII. Structure in Single-Stranded Viral DNAs

A. General Considerations

In addition to the types of structures described above for duplex DNAs, there is substantial evidence that the "single-stranded" DNAs from several small viruses (such as ϕX174, fd, and several animal viruses) possess a number of secondary and tertiary interactions. Denhardt (422) has written an exhaustive review on single-stranded DNA phages that provides the background up to 1975.

It is not unexpected that these DNAs contain substantial duplex structure since extensive amounts of base-pairing can occur even in random sequences. Such sequences of RNA generated by computers (423, 424) or by polynucleotide phosphorylase (425) can have 40–60% of their nucleotides in duplex structures. Moreover, even "random-coil" poly(U) has preferred conformations in solution and is not devoid of ordered structure (426).

Much more work has been done on the duplex regions in "single-stranded" RNAs than for DNAs, in part since RNA sequencing procedures predated DNA sequencing. Perhaps the best-known example of a single-stranded RNA with a high potential for secondary structure is MS2 RNA. The nucleotide sequence (427–429) provided the basis for a "flower" model with greater than 60% duplex structure. Limited nuclease digestion studies were the primary basis for the proposed structure. Also, electron micrographs of this RNA (430) showed several looped structures; the size and number of these loops were functions of the ionic conditions.

The 16 S and 23 S ribosomal RNAs of *E. coli* also have extensive secondary structure (431–433). Moreover, Perdue *et al.* (434) have isolated and characterized a large "hairpin" structure from the center of Rous sarcoma virus RNA, and electron microscopy of f2, Qβ, and PP7 bacteriophage RNAs (435) showed similar secondary structures for all three phages. Many other RNAs also have regions of duplex structure including 50–70 S RNA of Moloney sarcoma virus (436, 437), eukaryotic 5 S RNA (438), *Drosophila melanogaster* 26 S and 18 S rRNA (439), and HeLa hnRNA (440).

B. Restriction Endonuclease Susceptibility

There were hints for a number of years (422) that ϕX174 and related DNAs contained regions of ordered structure; these hints derived mainly from nuclease (such as *E. coli* Exo I) susceptibility data. However, the chance observation in 1975 that certain restriction endonucleases cleave the "single-stranded" DNA of ϕX174, M13 (441), and f1 (442), as well as other findings that soon followed, rapidly solidified and extended these hints.

Unexpectedly, *Hae*III was observed to cut the "single-strand" DNA of ϕX174 at specific sites, yielding 11 fragments of the same size as those obtained by a digestion of ϕX174 RF (the double-stranded replicative form of ϕX174 DNA) (441), and it was subsequently shown that conditions that should perturb the duplex structure in single-stranded DNA affected the

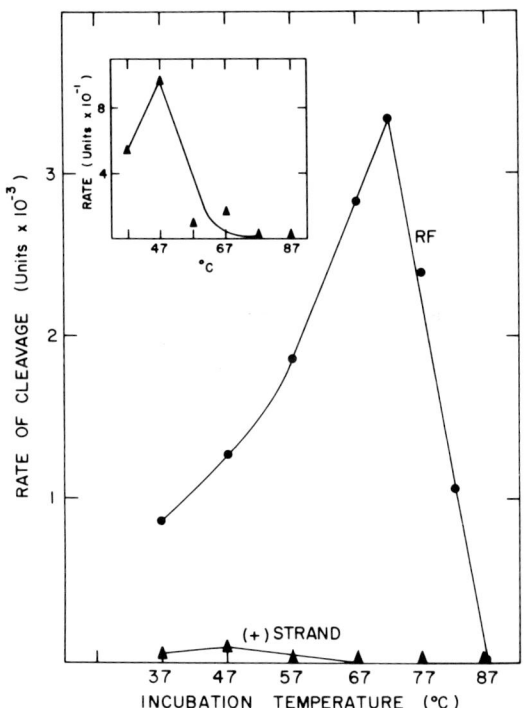

FIG. 17. Effect of temperature on cleavage of ϕX174 (+)-strand and RF DNAs. ^{32}P-labeled ϕX174 (+)-strand or RF DNA was incubated with *Haemophilus aegyptius* III at various temperatures. The initial rate of cleavage at each temperature was measured by following the degradation of RF I, II, and III DNAs into smaller fragments. Aliquots were removed from the reaction at 30-second intervals during the 3-minute period following enzyme addition. Quantitation of the unrestricted DNA was essentially as described. The inset is a replot of the data for the (+)-strand. Reprinted, with permission, from Blakesley *et al.* (443).

cleavage by HaeIII (443). As the temperature of the digestion is increased, two effects might be expected. The DNA substrate should lose duplex structure as regions of lesser stability "melt out," and the enzyme should lose activity by denaturation. Figure 17 shows that with increasing temperature the rate of HaeIII digestion of ϕX174 (+)-DNA rapidly declines from an optimum rate at 47°C to zero at 72°C. However, the reaction rate on fully duplex ϕX174 RF DNA reaches its maximum at 72°C. Since the enzyme is still active on duplex sites at 72°C, its lack of cleavage at lower temperature on ϕX174 (+)-DNA must be due to loss of duplex DNA structure in regions of lesser stability rather than to enzyme denaturation.

A second line of evidence comes from the use of actinomycin to inhibit the cleavage. This drug intercalates only into duplex DNA (444). When increasing amounts of actinomycin are added to a DNA solution, HaeIII digestion of that DNA is increasingly inhibited (443). At a molar ratio of actinomycin to nucleotide of 1.2, there is a 50% inhibition of HaeIII cleavage of ϕX174 RF. However, 50% inhibition of cleavage of the (+)-strand occurs at a molar ratio of only 0.04. This suggests that there is a small amount of duplex structure in the single-stranded DNA that is quickly saturated by the actinomycin.

Other workers (442) originally challenged the notion (441) that HaeIII was recognizing duplex regions in the "single-stranded" viral DNAs, however the idea that a truly random coil—GGCC—was the substrate has not been independently confirmed.

Several other restriction endonucleases, including EcoRI (445) and HPAII (446-448), have also been studied for their capacity to cleave single-stranded substrates. In all cases, it was concluded that a duplex region (the binary complex of two complementary strands) was required for cleavage.

C. Single-Strand-Specific Nuclease Reactions

Another line of evidence indicating duplex structures in single-stranded DNAs is their partial resistance to endo- or exonucleases specific for single strands. Schaller et al. (449) treated fd viral DNA with *Neurospora crassa* endonuclease and *E. coli* exonuclease I, and purified a portion of the DNA resistant to these nucleases. This core DNA retained nuclease resistance after alkaline pH or heat treatment, and it melted cooperatively with a t_m of 70°C, indicating a duplex structure. They estimated the chain length to be 40 to 50 nucleotides by end-labeling, G-75 gel filtration, and sucrose gradient sedimentation.

Similarly, the endo- and exonucleases from *N. crassa* have been employed to digest ϕX174(+) DNA; from 1.5 to 3.0% of the genome was resistant to these single-strand-specific nucleases, and two fragments of about 16 and 24 base-pairs and many shorter fragments were found.

Lavelle and Mitra (451) found linear single-stranded DNA from Kilham rat virus (KRV) to be only 5% resistant to S1 nuclease. After treatment with a combination of S1 and *E. coli* exonuclease I, two fragments of 135 and 110 base-pairs were isolated by hydroxyapatite column chromatography.

Niyogi and Mitra (452, 453) used similar techniques to isolate double-stranded regions from M13 (+)-DNA. Digestion with *E. coli* exonuclease I gave two major fragments about 60 and 44 nucleotides long, by polyacrylamide gel electrophoresis under denaturing conditions, that retained nuclease resistance after heating and quick-cooling, suggesting hairpin-type structures.

The minute virus of mice (MVM) DNA is 6% resistant to S1 nuclease or *E. coli* exonuclease I (454). The resistant region is at the 5' end of the molecule and is about 130 base-pairs long, as analyzed on polyacrylamide gels. The involvement of this duplex region in initiation of DNA synthesis is discussed in the following section.

Whereas some useful information has been obtained from partial digestion studies as described above, we feel that this approach suffers from the fact that a variety of lengths and types of "resistant cores" are derived from RNAs or DNAs depending on the reaction conditions ($MgCl_2$ or salt concentrations). Success is highly dependent on the nuclease specificity and in the degree of uniqueness and stability of the duplex region relative to the nonhelical regions (Grant and Wells, unpublished work).

D. DNA Polymerase Reactions *in Vitro*

Experiments with DNA polymerases also give indications of duplex structure in single-strand DNAs. Sherman and Gefter (455) used *E. coli* DNA polymerase II to extend a DNA primer annealed to fd viral DNA. It was expected that the polymerase would extend the primer completely around the genome, but instead it stopped at specific locations. The polymerase proceeded through these "barriers" if *E. coli* helix-destabilizing protein (see Section VI,C) was present. The most persistent barrier was near the origin of complementary strand synthesis (456). It is probable that these barriers are regions of duplex structure in the viral genome.

The same type of result occurs when vaccinia virus DNA polymerase acts on primed ϕX174 viral DNA (457). The major barriers in this case coincided with regions considered to have large hairpin structures. It is presumed that these barriers in the viral genomes are the same folded-back duplex structures recognized by the DNA restriction enzymes.

A somewhat different type of duplex structure in single-stranded DNAs comes from sequencing studies on parvovirus genomes. The 3' ends of KRV, H-1, H-3, and MVM DNAs were sequenced (458–460) and Y-shaped secon-

KRV DNA

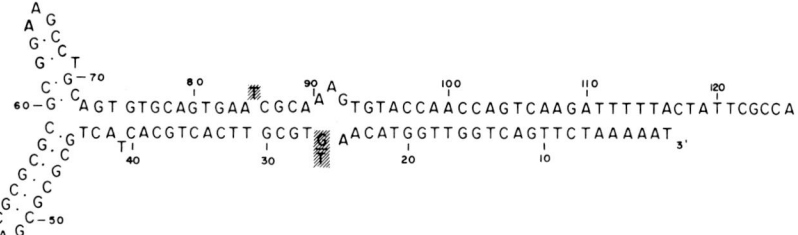

H-1 DNA

H-3 DNA

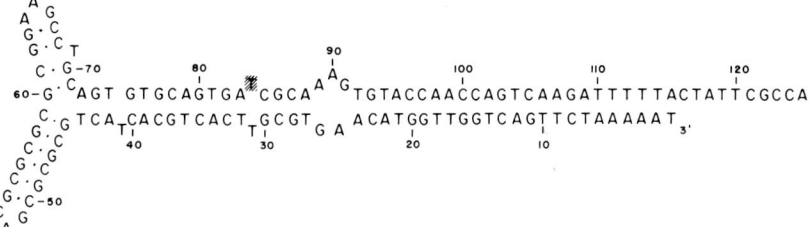

MVM DNA

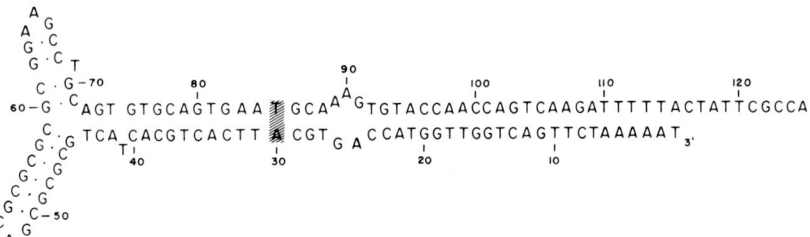

FIG. 18. Nucleotide sequence of the 3' hairpin of KRV, H-1, H-3, and MVM DNAs. Shaded areas indicate differences between these DNAs. Reprinted, with permission, from Astell *et al.* (458).

dary structures were proposed (Fig. 18). Such duplex structures would explain the resistance of KRV DNA to *E. coli* exonuclease I (*451*) and its ability to act as a self-priming template for DNA polymerases (*454, 461*). The extent of duplex structure in other regions of these animal virus genomes is uncertain at present and is somewhat complicated by the difficulty in preparing the viruses in quantity.

It is interesting to note that an alternating dG-dC duplex sequence is found for all four DNAs for nucleotides 45 to 60; this is the type of sequence that generates a Z-DNA conformation (see Section II). Also, a $(dA \cdot dT)_5$ sequence is found near the 3' end; this type of sequence also has unusual properties (Section II).

E. Electron Microscopy

Electron microscopy has also been used to examine single-strand DNA for duplex structure. Viral fd DNA was cross-linked with trioxsalen and viewed under the electron microscope (*462*). As the ionic strength was increased, more hairpins were observed. Mapping studies showed that the hairpins appeared at specific sites. Under certain conditions, more than 90% of the molecules had a hairpin of 150 base-pairs, localized to the region where the origin of complementary strand replication occurs (*456*). When the DNA was cross-linked in the virion (*463*), the same hairpin was found.

These cross-linking agents show great promise for the future because of their high cross-linking capacity and their permeability into cells and particles such as viruses.

F. Biological Importance

From the evidence cited, there is good reason to believe that single-strand DNAs contain some secondary structure, shown in some cases to be at specific sites. These duplex regions could be important for many biological processes. For example, RNA polymerase may require double-stranded DNA to synthesize an RNA primer for DNA replication (see Section V). When fd DNA is complexed with *E. coli* helix-destabilizing protein, which binds single-stranded DNA, RNA polymerase is directed to a specific site on the genome (*464*). This site is the origin of complementary strand synthesis. Also, Niyogi and Mitra found that the nuclease-resistant cores isolated from M13 (+)-DNA (*452*) were in the same region of the genome as the site to which RNA polymerase is directed in the presence of *E. coli* helix-destabilizing protein (*453*). Thus, the RNA polymerase seems to be binding to a duplex region of DNA. Similarly, the double-stranded region at the 3' end of KRV DNA can act as a self-primer for DNA synthesis in the presence of *E. coli* polymerase *in vitro* (*461*), and this may be its function *in vivo*.

Another possible biological function for duplex structure in fd viral DNA is phage adsorption to the bacteria. The filamentous phages are known to adsorb to the F-pili with only one end of the virion (*422*), the end that contains a gene 3 protein. The fd viral DNA has a duplex structure at only one end of the virion, and it is suggested that gene 3 protein may bind to the duplex region of DNA to form an adsorption complex (*463*).

Also, the folding of single-stranded genomes may be involved in, or necessitated by, assembly of the viral proteins to give the mature virus particle. In addition, certain hairpin structures in intercistronic regions have been implicated as transcription terminators (see following section).

However, it is possible that these duplex regions in the single-strand DNAs are not functional, but rather are reflections of functional structures in their polycistronic transcripts. These structured RNA regions may be involved in translational regulation of the RNA transcripts, as has been proposed for the single-strand RNA phages (*465, 466, 427, 428*). In either case, the duplex regions in single-strand DNAs may be of biological importance, not just randomly paired structures.

VIII. DNA Secondary Structures in Intercistronic Regions

A. General Considerations and Definitions

Sequence determinations of the complete genomes of several small bacteriophages and viruses as well as parts of larger genomes have given insight into the arrangement of genes and various regulatory elements. In several instances, the existence of untranslated sequences, located between the regions known to code for a protein or an RNA (other than mRNA) has been established. These sequences have been referred to as "intergenic space," "intercistronic region," "noncoding region," or "spacer regions."

The following requirements for an "intercistronic region" are defined for the sake of clarity:

1. The start of an intercistronic region is the nucleotide after the last of at least three termination codons (one in each reading frame), preceding the intercistronic region.
2. The end of the intercistronic region is the last nucleotide in front of the first initiation codon (which is preceded by a ribosome binding site active *in vivo*) and which is not immediately followed by a stop codon.
3. The intercistronic sequence may not code for the synthesis of a tRNA or rRNA (or a related precursor molecule).
4. The intercistronic region may be transcribed as part of a messenger RNA but may not be translated even in the event of gene splicing.
5. These requirements must be fulfilled for both strands.

It is obvious that this definition requires more information than the primary sequence of a DNA before a region may be classified as an intercistronic region. Most reports of intercistronic regions in sequenced DNAs or RNAs are necessarily based on much less information, but are accepted as such for this review. With these reservations in mind, we have attempted to compare the DNA sequences of several intercistronic regions in various genomes with emphasis on their proposed secondary structure in relation to their known or proposed function.

B. Single-Stranded DNA Phages

1. REPLICATION ORIGIN REGIONS

The best-characterized intercistronic regions are those in the genomes of icosahedral and filamentous single-stranded DNA phages (467–469). The genomes of the closely related icosahedral bacteriophages ϕX174 and G4 have been sequenced (467, 468, 470); four intercistronic regions have been found in ϕX and six in G4 (471). The nucleotide sequences around the negative strand initiation sites of the related phages St-1, ϕK, and α3 also contain an intercistronic region (472, 294). Of the filamentous phages M13, f1, and fd, only fd has been completely sequenced (469), and six intercistronic regions were identified. [The region between genes VII and VIII may code for a protein (473).] Sequences around the origins of replication of viral and complementary strands are also available for M13 and f1 (474–476, 317) and are located in the intercistronic region between genes II and IV, as also found for fd.

The structural and functional similarities of the intercistronic regions within the icosahedral phage group have been discussed in detail (471, 294), and the structure–function relationship of the largest intercistronic region in the filamentous phage group has been described (476a). We limit our discussion therefore to an overview of the significant features of these regions.

Figure 19 shows the largest intercistronic region in the genomes of the icosahedral phages ϕX, G4, St-1, ϕK, α3 and the filamentous phages fd, f1, and M13. The potential for secondary structure in all of these regions is readily apparent, but it must be stressed that these, like most other proposed secondary structures, are only computer-derived models with maximal base-pairing within the single-stranded DNA (317, 476a); other or additional secondary structures may also be drawn (294, 477, Müller, Fitch, and Wells, unpublished). However, several lines of evidence (reviewed in Section VII) suggest that at least some of these structures may exist under certain conditions. Also, indirect evidence comes from energetic considerations, since formation of all these structures in the single-stranded DNA molecule is favored by a $\triangle G$ varying from -6 to -54 kcal/mol [calculated

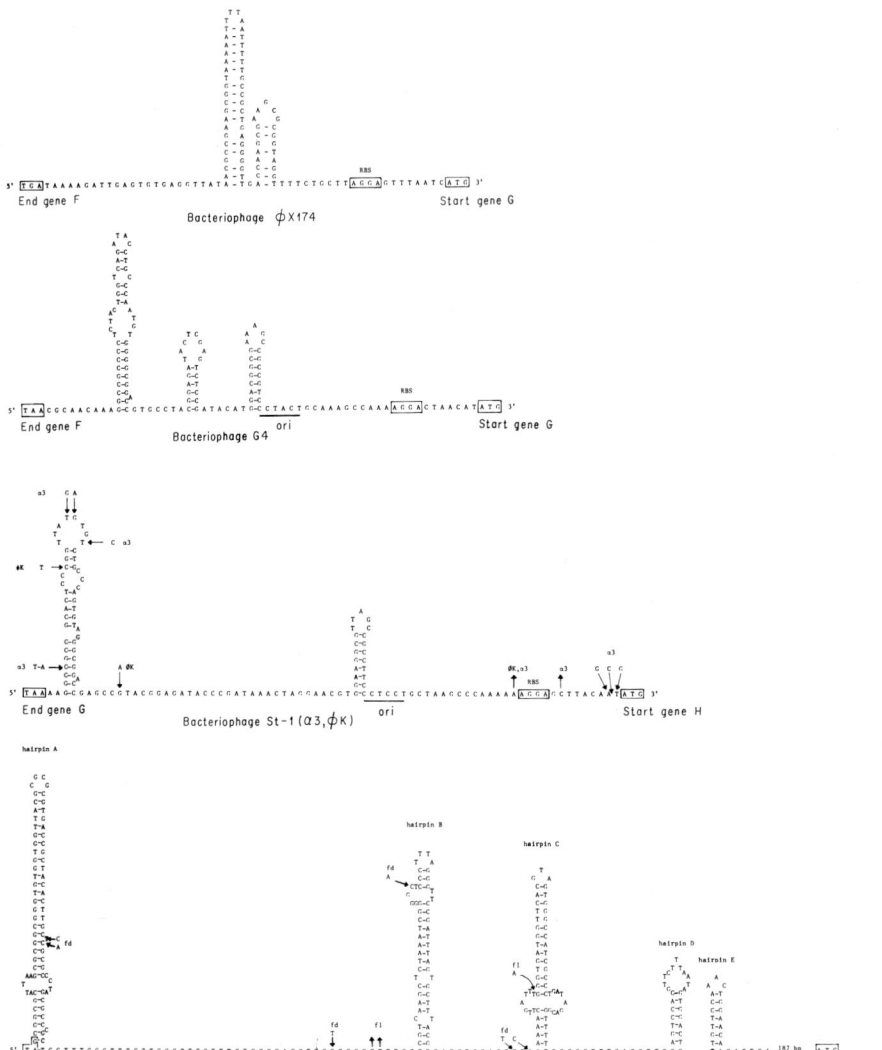

FIG. 19. Sequence of the largest intercistronic region of icosahedral and filamentous bacteriophages. Figures were redrawn from Godson et al. (468) (φX, G4), Sims and Dressler (294) (St-1, α3, φK), and Suggs and Ray (317) (M13, fd, f1). The sequences of bacteriophages α3 and φK are very similar to that of St-1, and variations from the St-1 sequence have been indicated by arrows. The sequence of f1 and fd is represented by the M13 sequence, and variations are indicated by arrows. Arrows pointing away from the sequence indicate deletions; arrows pointing toward a nucleotide indicate substitutions, and arrows pointing between two nucleotides indicate insertions. The origin of complementary strand synthesis (ori) is underlined. Proposed ribosome binding sites (RBS), initiation codons (ATG), and termination codons (TAA) are indicated by boxes.

according to Tinoco et al. (478)]. Moreover, the recurrence of these potential secondary structures in the same physical position in the genomes of closely related phages (Fig. 19) and in functionally similar position in the genomes of less related organisms (294, 479) is perhaps the best indirect evidence for their existence.

Since the F-G intercistronic regions of phages G4, α3, φK, and St-1 are known to contain the origin of complementary strand synthesis (Fig. 19) (479, 480, 294), and the intercistronic region between genes II and IV in the filamentous phage groups has been shown to contain the origin for both viral and complementary strand synthesis (Fig. 19) (313, 476a, 474, 481), it is tempting to correlate these functions with the potential secondary structure of these regions.

Several arguments may be raised in support of this hypothetical structure–function relationship. One of the strongest is the low probability with which sequences occur in a small genome (5 to 6 × 10^3 bases in length) that allows formation of such secondary structures. It has been pointed out in Section VII,A that any random sequence of nucleotides has the capability of forming secondary structures with approximately 60% base-pairing. The probability of formation of a hairpin structure with a relatively small loop size (3–10 bases) and a "perfect" (no oppositions of dG and dT: uninterrupted) stem of greater than 5 base-pairs is small. A computer search of the φX sequence for an example revealed only two perfect hairpins of small loop size (< 100 nucleotides) with a stem length of 8 base-pairs (Müller, Wells, and Fitch, unpublished); one is located in gene H and the other is in the intercistronic region H-A (471) which is discussed below. Despite this theoretically low probability of occurrence, the largest intercistronic region in the filamentous phage genomes has the potential for two such structures, hairpin D and E (Fig. 19); also, the equivalent region in the icosahedral phages has one.

In the F-G intercistronic region of φX, a perfect 8-base-pair hairpin stem cannot be drawn, though the potential for a large, less perfect hairpin is present. This structural difference is matched by a functional one, since φX is the only phage of this group that does not have a unique origin of complementary strand synthesis in this region. Instead, φX has multiple sites along its genome where synthesis of (−)-strand DNA is initiated (482). Of course, this does not by itself imply that an 8-base-pair hairpin stem is involved in the initiation of DNA replication, especially since the generally high (67%) sequence homology between the coding regions of the φX and G4 genomes is not maintained in this intercistronic region (468). Thus, the primary arrangement of nucleotides rather than the secondary structure may be of significance in this context.

On the other hand, the equivalent intercistronic regions in phages G4, α3, φK, and St-1 have maintained a high degree of sequence homology in addition to preservation of the secondary structure, while only little sequence homology is observed within the genes flanking this region (*294*). Again this does not allow the conclusion that only the secondary structure is of consequence to the crucial process of DNA replication, but may even point to an essential function involving the whole intercistronic region.

The sequences of the filamentous phage group suggest an important function of the structure other than the sequence of DNA in the initiation of DNA replication. Not enough sequence data for f1 and M13 are available yet to draw comparisons of sequence variation within coding and noncoding regions. However, it is intriguing that, around the origin of replication, sequence differences occur almost exclusively in the nonpaired region or outside the putative hairpin structures (Fig. 19) (*317*). More direct evidence for the functional role of these hairpins has been derived from the isolation of mutants carrying either insertions or deletions in this intercistronic region (*476a, 483, 484*). These studies prove that (at least under laboratory conditions) some sequences of this region and the continuity of others are not essential for phage growth. These nonessential regions include parts of the less perfect hairpins A and B and the sequence between them as well as sequences to the right of hairpin E (Fig. 19) (*476a*).

In summary, it appears that for single-stranded DNA phages, a structural recognition site has evolved that allows DNA replication with a minimum of host enzymes (*485, 472, 294*). A requirement for this secondary structure may have imposed an evolutionary constraint so great that it has been placed into a region separate from those coding for other functions. This intercistronic region may exert its function by assuming one or more specific secondary structures that act by binding other macromolecules involved in initiating or catalyzing reactions.

If this hypothesis is true, one may expect to find a high potential for secondary structures in other intercistronic regions. For comparison, we may consider the untranslated region in the genome of the eukaryotic virus SV40 (*486, 487*), though the intercistronic nature of this region has not been established. The region between gene A (T-antigen) and virion protein 2 is endowed with several unusually large inverted repeat sequences, and has the potential for a perfect 12-base-pair hairpin stem (*488*); this is a rather improbable structure for a genome smaller than φX. The location of this structure coincides with the origin of DNA replication and has been shown to be one of the binding sites for T-antigen (*489*).

2. Other Intercistronic Regions

Further searches for unusual secondary structures in intercistronic regions are limited at this time because of very few available sequences. In φX there are only two other intercistronic regions besides the one discussed above that are large enough to allow for any significant secondary structures. These are located between genes H and A (63 base-pairs) and J and F (36 base-pairs) (467) (Fig. 20). Both of these exist in the G4 genome in similar size (59 and 45 base-pairs), but an additional sizable intercistronic region is located in G4 between genes E and J (32 base-pairs) (468).

In the filamentous phage fd, there appears to be only one additional intercistronic region of comparable size (53 base-pairs), while the others, as in ϕX or G4, are less than 12-base-pairs long (469, 473).

With the exception of the E-J intercistronic region of G4, all the above-mentioned regions have the possibility of perfect hairpins with a stem of 6 to 9 pairs and a small loop (Fig. 20) (471, 473). A comparative analysis reveals that all these structures share common features. First, their stem is quite rich in dG and dC, which favors a relatively stable stem with a free energy of formation of -14 to -18 kcal/mol when the parent DNA is in the single-stranded form [calculated according to Tinoco et al. (478)]. Second, in each case the stem is flanked by a sequence rich in dA and dT, which consists mostly of a stretch of thymidines on the 3' side.

We have pointed out already that the probability of finding a perfect hairpin with a small loop size is low per se. Alternating blocks of $dG \cdot dC$ or $dA \cdot dT$ have an equally low or lower probability of occurrence (490), and the two together should make these structures unique for a small genome. This theory holds true for the ϕX genome, since a computer search of the ϕX sequence revealed only 9 potential perfect hairpins with not more than 10 base-pairs in the loop and 6 or more base-pairs in the stem (Table II) (Müller, Wells, and Fitch, unpublished). Three of these are located in intercistronic regions, which together comprise only 4% of the genome, and another three are located either at the end or at the beginning of a gene. At least for ϕX, it appears therefore that the distribution of these potential structures is not random along the sequence.

Several *in vivo* and *in vitro* transcription experiments suggest an association of at least some of these hairpins with a regulatory function in transcription by providing termination and/or initiation signals (491, 492, 468). The strongest terminator in the ϕX sequence is known to be located in front of gene A (493–496, 468) and coincides with the 8-base-pair stem in the hairpin in the intercistronic region H-A (Fig. 20) (467, 468). A similar hairpin structure can be formed in the analogous intercistronic region of phage G4 (Fig. 20), although evidence for its function as terminator of transcription is not yet available (468). Both of these hairpins also contain a promoter sequence immediately to the left of their stem (467, 468). *It is very improbable that overlap of two strong regulatory sites with one of the two potentially perfect 8-base-pair stems in the hairpins in the ϕX sequence is merely coincidental.*

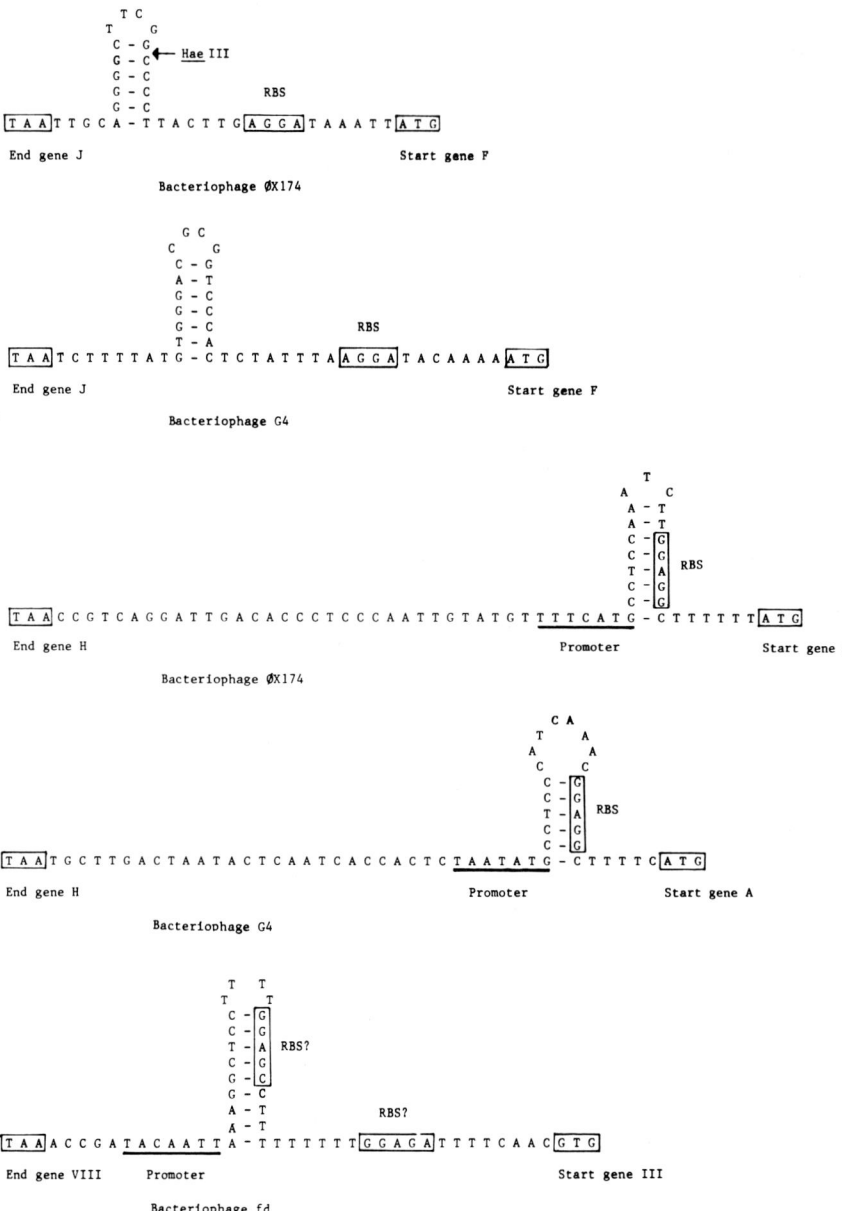

FIG. 20. Intercistronic regions in φX, G4, and fd containing a termination site of transcription. Sequences are redrawn from Fiddes and Godson (497), Godson et al. (471), and Schaller et al. (449). Proposed ribosome binding sites (RBS), initiation codons (ATG or GTG), and termination codons (TAA) are boxed in. Promoter sequences are underlined.

TABLE II

"Perfect Hairpins" in φX174 Viral DNA with up to Six Base-Pairs in the Stem and No More Than Ten Bases in the Loop[a]

Location		Stem length (No. of base-pairs)	Loop length (No. of nucleotides)	Regulatory site in close proximity[c]
Number of first nucleotide[b]	Gene			
	End B			
75	End A	6	9	T_C/P_G/RBS
797	End E	7	10	T_E/RBS
969	J-F intercistronic	6	4	T/RBS[d]
2316	F-G intercistronic	6	4	T_F/RBS
2575	G	6	4	T_G
2997	H	8	8	—
3571	H	6	9	—
3956	H-A intercistronic	8	3	$T_H/P_{A'}$/RBS
5042	Start B	6	7	P_A/RBS

[a] The data were derived from a computer search (Müller, Fitch, and Wells, unpublished results) of the φX174 sequence (467,470). Only hairpins with uninterrupted perfectly base-paired stems (dG · dT pairs not allowed) with loop size of 10 or fewer nucleotides (chosen arbitrarily) are listed.

[b] Numbering of nucleotides is according to Sanger et al. (470).

[c] Regulatory sites refer to proposed promoters (P) or terminators (T) of transcription or ribosome binding sites (RBS). The maximum distance of the potential secondary structure to the suspected site is approximately 60 base-pairs. The designation of terminators and promoters is according to Fujimura and Hayashi (491).

[d] Terminator proposed by Godson et al. (468).

The hairpin structures in the J-F intercistronic region of phages φX and G4 have also been inferred to be terminators of transcription (497, 468), although there is no direct evidence for this for the G4 sequence. Müller and Wells constructed a series of φX mutants with modified J-F intercistronic regions (498) in an effort to evaluate the biological role of hairpin loops in this region. One set of mutants contained inserts of pBR322 DNA, which ranged in size from 40 to 160 base-pairs into the HaeIII site of this region (Fig. 20). A second set carried deletions starting at this HaeIII site and proceeding in one or both directions, deleting part or all of the potential hairpin and up to 75% of the J-F intercistronic region (498, 499). Since all mutants were capable of growth in the absence of helper phage, formation of this hairpin and most of the sequence in the J-F intercistronic region is apparently not essential for lytic growth under laboratory conditions. Two of these mutants were characterized with respect to their biological properties; a small reduction in growth efficiency was found for the insertion mutant whereas no growth deficiency was found for the large deletion mutant (499).

An *in vivo* termination site has been associated with the φX intercistronic region F-G (494–496, 491), which has the potential for a perfect 6-base-pair stem in the hairpin (top of left loop). This may be extended in length if one allows dG·dT base-pairs (477) (Fig. 19).

For fd, only one so-called central terminator exists at which all *in vitro* transcription stops. The sequence at the 3' end of some transcripts shows that this terminator is located in the intercistronic region between genes VIII and III (473, 492) and coincides with the sequence of the potential hairpins shown in Fig. 20.

C. Phage Lambda

For the larger double-stranded DNA of phage λ, an intercistronic region 118 base-pairs long has been identified between genes *cro* and *cII* (500, 501) (Fig. 21). Like the intercistronic regions of the single-stranded DNA phages, this sequence exhibits the potential for various secondary structures, including a perfect 8-pair hairpin stem. This structure, which has dA·dT-rich region on its 3' side, was also shown to act as a terminator of transcription with 80% efficiency in the presence of rho factor (500, 501). The availability of mutants with base changes in the stem of this hairpin has allowed comparative *in vitro* and *in vivo* transcription studies, which substantially increases the confidence in the proposed structure–function relationship. CNC mutations, which introduce base mismatches in the stem (Fig. 21) of this hairpin, almost completely abolish the termination function. On the other hand, increases in termination efficiency are observed when the stem length was extended by a dA·dT pair (CIN mutation), even though it is separated from the stem by a dG·dT pair (500, 501).

Other regulatory functions, such as interaction with the N protein (Nut R), ribosome binding, transcription initiation and/or attenuation, have been

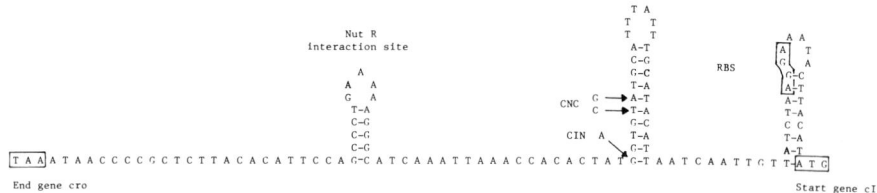

FIG. 21. Intercistronic region between genes *cro* and *cII* in bacteriophage λ. Redrawn from Rosenberg *et al.* (500, 501). Sequence changes by CNC and CIN mutations are indicated by arrows. Other recognition sites are indicated as in Fig. 19. See Fig. 15 for a genetic map of this region.

associated with this intercistronic region and with its potential secondary structure (*500, 501*).

A comparison of these viral intercistronic regions with those in bacterial or even eukaryotic DNAs is not productive at this time owing to the limited amount of information. Only a few bacterial intercistronic regions are known. One consists of two untranslated base pairs of the *trp* B gene termination codon of *E. coli*, and would not satisfy our definition (*502*). Two others are too short (9 base-pairs) to contain significant secondary structure (*503, 504*), although they may be part of a hairpin structure thought to be involved in ribosome binding (*503*).

More significant in this context are two sequenced regions of the *E. coli* chromosome containing the genes for RNA polymerase α and β subunits as well as several ribosomal proteins. The gene cluster adjacent to subunit β gene contains several intercistronic regions with the potential for relatively large but imperfect hairpins (*507*). Though not specifically indicated by the authors, a perfect 9-base-pair stem in the hairpin (3 pairs in a loop) can be drawn at the end of the gene coding for polymerase subunit α. Part of this hairpin (starting at base-pair 167) reaches into the adjacent intercistronic region that separates this gene from ribosomal protein gene S4 (*508*).

D. Summary: Secondary Structure in Intercistronic Regions

It appears that almost all the larger intercistronic regions described so far have the potential for secondary structures. The hairpins with small loop size and perfectly base-paired, uninterrupted stems ($\geq$ 6 base-pairs in length) are the most significant. These seem to coincide in every case with known regulatory loci, and, at least for the putative terminators of transcription, there seems to be direct correlation between the biological activity and the stem length of these structures. Sequencing of other genomes and other studies will be necessary to determine if the proposed correlation of structure and function is merely coincidental or is biologically relevant.

IX. Conclusions and Prospects for the Future

A complete knowledge of the kinetic and equilibrium properties of the interaction of specific DNA target sites with important regulatory proteins is fundamental for the eventual comprehension of gene regulation. Cellular differentiation is the orderly and programmed expression of a family of genes. Alternatively, the malignant transformation of a cell will eventually be recognized as a faulty interaction between one or more key proteins and their DNA receptor sites.

In spite of the importance of this area of research, this review exposes our lack of knowledge. The field is in its infancy. Of the thousands of regulatory and metabolic proteins that act on DNA, fundamental genetic and biochemical (enzymological) characterization has been reported on only about one hundred, and careful studies on the mechanism of the interaction between a defined DNA target site and a specific protein for ten or fewer cases. Furthermore, no X-ray crystal structure is known for any duplex DNA–protein complex, and there is but a single example of work on a complex of protein and random-coil DNA.

There is good reason to believe that progress will rapidly accelerate in the next decade, since it is now possible to obtain large amounts of important regulatory and metabolic proteins in pure form, and to isolate milligram amounts of small defined DNA fragments in pure form by gene cloning in conjunction with liquid chromatography (*164*, *171*). A wide array of DNA restriction enzymes facilitates the excision of DNA fragments at precise sites, rapid DNA sequencing techniques permit the characterization as defined chemical entities, and new genetic techniques will provide a broad spectrum of mutant DNAs and mutant proteins.

However, in the face of this optimism regarding biochemical and genetic advances, a note of pessimism must be expressed regarding physical determinations on these molecules. In order to determine DNA conformation, it is necessary to employ physical or spectroscopic techniques. In the past, a number of these techniques have yielded equivocal results. Moreover, most physical or spectroscopic methods cannot be interpreted to give absolute conformational information (i.e., absolute orientation of atoms in space) at present. These methods usually permit only the conclusion that the conformation of a large molecule was altered by a particular perturbant without revealing the details of the alteration in absolute terms. This note of pessimism is expressed as a plea for the exploration of new and improved physical and spectroscopic probes for the detailed analysis of macromolecular conformations.

Acknowledgments

We are grateful to the many people who generously contributed unpublished manuscripts to assist in the preparation of this review and to our colleagues for numerous enlightening discussions. We wish especially to thank Michael Chamberlin, Joseph Coleman, David Dressler, Olga Kennard, Alexander Rich, Akiyoshi Wada, and David Ward for providing prints of figures. Also, Walter Fitch's help with both computer searches and editorial matters (on Section VIII) is gratefully noted. The financial support of the N.I.H. (CA 20279) and the N.S.F. (PCM 77-15033) is gratefully acknowledged; W. H. was supported by the Max Kade Foundation and the Deutsche Forschungsgemeinschaft, and N. P. by an N.I.H. Postdoctoral Training Grant (T32 CA09075).

References

1. R. D. Wells, R. W. Blakesley, J. F. Burd, H. W. Chan, J. B. Dodgson, S. C. Hardies, G. T. Horn, K. F. Jensen, J. Larson, I. F. Nes, E. Selsing, and R. M. Wartell, *Crit. Rev. Biochem.* **4**, 305 (1977).
2. P. H. von Hippel and J. D. McGhee, *ARB* **41**, 231 (1972).
3. M. Rosenberg and D. Court, *Annu. Rev. Genet.* **13**, 319 (1979).
4. P. H. von Hippel, in "Biological Regulation and Development" (R. F. Goldberger, ed.), Vol. 1, p. 279. Plenum, New York, 1979.
5. S. Arnott, R. Chandrasekaran, and E. Selsing, in "Structure and Conformation of Nucleic Acids and Protein-Nucleic Acid Interactions" (M. Sundaralingam and S. T. Rao, eds.), p. 577. Univ. Park Press, Baltimore, Md., 1975.
6. S. Arnott and E. Selsing, *JMB* **98**, 265 (1975).
7. R. Langridge, D. A. Marvin, W. E. Seeds, H. R. Wilson, C. W. Hooper, M. H. F. Wilkins, and C. D. Hamilton, *JMB* **2**, 38 (1960).
8. S. Arnott and D. W. L. Hukins, *BBRC* **47**, 1504 (1972).
9. B. Wolf and S. Hanlon, *Bchem* **14**, 1661 (1975).
10. S. Hanlon, S. Brudno, T. T. Wu, and B. Wolf, *Bchem* **14**, 1648 (1975).
11. A. Chan, R. Kilkuskie, and S. Hanlon, *Bchem* **18**, 84 (1979).
12. V. I. Ivanov, L. E. Minchenkova, E. E. Minyat, M. D. Frank-Kamenetskii, and A. K. Schyolkina, *JMB* **87**, 817 (1974).
13. C. A. Sprecher, W. A. Baase, and W. C. Johnson, Jr., *Biopolymers* **18**, 1009 (1979).
14. S. K. Zavriev, L. E. Minchenkova, M. D. Frank-Kamenetskii, and I. V. Ivanov, *NAR* **5**, 2651 (1978).
15. E. E. Minyat, V. I. Ivanov, A. M. Kritzyn, L. E. Minchenkova, and A. K. Schoylkina, *JMB* **128**, 397 (1978).
16. N. R. Kallenbach and H. M. Berman, *Q. Rev. Biophys.* **10**, 138 (1977).
17. E. Selsing, R. D. Wells, T. A. Early, and D. R. Kearns, *Nature* **275**, 149 (1978).
18. E. Selsing and R. D. Wells, *JBC* **254**, 5410 (1979).
19. E. Selsing, R. D. Wells, C. J. Alden, and S. Arnott, *JBC* **254**, 5417 (1979).
20. R. D. Wells and R. M. Wartell, in "Biochemistry of Nucleic Acids" (K. Burton, ed.), Vol. 6, p. 41. Butterworths, London, 1974.
21. F. H. C. Crick and A. Klug, *Nature* **255**, 530 (1975).
22. H. M. Sobell, C. Tsai, S. G. Gilloy, S. C. Jain, and T. D. Sakote, *PNAS* **75**, 3068 (1976).
23. J. Feigon and D. R. Kearns, *NAR* **6**, 2327 (1979).
24. J. T. Finch, C. Lutter, D. Rhodes, R. Brown, B. Rushton, M. Levitt, and A. Klug, *Nature* **269**, 29 (1977).
25. A. Dugaiczyk, H. W. Boyer, and H. M. Goodman, *JMB* **96**, 171 (1975).
26. J. L. Sussman and E. N. Trifonov, *PNAS* **75**, 103 (1978).
27. M. Levitt, *PNAS* **75**, 640 (1978).
28. W. K. Olson, *Biopolymers* **14**, 1775 (1975).
29. N. Yathindra and M. Sundaralingam, *PNAS* **71**, 3325 (1974).
30. R. Teveari, K. Nanda, and G. Govil, *Biopolymers* **13**, 2015 (1974).
31. W. K. Olson, *Biopolymers* **18**, 1213 (1979).
32. W. K. Olson, *Biopolymers* **18**, 1235 (1979).
33. A. D. Mizzalukov and A. Rich, *PNAS* **76**, 1118 (1979).
34. *Nature* editorial **281**, 631 (1979).
35. J. D. Griffith, *Science* **201**, 525 (1978).
36. J. C. Wang, *PNAS* **76**, 200 (1979).
37. R. E. Depew and J. C. Wang, *PNAS* **72**, 4275 (1975).

38. M. Nall, *JMB* **116**, 49 (1977).
39. E. N. Trifonov and T. Bettecken, *Bchem* **18**, 454 (1979).
40. S. B. Zimmerman and B. H. Pheiffer, *PNAS* **76**, 2705 (1979).
41. L. Klevan, I. M. Armitage, and D. M. Crothers, *NAR* **6**, 1607 (1979).
42. T. A. Early and D. R. Kearns, *PNAS* **76**, 4165 (1979).
43. M. A. Viswamitra, O. Kennard, Z. Shakked, P. G. Jones, G. M. Sheldrick, S. Salisbury, and C. Falvello, *Nature* **173**, 687 (1978).
44. A. Klug, A. Jack, N. A. Viswamitra, O. Kennard, Z. Shakked, and T. A. Steitz, *JMB* **131**, 669 (1979).
45. A. H. J. Wang, G. J. Quigley, F. J. Kolpak, J. L. Crawford, J. H. Van Boom, G. van der Marel, and A. Rich, *Nature* **282**, 680 (1979).
46. S. Arnott, R. Chandrasekaran, D. L. Birdsall, A. G. W. Leslie, and R. L. Ratliff, *Nature* **283**, 743 (1980).
47. R. D. Wells, J. E. Larson, R. C. Grant, B. E. Shortle, and C. R. Cantor, *JMB* **54**, 465 (1970).
48. R. C. Grant, S. J. Harwood, and R. D. Wells, *JACS* **90**, 4474 (1968).
49. R. C. Grant, M. Kodama, and R. D. Wells, *Bchem* **11**, 805 (1972).
50. R. D. Wells, E. Ohtsuka, and H. G. Khorana, *JMB* **14**, 221 (1965).
51. Y. Mitsui, R. Langridge, B. E. Shortle, C. R. Cantor, R. C. Grant, M. Kodama, and R. D. Wells, *Nature* **228**, 1166 (1970).
52. N. Yathindra and M. Sundaralingam, *NAR* **3**, 729 (1976).
53. M. Sundaralingam and N. Yathindra, *Int. J. Quantum Chem.* **4**, 285 (1977).
54. D. J. Patel, L. L. Canuel, and F. M. Pohl, *PNAS* **76**, 2508 (1979).
55. F. M. Pohl and T. M. Jovin, *JMB* **67**, 375 (1972).
56. F. Pohl, T. M. Jovin, W. Baehr, and J. Holbrook, *PNAS* **69**, 3805 (1972).
57. H. R. Drew, R. E. Dickerson, and K. Itahkura, *JMB* **25**, 535 (1978).
58. H. Shindo, R. T. Simpson, and J. S. Cohen, *JBC* **254**, 8125 (1979).
59. E. Selsing and R. D. Wells, *NAR* **6**, 3025 (1979).
60. E. Paleček, *This Series* **18**, 151 (1976).
61. P. Anderson and W. Bauer, *Bchem* **17**, 594 (1978).
62. J. Marmur, R. Rownd, and C. L. Schildkraut, *This Series* **1**, 232 (1965).
63. A. Wada, S. Yabuki, and Y. Husimi, *CRC Crit. Rev. Biochem.*, in press (1979).
64. B. B. Jones, H. W. Chan, S. Rothstein, R. D. Wells, and W. S. Reznikoff, *PNAS* **74**, 4914 (1977).
65. S. Dasgupta, D. P. Allison, C. E. Snyder, and S. Mitra, *JBC* **252**, 5916 (1977).
66. H. J. Vollenweider, M. Fiandt, and W. Szybalski, *Science* **205**, 508 (1979).
67. P. Botchan, *JMB* **105**, 161 (1976).
68. N. Panayotatos and R. D. Wells, *Nature* **280**, 35 (1979).
69. R. M. Wartell, *NAR* **4**, 2779 (1977).
70. M. J. Chamberlin, in "RNA Polymerase" (R. Losick and M. J. Chamberlin, eds.), p. 159. Cold Spring Harbor Lab., New York, 1976.
71. J. C. Wang, J. H. Jacobsen, and J. M. Saucier, *NAR* **4**, 1225 (1977).
72. A. Helnikova, R. Beabealashvilli, and A. D. Mirzavekov, *EJB* **84**, 301 (1978).
73. T. Hsieh and J. C. Wang, *NAR* **5**, 3337 (1978).
74. U. Siebenlist, *Nature* **279**, 651 (1979).
75. D. L. Vizard and A. T. Ansevin, *Bchem* **15**, 741 (1976).
76. O. Gotoh, Y. Husimi, S. Yaleuki, and A. Wada, *Biopolymers* **15**, 655 (1976).
77. G. S. Manning, *Qt. Rev. Biophys.* **11**, 179 (1978).
78. T. M. Record Jr., C. F. Anderson, and T. M. Lohman, *Qt. Rev. Biophys.* **11**, 103 (1978).
79. F. Michel, J. Lazowska, G. Faye, H. Fukuhara, and P. P. Slonimski, *JMB* **85**, 411 (1974).

80. F. Michel, *JMB* **89**, 305 (1974).
81. C. Reiss and T. Arpa-Gabarro, in "Progress in Molecular and Subcell Biology" (F. E. Hahn, ed.), Vol. 5. Springer-Verlag, Berlin and New York, 1977.
82. S. Yabuki, O. Gotoh, and A. Wada, *BBA* **395**, 258 (1975).
83. O. Gotoh, Y. Husimi, S. Yabuki, and A. Wada, *Biopolymers* **15**, 655 (1976).
84. C. Takiyama, O. Gotoh, and A. Wada, *Biopolymers* **16**, 427 (1977).
85. O. Gotoh, A. Wada, and S. Yabuki, *Biopolymers* **18**, 805 (1979).
86. D. L. Vizard and A. T. Ansevin, *Bchem* **15**, 741 (1976).
87. J. Marmur and P. Doty, *JMB* **5**, 109 (1962).
88. H. Tachibana, A. Wada, O. Gotoh, and M. Takanami, *BBA* **517**, 319 (1978).
89. A. Wada, S. Ueno, H. Tachibana, and Y. Husimi, *J. Biochem.* **85**, 827 (1979).
90. B. Y. Tong and S. J. Battersby, *Biopolymers* **17**, 2933 (1978).
91. Y. L. Lyubchenko, A. V. Vologodskii, and M. D. Frank-Kamenetskii, *Nature* **271**, 28 (1978).
92. M. D. Frank-Kamenetskii and A. V. Vologodskii, *Nature* **269**, 729 (1977).
93. B. Y. Tong and S. J. Battersby, *Biopolymers* **18**, 1917 (1979).
94. G. I. Karataev, V. I. Permogorov, A. V. Vologodskii, and M. D. Frank-Kamenetskii, *NAR* **5**, 2493 (1978).
95. S. C. Hardies, W. Hillen, T. C. Goodman, and R. D. Wells, *JBC* **254**, 10128, (1979).
96. M. A. Grachev and M. P. Perelroyzen, *NAR* **5**, 2557 (1978).
97. S. C. Hardies, R. K. Patient, R. D. Klein, F. Ho, W. S. Reznikoff, and R. D. Wells, *JBC* **254**, 5527 (1979).
98. S. C. Hardies and R. D. Wells, *Gene* **7**, 1 (1979).
99. J. E. Larson, S. C. Hardies, R. K. Patient, and R. D. Wells, *JBC* **254**, 5535 (1979).
100. R. K. Patient, S. C. Hardies, and R. D. Wells, *JBC* **254**, 5542 (1979).
101. R. K. Patient, S. C. Hardies, J. E. Larson, R. B. Inman, L. E. Maquat, and R. D. Wells, *JBC* **254**, 5548 (1979).
102. J. F. Burd, R. M. Wartell, J. B. Dodgson, and R. D. Wells, *JBC* **250**, 5109 (1975).
103. R. C. Dickson, J. Abelson, W. M. Barnes, and W. S. Reznikoff, *Science* **187**, 27 (1975).
104. D. R. Kearns, *Annu. Rev. Biophys. Bioeng.* **6**, 477 (1977).
105. R. D. Blake and S. G. Lefoley, *BBA* **518**, 235 (1978).
106. D. J. Patel and L. L. Canuel, *PNAS* **73**, 674 (1976).
107. D. J. Patel and L. L. Canuel, *EJB* **96**, 267 (1979).
108. T. A. Early, D. R. Kearns, J. F. Burd, J. E. Larson, and R. D. Wells, *Bchem* **16**, 541 (1977).
109. J. F. Burd and R. D. Wells, *JBC* **249**, 7094 (1974).
110. J. F. Burd, J. E. Larson, and R. D. Wells, *JBC* **250**, 6002 (1975).
111. D. B. Davies and S. S. Danyluk, *Bchem* **14**, 543 (1975).
112. D. Patel, in "Geometry and Dynamics of Nucleic Acids" (R. Sarma, ed.). Pergamon, New York, 1980.
113. P. H. von Hippel and G. Felsenfeld, *Bchem* **3**, 27 (1964).
114. L. Wingert and P. H. von Hippel, *BBA* **157**, 114 (1968).
115. V. M. Vogt, *EJB* **33**, 192 (1973).
116. D. Kowalski, W. D. Kroeker, and M. Laskowski, *Bchem* **15**, 4457 (1976).
117. H. W. Chan and R. D. Wells, *Nature* **252**, 205 (1974).
118. H. W. Chan, J. B. Dodgson, and R. D. Wells, *Bchem* **16**, 2356 (1977).
119. B. B. Jones, H. Chan, S. Rothstein, R. D. Wells, and W. S. Reznikoff, *PNAS* **74**, 4914 (1977).
120. W. D. Kroeker and D. Kowalski, *Bchem* **17**, 3236 (1978).
121. A. E. Pritchard and M. Laskowski, *JBC* **253**, 7989 (1978).

122. A. E. Pritchard and M. Laskowski, *JBC* **253**, 6606 (1978).
122a. R. J. Legerski, H. B. Gray, and D. L. Robberson, *JBC* **252**, 8740 (1977).
122b. R. J. Legerski, J. L. Hodnett, and H. B. Gray, *NAR* **5**, 1445 (1978).
123. J. Kinberg-Calhoun, W. T. Ruyechan, and J. G. Wetmur, *FP* **38**, 1431 (1979).
124. J. B. Dodgson and R. D. Wells, *Bchem* **16**, 2367 (1977).
125. J. B. Dodgson and R. D. Wells, *Bchem* **16**, 2374 (1977).
126. J. R. Hutton and J. G. Wetmur, *BBRC* **66**, 942 (1975).
127. H. Chan, Ph.D. Thesis, University of Wisconsin (1976).
128. B. E. Funnell and R. B. Inman, *JMB* **131**, 331 (1979).
129. C. Brack, T. A. Bickle, and R. Yuan, *JMB* **96**, 693 (1975).
130. K. F. Jensen, I. F. Nes, and R. D. Wells, *NAR* **3**, 3143 (1976).
131. M. S. Flashner, M. A. Katopes, and J. Lebowitz, *NAR* **4**, 1713 (1977).
132. P. Hale and J. Lebowitz, *J. Virol.* **25**, 305 (1978).
133. P. Hale, R. S. Woodward, and J. Lebowitz, *Nature* **284**, 640 (1980).
134. H. Teitelbaum and S. W. Englander, *JMB* **92**, 79 (1975).
135. M. Y. Feldman, *This Series* **13**, 1 (1973).
136. J. D. McGhee and P. H. von Hippel, *Bchem* **14**, 1281 (1975).
137. J. D. McGhee and P. H. von Hippel, *Bchem* **14**, 1297 (1975).
138. W. M. Brown and J. Vinograd, *PNAS* **71**, 4617 (1974).
139. S. Dasgupta, D. P. Allison, C. E. Snyder, and S. Mitra, *JBC* **252**, 5916 (1977).
140. A. J. Lomant and J. R. Fresco, *This Series* **15**, 185 (1975).
141. R. M. Wartell and J. F. Burd, *Biopolymers* **15**, 1461 (1976).
142. J. Gabbarro-Arpa, P. Tougard, and C. Reiss, *Nature* **280**, 515 (1979).
143. J. Gabbarro-Arpa and C. Reiss, *C. R.*, submitted (1979).
144. R. D. Klein, E. Selsing, and R. D. Wells, *Inte. Congr. Bchem, XIth*, 01-1-H26 (1979).
145. J. F. Burd and R. D. Wells, *FP* **33**, 1128 (1974).
146. M. Hogan, N. Dattagupta, and D. M. Crothers, *Nature* **278**, 521 (1979).
147. D. J. Patel and L. L. Canuel, *PNAS* **74**, 5209 (1977).
148. D. J. Patel, *Acc. Chem. Res.* **12**, 118 (1979).
149. J. Kania and T. G. Fanning, *EJB* **67**, 367 (1976).
150. V. V. Nosikov, E. A. Braga, A. V. Karlishev, A. L. Zhuze, and O. L. Polyanovsky, *NAR* **3**, 2293 (1976).
151. E. Loucks, G. Chaconas, R. W. Blakesley, R. D. Wells, and J. H. van de Sande, *NAR* **6**, 1869 (1979).
152. T. R. Krugh and M. A. Young, *Nature* **269**, 627 (1977).
153. T. R. Krugh, J. W. Hook, S. Lin, and F. M. Chen, *In* "Sterochemistry of Molecular Systems" (R. H. Sarma, ed.), p. 423. Pergamon, New York, 1979.
154. R. M. Wartell, J. E. Larson, and R. D. Wells, *JBC* **250**, 2698 (1975).
155. E. Selsing, R. D. Wells, C. J. Alden, and S. Arnott, *JBC* **254**, 5417 (1979).
156. H. Asakura, M. Hori, and H. Umezawa, *J. Antibiot.* **28**, 537 (1975).
157. G. R. Keilman, B. Tanimoto, and R. H. Doi, *BBRC* **67**, 414 (1975).
158. T. J. Lindell, A. F. O'Malley, and B. Puglisi, *Bchem* **17**, 1154 (1978).
159. A. Gierer, *Nature* **212**, 1480 (1966).
160. H. M. Sobell, B. S. Reddy, K. K. Bhandary, S. C. Jain, T. D. Sakore, and T. P. Seshadri, *CSHSQB* **42**, 87 (1977).
161. C. J. Benham, *PNAS* **76**, 3870 (1979).
162. M. Gellert, K. Mizuuchi, M. H. O'Dea, H. Ohmori, and J. Tomizawa, *CSHSQB* **43**, 35 (1978).
163. J. R. Sadler, M. Tecklenburg, and J. L. Betz, *Gene* **8**, 279 (1980).
164. S. C. Hardies and R. D. Wells, *PNAS* **73**, 3117 (1976).

165. R. D. Wells, S. C. Hardies, G. T. Horn, B. Klein, J. E. Larson, S. K. Neuendorf, N. Panayotatos, R. K. Patient, and E. Selsing, *Methods Enzymol.* **65**, 327 (1980).
166. G. C. Walker, O. D. Uhlenbeck, E. Bedows, and R. I. Gumport, *PNAS* **72**, 122 (1975).
167. A. Landy, C. Foeller, R. Reszelbach, and B. Dudock, *NAR* **3**, 2575 (1976).
168. B. W.-K. Shum and D. M. Crothers, *NAR* **5**, 2297 (1978).
169. E. M. Trip and M. Smith, *NAR* **5**, 1529 (1978).
170. H.-J. Fritz, R. Belagaje, E. L. Brown, R. H. Fritz, R. A. Jones, R. G. Lees, and H. G. Khorana, *Bchem* **17**, 1257 (1978).
171. H. Eshaghpour and D. M. Crothers, *NAR* **5**, 13 (1978).
172. D. A. Usher, *NAR* **6**, 2289 (1979).
173. S. M. Tilghman, D. C. Tiemeier, F. Polsky, M. H. Edgell, J. G. Seidman, A. Leder, L. W. Enquist, B. Norman, and P. Leder, *PNAS* **74**, 4406 (1977).
174. J. G. Seidman, A. Leder, M. H. Edgell, F. Polsky, S. M. Tilghman, D. C. Tiemeier, and P. Leder, *PNAS* **75**, 3881 (1978).
175. A. Leder, H. I. Miller, D. H. Hamer, J. G. Seidman, B. Norman, M. Sullivan, and P. Leder, *PNAS* **75**, 6187 (1978).
176. D. C. Tiemeier, S. M. Tilghman, F. I. Polsky, J. G. Seidman, A. Leder, M. H. Edgell, and P. Leder, *Cell* **14**, 237 (1978).
177. W. McClements, H. Hanafusa, S. Tilghman, and A. Skalka, *PNAS* **76**, 2165 (1979).
178. F. Polsky, M. H. Edgell, J. G. Seidman, and P. Leder, *Anal. Bchem* **87**, 397 (1978).
179. J. Scaife and J. Beckwith, *CSHSQB* **31**, 403 (1966).
180. A. Schmitz and D. J. Galas, *NAR* **6**, 111 (1979).
181. "The Operon" (W. S. Reznikoff and J. Miller, eds.). Cold Spring Harbor Lab., New York, 1978.
182. A. Travers, *Cell* **3**, 97 (1974).
183. "RNA Polymerase" (R. Losick and M. J. Chamberlin, eds.). Cold Spring Harbor Lab., New York, 1976.
184. W. Gilbert, *in* "RNA Polymerase" (R. Losick and M. J. Chamberlin, eds.), p. 193. Cold Spring Harbor Lab., New York, 1976.
185. D. Pribnow, *in* "Biological Regulation and Development. Vol. I. Gene Expression" (R. Goldberger, ed.), p. 219. Plenum, New York, 1979.
186. W. S. Reznikoff and J. Abelson, *in* "The Operon" (W. S. Reznikoff and J. Miller, eds.), p. 221. Cold Spring Harbor Lab., New York, 1978.
187. M. J. Chamberlin, *ARB* **43**, 721 (1974).
188. M. J. Chamberlin, W. C. Nierman, J. Wiggs, and N. Neff, *JBC* **254**, 10061 (1979).
189. A. Travers, *Mol. Gen. Genet.* **147**, 225 (1976).
190. A. Travers, *Abstr. Int. Congr. Biochem. XIth*, p. 61 (1979).
191. P. Primakoff and S. W. Artz, *PNAS* **76**, 1726 (1979).
192. L. Johnsrud, *PNAS* **75**, 5314 (1978).
193. R. B. Simpson, *PNAS* **76**, 3233 (1979).
194. L. M. Maquat and W. S. Reznikoff, *JMB* **125**, 467 (1979).
195. H. Schaller, C. Gray, and K. Herrman, *PNAS* **72**, 737 (1975).
196. D. Pribnow, *PNAS* **72**, 784 (1975).
197. D. Pribnow, *JMB* **99**, 419 (1975).
198. T. M. Maniatis, M. Ptashne, K. Backman, D. Kleid, S. Flashman, A. Jeffrey, and R. Maurer, *Cell* **5**, 109 (1975).
199. G. E. F. Scherer, M. D. Walkinshaw, and S. Arnott, *NAR* **5**, 3759 (1978).
200. S. F. Gilbert, H. deBoer, and M. Nomura, *Cell* **17**, 211 (1979).
201. R. A. Young and J. A. Steitz, *Cell* **17**, 225 (1979).
202. H. Schimatake and M. Rosenberg, *FP* **37**, 1499 (1978).

203. W. S. Reznikoff, in "RNA Polymerase" (R. Losick and M. J. Chamberlin, eds.), p. 441. Cold Spring Harbor Lab., New York, 1976.
204. J. Majors, PNAS 72, 4394 (1975).
205. M. L. Berman and J. Beckwith, JMB 130, 303 (1979).
206. W. R. McClure, C. L. Cech, and D. E. Johnston, JBC 253, 8941 (1978).
207. H. S. Strauss, R. R. Burgess, and M. T. Record, Bchem, in press (1979).
208. J. C. Wang and L. F. Liu, in "Molecular Genetics" (J. H. Taylor, ed.), Vol. III Academic Press, New York, 1979.
209. S. Nakanishi, S. Adhya, M. Gottesman, and I. Pastan, JBC 249, 4050 (1974).
210. S. Nakanishi, S. Adhya, M. Gottesman, and I. Pastan, JBC 250, 8202 (1975).
211. B. Y. Tong and S. J. Battersby, NAR 6, 1073 (1979).
212. K. Nakamura and M. Inouye, Cell 18, 1109 (1979).
213. K. Mueller, C. Oebbecke, and G. F. Föster, Cell 10, 121 (1977).
214. H. A. deBoer, S. F. Gilbert, and M. Nomura, Cell 17, 201 (1979).
215. J. Vinograd, J. Lebowitz, and R. Watson, JMB 33, 173 (1968).
216. P. Botchan, J. C. Wang, and H. Echols, PNAS 70, 3077 (1973).
217. B. Sanzey, J. Bact. 138, 40 (1979).
218. H.-L. Yang, K. Heller, M. Gellert, and G. Zubay, PNAS 76, 3304 (1979).
219. W. R. Bauer, Annu. Rev. Biophys. Bioeng. 7, 287 (1978).
220. H. Kasamatsu and J. Vinograd, ARB 43, 695 (1974).
221. D. R. Helinski and D. B. Clewell, ARB 40, 899 (1971).
222. J. J. Champoux, ARB 47, 449 (1978).
223. C. L. Peebles, N. P. Higgins, K. N. Kreuzer, A. Morrison, P. O. Brown, A. Sugino, and N. R. Cozzarelli, CSHSQB 43, 41 (1978).
224. M. Gellert, K. Mizuuchi, M. H. O'Dea, and H. A. Nash, PNAS 73, 3872 (1976).
225. N. P. Higgins, C. L. Peebles, A. Sugino, and N. R. Cozzarelli, PNAS 75, 1773 (1978).
226. A. Sugino, C. L. Peebles, K. N. Kreuser, and N. R. Cozzarelli, PNAS 74, 4767 (1977).
227. M. Gellert, K. Mizuuchi, M. H. O'Dea, T. Itoh, and J. Tomizawa, PNAS 74, 4772 (1977).
228. L. F. Liu and J. C. Wang, PNAS 75, 2098 (1978).
229. L. F. Liu and J. C. Wang, Cell 15, 979 (1978).
230. P. O. Brown and N. R. Cozzarelli, Science 206, 1081 (1979).
231. M. Gellert, M. H. O'Dea, T. Itoh, and J. Tomizawa, PNAS 73, 4474 (1976).
232. A. Sugino, N. P. Higgins, P. O. Brown, C. L. Peebles, and N. R. Cozzarelli, PNAS 75, 4838 (1978).
233. K. N. Kreuzer and N. R. Cozzarelli, J. Bact. 140, 424 (1979).
234. A. Morrison and N. R. Cozzarelli, Cell 17, 175 (1979).
235. M. J. Ryan, in "The Antibiotics" (F. E. Hahn, ed.), Vol. V, Part 1, p. 214. Springer-Verlag, Berlin and New York, 1979.
236. M. J. Ryan, Bchem 15, 3769 (1976).
237. M. J. Ryan and R. D. Wells, Bchem 15, 3778 (1976).
238. T. Itoh and J. Tomizawa, Nature 270, 78 (1977).
239. M. A. DeWyngaert and D. C. Hinkle, J. Virol. 29, 529 (1979).
240. J. D. Watson, Nature NB 239, 197 (1972).
241. K. Drilca and M. Snyder, JMB 120, 145 (1978).
242. W. A. Gross and T. M. Cook, in "Antibiotics" (J. W. Corcoran and F. E. Hahn, eds.), Vol. III, p. 174. Springer-Verlag, Berlin and New York, 1975.
243. A. Puga and I. Tessman, JMB 75, 99 (1973).
244. S. C. Falco, R. Zivin, and L. B. Rothman-Denes, PNAS 75, 3220 (1978).
245. H. Shuman and M. Schwartz, BBRC 64, 204 (1975).
246. J. C. Wang, JMB 87, 797 (1974).

247. T.-S. Hsieh and J. C. Wang, *Bchem* **14**, 527 (1975).
248. Y. Hayashi and M. Hayashi, *Bchem* **10**, 4212 (1971).
249. J. P. Richardson, *JMB* **91**, 447 (1975).
250. P. H. Seeburg, C. Nüsslein, and H. Schaller, *EJB* **74**, 107 (1977).
251. A. Sankar and W. Miller, *Nature* **279**, 492 (1979).
252. C. L. Smith, M. Kubo, and F. Imamoto, *Nature* **275**, 420 (1978).
253. M. Kubo, Y. Kano, H. Nakamura, A. Nagata, and F. Imamoto, *Gene* **7**, 153 (1979).
254. A. D. Levine and W. D. Rupp, *in* "Microbiology 1978" (D. Schlessinger, ed.), p. 163. Am. Soc. Microbiology, Washington D.C., 1978.
255. P. T. Chan, J. Lebowitz, and D. Bastia, *NAR* **7**, 1247 (1979).
256. S. C. Conrad and J. L. Campbell, *Cell* **18**, 61 (1979).
257. F. W. Studier, *Science* **176**, 367 (1972).
258. F. W. Studier, *in* "Proceedings, Tenth FEBS Meeting," p. 45, 1975.
259. R. H. Doi, *Bacteriol. Rev.* **41**, 568 (1977).
260. W. C. Summers, *Annu. Rev. Genet.* **6**, 191 (1972).
261. M. W. McDonell, M. N. Simon, and F. W. Studier, *JMB* **110**, 119 (1977).
262. J. L. Oakley and J. E. Coleman, *PNAS* **74**, 4266 (1977).
263. G. A. Kassavetis and M. J. Chamberlin, *J. Virol.* **29**, 196 (1979).
264. N. Panayotatos and R. D. Wells, *JMB* **135**, 91 (1979).
265. M. Rosa, *Cell* **16**, 815 (1979).
266. J. L. Oakley, R. E. Strothkamp, A. H. Sarris, and J. E. Coleman, *Bchem* **18**, 528 (1979).
267. J. J. Dunn and F. W. Studier, *Brookhaven Symp. Biol.* **26**, 267 (1975).
268. M. Golomb and M. Chamberlin, *PNAS* **71**, 760 (1974).
269. E. G. Niles and R. C. Condit, *JMB* **98**, 57 (1975).
270. H. Beier, M. Golomb, and M. Chamberlin, *J. Virol.* **21**, 753 (1977).
271. W. T. McAllister and R. J. McCarron, *Virology* **82**, 288 (1977).
272. T. Platt *in* "The Operon" (J. H. Miller and W. S. Reznikoff, eds.), p. 263. Cold Spring Harbor Lab., New York, 1978.
273. S. Adhya and M. Gottesman, *ARB* **27**, 967 (1978).
274. J. W. Roberts, *in* "RNA Polymerase" (R. Losick and M. Chamberlin, eds.), p. 247. Cold Spring Harbor Lab., New York, 1976.
275. S. Adhya, M. Gottesman, B. de Crombrugghe, and D. Court, *in* "RNA Polymerase" (R. Losick and M. Chamberlin, eds.), p. 719. Cold Spring Harbor Lab., New York, 1976.
276. L. P. Guarente, J. Beckwith, A. M., Wu, and T. Platt, *JMB* **133**, 189 (1979).
277. F. Lee and C. Yanofsky, *PNAS* **74**, 4365 (1977).
278. S. Adhya, P. Sarkar, D. Valenzuela, and U. Maitra, *PNAS* **76**, 1613 (1979).
279. J. F. Gardner, *PNAS* **76**, 1706 (1979).
280. R. M. Gemmill, S. R. Wessler, E. B. Keller, and J. M. Calvo, *PNAS* **76**, 4941 (1979).
281. W. M. Barnes, *PNAS* **75**, 4281 (1978).
282. G. Zurawski, K. Brown, D. Killingly, and C. Yanofsky, *PNAS* **75**, 4271 (1978).
283. A. Kornberg, *in* "DNA Synthesis." Freeman, San Francisco, 1974.
284. D. Dressler, *Annu. Rev. Microbiol.* **29**, 525 (1975).
285. M. L. Gefter, *ARB* **44**, 45 (1975).
286. K. Geider, *Curr. Top. Microbiol. Immunol.* **74**, 55 (1976).
287. T. M. Jovin, *ARB* **45**, 889 (1976).
288. A. Kornberg, *in* "RNA Polymerase" (R. Losick and M. Chamberlin, eds.), p. 331. Cold Spring Harbor Lab., New York, 1976.
289. B. Alberts and R. Sternglanz, *Nature* **269**, 655 (1977).
290. J. A. Wechsler, *in* "DNA Synthesis, Present and Future" (I. Molineux and M. Kohiyama, eds.), p. 49. Plenum, New York, 1977.

291. W. H. Wickner, *ARB* **47**, 1163 (1978).
292. J.-I. Tomizawa and G. Selzer, *ARB* **48**, 999 (1979).
293. R. Kolter and D. R. Helinski, *Annu. Rev. Genet.* **13**, 355 (1979).
294. J. Sims, D. Capon, and D. Dressler, *JBC* **254**, 12615 (1979).
295. J. Ikeda, A. Yudelevich, and J. Hurwitz, *PNAS* **73**, 2669 (1976).
296. K. Denniston-Thompson, D. D. Moore, K. E. Kruger, M. E. Furth, and F. R. Blattner, *Science* **198**, 1051 (1977).
297. G. Hobom, M. Lusky, R. Grosschedl, and G. Scherer, *CSHSQB* **43**, 165 (1978).
298. D. D. Moore, K. Denniston-Thompson, K. E. Kruger, M. E. Furth, B. G. Williams, D. L. Daniels, and F. R. Blattner, *CSHSQB* **43**, 155 (1978).
299. D. M. Stalker, R. Kolter, and D. R. Helinski, *PNAS* **76**, 1150 (1979).
300. D. Dressler, J. Wolfson, and M. Magazin, *PNAS* **69**, 499 (1972).
301. N. Panayotatos and R. D. Wells, *JBC* **254**, 5555 (1979).
302. J.-P. Bouché, K. Zechel, and A. Kornberg, *JBC* **250**, 5995 (1975).
303. J.-P. Bouché, L. Rowen, and A. Kornberg, *JBC* **253**, 765 (1978).
304. S. Wickner and J. Hurwitz, *PNAS* **71**, 4120 (1974).
305. R. Schekman, A. Weiner, and A. Kornberg, *Science* **186**, 987 (1974).
306. A. M. Skalka, *Curr. Top. Mol. Biol. Immunol.* **78**, 201 (1978).
307. R. B. Wickner, M. Wright, S. Wickner, and J. Hurwitz, *PNAS* **69**, 3233 (1972).
308. D. Brutlag, R. Schekman, and A. Kornberg, *PNAS* **68**, 2826 (1971).
309. F. Jacob, S. Brenner, and F. Cuzin, *CSHSQB* **28**, 329 (1963).
310. E. Goldberger, *Science* **183**, 810 (1974).
311. L. Sompayrac and O. Maaloe, *Nature NB* **241**, 133 (1973).
312. N. Sigal, H. Delius, A. Kornberg, M. Gefter, and B. Alberts, *PNAS* **69**, 3537 (1972).
313. K. Geider, E. Beck, and H. Schaller, *PNAS* **75**, 645 (1978).
314. C. P. Gray, R. Sommer, C. Polke, E. Beck, and H. Schaller, *PNAS* **75**, 50 (1978).
315. K. Geider and A. Kornberg, *JBC* **249**, 3999 (1974).
316. J. V. Ravetch, K. Horiuchi, and N. D. Zinder, *PNAS* **74**, 4219 (1977).
317. S. V. Suggs and D. S. Ray, *CSHSQB* **43**, 379 (1978).
318. D. M. Martin and G. N. Godson, *JMB* **117**, 321 (1977).
319. K. Zechel, J.-P. Bouché, and A. Kornberg, *JBC* **250**, 4684 (1975).
320. R. Groschedl and G. Hobom, *Nature* **277**, 621 (1979).
321. M. Meijer, E. Beck, F. G. Hansen, H. E. N. Bergmans, W. Messer, K. von Meyenburg, and H. Schaller, *PNAS* **76**, 580 (1979).
322. K. Sugimoto, A. Oka, H. Mugisaki, M. Takanami, A. Nishimura, Y. Yasuda, and Y. Hirota, *PNAS* **76**, 575 (1979).
323. M. Lusky and G. Hobom, *Gene* **6**, 137 (1979).
324. M. E. Furth, J. L. Yates, and W. F. Dove, *CSHSQB* **43**, 147 (1978).
325. A. Johnson, B. J. Meyer, and M. Ptashne, *PNAS* **75**, 1783 (1978).
326. W. F. Dove, E. Hargrove, M. Ohashi, F. Hougli, and A. Guha, *Jpn J. Genet.* **44**, *Suppl.* 1 11 (1969).
327. M. Lusky and G. Hobom, *Gene* **6**, 173 (1979).
328. T. Itoh and J.-I. Tomizawa, *Nature* **270**, 78 (1977).
329. L. F. Liu, C.-C. Liu, and B. M. Alberts, *Nature* **281**, 456 (1979).
330. J. W. Zyskind and D. W. Smith, *PNAS* **77**, 2460 (1980).
331. M. T. Record, Jr., T. M. Lohman, and P. L. Haseth, *JMB* **107**, 145 (1976).
332. E. J. Gabbay, K. Sanford, C. S. Baxter, and L. Kapicak, *Bchem* **12**, 4021 (1973).
333. M. Durand, J.-C. Maurizot, H. N. Barazan, and C. Héléne, *Bchem* **14**, 563 (1975).
334. F. Brun, J.-J. Toulme, and C. H. Héléne, *Bchem* **14**, 558 (1975).
335. J.-L. Dimicoli and C. Héléne, *Bchem* **13**, 714 (1974).

336. J.-L. Dimicoli and C. Héléne, *Bchem* **13**, 724 (1974).
337. K. Wehling, H.-A. Arfmann, G. Seipke, and K. G. Wagner, *NAR* **4**, 413 (1977).
338. J.-J. Toulme and C. Héléne, *JBC* **252**, 244 (1977).
339. C. Chang, M. Weiskopf, and H. J. Li, *Bchem* **12**, 3028 (1973).
340. C. Helene, T. Montenay-Garestier, and J.-L. Dimicoli, *BBA* **254**, 349 (1971).
341. J.-C. Maurizot, G. Boubault, and C. Héléne, *Bchem* **17**, 2096 (1978).
342. B. M. Alberts, F. J. Amadia, M. Jenkins, E. D. Gutmann, and F. L. Ferris, *CSHSQB* **33**, 289 (1968).
343. B. M. Alberts and L. Frey, *Nature* **227**, 1313 (1970).
344. D. E. Jensen and P. H. von Hippel, *JBC* **251**, 7198 (1976).
345. R. H. Epstein, A. Balle, C. M. Steinberg, E. Kellenberger, E. Boy de la Tour, R. Chevallez, R. S. Edgar, M. Susman, G. H. Denhardt, and A. Lielausis, *CSHSQB* **28**, 375 (1963).
346. J. Huberman, A. Kornberg, and B. M. Alberts, *JMB* **62**, 39 (1971).
347. J. Tomizawa, N. Aniakev, and Y. Iwama, *JMB* **21**, 247 (1966).
348. H. Berger, A. J. Warren, and K. E. Fry, *J. Virol.* **3**, 171 (1969).
349. G. Mosig and S. Bock, *J. Virol.* **17**, 756 (1976).
350. S. R. Wu and Y. C. Yeh, *J. Virol.* **12**, 758 (1973).
351. K. R. Williams and W. Konigsberg, *JBC* **253**, 2463 (1978).
352. R. C. Kelly, D. E. Jensen, and P. H von Hippel, *JBC* **251**, 7240 (1976).
353. D. E. Jensen, R. D. Kelly, and P. H. von Hippel, *JBC* **251**, 7215 (1976).
354. R. A. Anderson and J. E. Coleman, *Bchem* **14**, 5485 (1975).
355. R. C. Kelly and P. H. von Hippel, *JBC* **251**, 7229 (1976).
356. H. Delius, N. J. Mantell, and B. M. Alberts, *JMB* **67**, 341 (1972).
357. H. Moise and J. Hosada, *Nature* **259**, 455 (1976).
358. E. K. Spicer, K. R. Williams, and W. H. Konigsberg, *JBC* **254**, 6433 (1979).
359. K. R. Williams, L. O. Sillerud, D. E. Schafer, and W. H. Konigsberg, *JBC* **254**, 6426 (1979).
360. J. Greve, M. F. Maestre, H. Moise, and J. Hosoda, *Bchem* **17**, 887 (1978).
361. J. L. Oey and R. Knippers, *JMB* **68**, 125 (1972).
362. B. M. Alberts, L. Frey, and H. Delius, *JMB* **68**, 139 (1972).
363. D. Pratt and W. S. Erdahl, *JMB* **37**, 181 (1968).
364. J. S. Salstrom and D. Pratt, *JMB* **61**, 489 (1971).
365. W. J. Staudenbauer and P. H. Hofschneider, *EJB* **34**, 569 (1973).
366. B. J. Mazur and P. Model, *JMB* **78**, 285 (1973).
367. Y. Nakashima, A. K. Dunker, D. A. Marvin, and W. Konigsberg, *FEBS Lett.* **40**, 290 (1974).
368. Y. Nakashima, A. K. Dunker, D. A. Marvin, and W. Konigsberg, *FEBS Lett.* **43**, 125 (1974).
369. S. J. Cavalieri, K. E. Neet, and D. A. Goldthwait, *JMB* **102**, 697 (1976).
370. L. A. Day, *Bchem* **12**, 5329 (1973).
371. R. A. Anderson, Y. Nakashima, and J. E. Coleman, *Bchem* **14**, 907 (1975).
372. J. E. Coleman, R. A. Anderson, R. G. Ratcliffe, and I. M. Armitage, *Bchem* **15**, 5419 (1976).
373. D. Pratt, D. Laws, and J. Griffith, *JMB* **82**, 425 (1974).
374. P. M. D. Fitzgerald, A. H. J. Wang, A. McPherson, F. A. Jurnak, I. Molineux, F. Kolpak, and A. Rich, *J. Sup Str.* **10**, 479 (1979).
375. A. McPherson, F. Jurnak, A. Wang, I. Molineux, and A. Rich, *J. Sup. Str.* **10**, 457 (1979).
376. N. Sigal, H. Delius, T. Kornberg, M. L. Gefter, and B. M. Alberts, *PNAS* **69**, 3537 (1972).
377. S. Eisenberg, J. F. Scott, and A. Kornberg, *PNAS* **73**, 3151 (1976).
378. S. Wickner and J. Hurwitz, *PNAS* **71**, 4120 (1974).
379. R. Shekman, J. H. Weiner, A. Weiner, and A. Kornberg, *JBC* **250**, 5859 (1975).
380. I. J. Molineux, S. Friedman, and M. L. Gefter, *JBC* **249**, 6090 (1974).
381. I. J. Molineux and M. L. Gefter, *PNAS* **71**, 3858 (1974).
382. I. J. Molineux and M. L. Gefter, *JMB* **98**, 811 (1975).

383. J. H. Weiner, L. L. Bertsch, and A. Kornberg *JBC* **250**, 1972 (1975).
384. I. J. Molineux and M. L. Gefter, *JBC* **249**, 6090 (1974).
385. W. T. Ruyechan and J. G. Wetmur, *Bchem* **15**, 5057 (1976).
386. P. K. Bandyapodhyay and C. W. Wu, *Bchem* **17**, 4078 (1978).
387. M. D. Barkley and S. Bourgeois, in "The Operon" (J. H. Miller and W. S. Reznikoff, eds.), p. 177. Cold Spring Harbor Lab., New York, 1978.
388. N. Geisler and K. Weber, *Bchem* **16**, 938 (1977).
389. R. Ogata and W. Gilbert, submitted (1979).
390. J. H. Miller, C. Coulandre, M. Hofer, U. Schmeissner, H. Sommer, A. Schmitz, and P. Lu, *JMB* **131**, 191 (1979).
391. A. P. Butler, A. Revzin, and P. H. von Hippel, *Bchem* **16**, 4757 (1977).
392. S. A. Narang, C. P. Bahl, and R. Wu, *Can. J. Biochem.* **55**, 1125 (1977).
393. P. L. de Haseth, T. M. Lohman, and M. T. Record, Jr., *Bchem* **16**, 4783 (1977).
394. D. V. Goeddel, D. G. Yansura, C. Winston, and M. H. Caruthers, *JMB* **123**, 661 (1978).
395. A. Revzin and P. H. von Hippel, *Bchem* **16**, 4769 (1977).
396. M. T. Record, Jr., P. L. de Haseth, and T. M. Lohman, *Bchem* **16**, 4791 (1977).
397. M. E. Alexander, A. A. Burgum, A. Noall, M. D. Shaw, and K. S. Matthews, *BBA* **493**, 367 (1977).
398. A. D. Riggs, H. Suzuki, and S. Bourgeois, *JMB* **48**, 67 (1970).
399. A. C. Wang, A. Revzin, A. P., Butler, and P. H. von Hippel, *NAR* **4**, 1579 (1977).
400. T. J. Richmond and T. A. Steitz, *JMB* **103**, 25 (1976).
401. A. M. Kolchinsky, A. D. Mirzabekov, W. Gilbert, and L. Li, *NAR* **3**, 11 (1976).
402. R. Ogata and W. Gilbert, *PNAS* **74**, 4973 (1977).
403. W. Gilbert, A. Maxam and A. Mirzabekov, in "Control of Ribosome Synthesis." Alfred Benzon Symp. IX, p. 139. Academic Press, New York, 1976.
404. C. P. Bahl, R. Wu, J. Stavinsky, and S. A. Narang, *PNAS* **74**, 966 (1977).
405. D. V. Goeddel, D. G. Yansura, and M. H. Caruthers, *PNAS* **74**, 3292 (1977).
406. J. C. Wang, M. D. Barkley, and S. Bourgeois, *Nature* **251**, 247 (1974).
407. H. L. Heyneker, J. Shine, H. M. Goodman, H. W. Boyer, J. Rosenberg, R. E. Dickerson, S. A. Narang, K. Itakwa, S.-Y. Lin, and A. D. Riggs, *Nature* **263**, 748 (1976).
408. M. Pfahl, *BBA* **520**, 285 (1978).
409. J.-C. Maurizot, M. Charlier, and C. Héléne, *BBRC* **60**, 951 (1974).
410. J.-C. Maurizot and M. Charlier, *FEBS Lett.* **93**, 107 (1977).
411. A. F. Melnikova, R. Beabealashvilli, and A. D. Mirzalveskov, *EJB* **84**, 301 (1978).
412. J. P. Richardson, *JMB* **21**, 83 (1966).
413. P. L. de Haseth, T. M. Lohman, R. R. Burgess, and M. T. Record, Jr., *Bchem* **17**, 1612 (1978).
414. J. S. Krakow, G. Rhodes, and T. M. Jovin, in "RNA Polymerase" (R. Losick and M. Chamberlin, eds.), p. 127. Cold Spring Harbor Lab., New York, 1976.
415. Z. Hillel and C. W. Wu, *Bchem* **17**, 2954 (1978).
416. C. W. Wu, *FP* **37**, 787 (1979).
417. J.-M. Saucier and J. C. Wang, *Nature NB* **239**, 167 (1972).
418. R. S. Beabealashvily, V. I. Ivanov, L. E. Menchenkova, and L. P. Savochkina, *BBA* **259**, 35 (1972).
419. S. A. Saxe and A. Revzin, *Bchem* **18**, 255 (1979).
420. L. K. Miller and R. D. Wells, *JBC* **247**, 2667 (1972).
421. J. N. Raquet and J. Brahms, *Biochimie* **55**, 111 (1973).
422. D. T. Denhardt, *CRC Crit. Rev. Microbiol.* **4**, 161 (1975).
423. J. Gralla and C. DeLisi, *Nature* **248**, 330 (1974).
424. W. M. Fitch, *J. Mol. Evol.* **3**, 279 (1974).

425. B. Ricard and W. Salser, *BBRC* **63**, 548 (1975).
426. L. D. Inners and G. Felsenfeld, *JMB* **50**, 373 (1970).
427. W. Fiers, R. Contreras, F. Duerinck, G. Haegeman, J. Merregaert, W. Min-Jou, A. Raeymakers, G. Volckaert, M. Ysebaert, J. Van de Kerckhove, F. Nolf, and M. Van Montagu, *Nature* **256**, 273 (1975).
428. W. Fiers, R. Contreras, F. Duerinck, G. Haegeman, D. Iserentant, J. Merregaert, W. Min-Jou, F. Molemans, A. Raeymaekers, A. Van den Berghe, G. Volckaert, and M. Ysebaert, *Nature* **260**, 500 (1976).
429. W. Min-Jou, G. Haegeman, M. Tsebaert, and W. Fiers, *Nature* **237**, 82 (1972).
430. A. B. Jacobson, *PNAS* **73**, 307 (1976).
431. T. D. Edlind and A. R. Bassel, *J. Bact.* **141**, 365 (1980).
432. P. Thammana, C. R. Cantor, P. L. Wollenzien, and J. E. Hearst, *JMB* **135**, 271 (1979).
433. P. Wollenzien, J. E. Hearst, P. Thammana, and C. R. Cantor, *JMB* **135**, 255 (1979).
434. M. L. Perdue, W. Wunderli, and W. K. Joklik, *Virology* **95**, 24 (1979).
435. T. D. Edlind and A. R. Bassel, *J. Virol.* **24**, 135 (1977).
436. A. M. Q. King, *JBC* **251**, 141 (1976).
437. J. Maisel, W. Bender, S. Hu, P. H. Duesberg, and N. Davidson, *J. Virol.* **25**, 384 (1978).
438. G. A. Luoma and A. G. Marshall, *JMB* **125**, 95 (1978).
439. P. L. Wollenzien, D. C. Youvan, and J. E. Hearst, *PNAS* **75**, 1642 (1978).
440. J. P. Calvet and T. Pederson, *NAR* **6**, 1993 (1979).
441. R. W. Blakesley and R. D. Wells, *Nature* **257**, 421 (1975).
442. K. Horiuchi and N. D. Zinder, *PNAS* **72**, 2555 (1975).
443. R. W. Blakesley, J. B. Dodgson, I. F. Nes, and R. D. Wells, *JBC* **252**, 7300 (1977).
444. R. D. Wells and J. E. Larson, *JMB* **49**, 319 (1970).
445. P. J. Greene, M. S. Poonian, A. L. Nussbaum, L. Tobias, D. E. Garfin, H. W. Boyer, and H. M. Goodman, *JMB* **99**, 237 (1975).
446. B. R. Baumstark, R. J. Roberts, and U. L. RajBhandary, *JBC* **254**, 8943 (1979).
447. G. N. Godson and R. J. Roberts, *Virology* **73**, 561 (1976).
448. R. D. Wells and S. K. Neuendorf, in "The Restriction Endonucleases" (J. Chirikjian, ed.). Elsevier, Amsterdam, 1980.
449. H. Schaller, H. Voss, and S. Gucker, *JMB* **44**, 445 (1969).
450. K. Bartok, B. Harbers, and D. T. Denhardt, *JMB* **99**, 93 (1975).
451. G. Lavelle and S. Mitra, in "Replication of Mammalian Proviruses" (D. Ward and P. Tattersall, eds.), p. 219. Cold Spring Harbor Lab., New York, 1978.
452. S. K. Niyogi and S. Mitra, *BBRC* **79**, 1037 (1977).
453. S. K. Niyogi and S. Mitra, *JBC* **253**, 5562 (1978).
454. G. J. Bourguignon, P. J. Tattersall, and D. C. Ward, *J. Virol.* **20**, 290 (1976).
455. L. A. Sherman and M. L. Gefter, *JMB* **103**, 61 (1976).
456. H. F. Tabak, J. Griffith, K. Geider, H. Schaller, and A. Kornberg, *JBC* **249**, 3049 (1974).
457. M. D. Challberg and P. T. Englund, *JBC* **254**, 7820 (1979).
458. C. R. Astell, M. Smith, M. B. Chow, and D. C. Ward, *Cell* **17**, 691 (1979).
459. C. R. Astell, M. Smith, M. B. Chow, and D. C. Ward, *Virology* **96**, 669 (1979).
460. L. A. Salzman and P. Fabisch, *J. Virol.* **30**, 946 (1979).
461. L. A. Salzman, P. Fabisch, R. Parr, C. Garon, and T. Wali, *J. Virol.* **27**, 784 (1978).
462. C. J. Shen and J. E. Hearst, *PNAS* **73**, 2649 (1976).
463. C. J. Shen, A. Ikoku, and J. E. Hearst, *JMB* **127**, 163 (1979).
464. H. Schaller, A. Uhlmann, and K. Geider, *PNAS* **73**, 49 (1976).
465. M. Kozak and D. Nathans, *Bact. Rev.* **36**, 109 (1972).
466. C. Weissmann, M. A. Billeter, H. M. Goodman, J. Hindley, and H. Weber, *ARB* **42**, 303 (1973).

467. F. Sanger, G. M. Air, B. G. Barrell, N. L. Brown, A. R. Coulson, J. C. Fiddes, C. A. Hutchison, III, P. M. Slocombe, and M. Smith, *Nature* **265,** 687 (1977).
468. G. N. Godson, B. G. Barrell, R. Staden, and J. C. Fiddes, *Nature* **276,** 236 (1978).
469. E. Beck, R. Sommer, E. A. Auerswald, C. Kurz, B. Zink, G. Osterburg, and H. Schaller, *NAR* **5,** 4495 (1978).
470. F. Sanger, A. R. Coulson, T. Friedmann, C. M. Air, B. G. Barrell, N. L. Brown, J. C. Fiddes, C. A. Hutchison, III, P. M. Slocombe, and M. Smith, *JMB* **125,** 225 (1978).
471. G. N. Godson, J. C. Fiddes, B. G. Barrell, and F. Sanger, *in* "The Single-Stranded DNA Phages" (D. T. Denhardt, D. Dressler and D. S. Ray, eds.), p. 51. Cold Spring Harbor Lab., New York, 1978.
472. J. Sims, K. Koths, and D. Dressler, *CSHSQB XLIII*, p. 349 (1978).
473. H. Schaller, E. Beck, and M. Takanami, *in* "The Single-Stranded DNA Phages" (D. T. Denhardt, D. Dressler and D. S. Ray, eds.), p. 139. Cold Spring Harbor Lab., New York, 1978.
474. J. V. Ravetch, K. Horiuchi, and N. D. Zinder, *PNAS* **74,** 4219 (1977).
475. J. V. Ravetch, K. Horiuchi, and N. D. Zinder, *JMB* **128,** 305 (1979).
476. K. Horiuchi, J. V. Ravetch, and N. D. Zinder, *CSHSQB* **XLIII,** 389 (1978).
476a. H. Schaller, *CSHSQB* **43,** 401 (1978).
477. J. C. Fiddes, *JMB* **107,** 1 (1976).
478. I. Tinoco, P. N. Borer, B. Dengler, and M. D. Levine, *Nature NB* **246,** 40 (1973).
479. J. C. Fiddes, B. G. Barrell, and G. N. Godson, *PNAS* **75,** 1081 (1978).
480. D. Hourcade and D. Dressler, *PNAS* **75,** 1652 (1978).
481. C. P. Gray, R. Sommer, C. Polke, E. Beck, and H. Schaller, *PNAS* **75,** 50 (1978).
482. R. McMacken, L. Rowen, K. Ueda, and A. Kornberg, *in* "The Single-Stranded DNA Phages" (D. T. Denhardt, D. Dressler and D. S. Ray, eds.), p. 273. Cold Spring Harbor Lab., New York, 1978.
483. R. Herrmann, K. Neugebauer, H. Schaller, and H. Zentgraf, *in* "The Single-Stranded DNA Phages" (D. T. Denhardt, D. Dressler and D. S. Ray, eds.), p. 473. Cold Spring Harbor Lab., New York, 1978.
484. J. Messing, B. Gronenborn, B. Müller-Hill, and P. H. Hofschneider, *PNAS* **74,** 3642 (1977).
485. G. N. Godson, *CSHSQB* **XLIII,** 367 (1978).
486. V. B. Reddy, B. Thimmappaya, R. Dhar, K. N. Subramanian, B. S. Zain, J. Pan, P. K. Ghosh, M. L. Celma, and S. M. Weissman, *Science* **200,** 494 (1978).
487. W. Fiers, R. Contreras, G. Haegeman, R. Rogiers, A. Van de Voorde, H. Van Heuverswyn, J. Van Herreweghe, G. Volckaert, and M. Ysebaert, *Nature* **273,** 113 (1978).
488. K. N. Subramanian, R. Dhar, and S. M. Weissman, *JBC* **152,** 355 (1977).
489. R. Tjian, *CSHSQB* **XLIII,** 655 (1978).
490. G. Dykes, R. Bambara, K. Marians, and R. Wu, *NAR* **2,** 327 (1974).
491. F. K. Fujimura and M. Hayashi, *in* "The Single-Stranded DNA Phages" (D. T. Denhardt, D. Dressler and D. S. Ray, eds.), p. 485. Cold Spring Harbor Lab. New York, 1978.
492. R. N. H. Konings and J. G. G. Shoenmakers, *in* "The Single-Stranded DNA Phages" (D. T. Denhardt, D. Dressler and D. S. Ray, eds.), p. 507. Cold Spring Harbor Lab., New York, 1978.
493. M. Hayashi, F. K. Fujimura, and M. Hayashi, *PNAS* **73,** 3519 (1976).
494. N. Axelrod, *JMB* **108,** 753 (1976).
495. N. Axelrod, *JMB* **108,** 771 (1976).
496. J. E. McMahon and I. Tinoco, *Nature* **271,** 275 (1978).
497. J. C. Fiddes and G. N. Godson, *Cell* **15,** 1045 (1978).
498. U. R. Müller and R. D. Wells, *JMB*, in press (1980).
499. U. R. Müller and R. D. Wells, *JMB*, in press (1980).

500. M. Rosenberg, D. Court, H. Shimatake, C. Brady, and D. L. Wulff, *Nature* **272**, 414 (1978).
501. M. Rosenberg, D. Court, H. Shimatake, C. Brady, and D. L. Wulff *in* "The Operon" (W. S. Reznikoff and J. Miller, eds), p. 345. Cold Spring Harbor Lab., New York, 1978.
502. T. Platt and C. Yanofsky, *PNAS* **72**, 2399 (1975).
503. E. Selker and C. Yanofsky, *JMB* **130**, 135 (1979).
504. N. D. F. Grindley, *Cell* **13**, 419 (1978).
505. D. A. Marvin, M. Spencer, M. H. F. Wilkins, and L. D. Hamilton, *JMB* **3**, 547 (1961).
506. S. Arnott and E. Selsing, *JMB* **98**, 265 (1975).
507. L. E. Post, G. D. Strycharz, M. Nomura, H. Lewis, and P. P. Dennis, *PNAS* **76**, 1697 (1979).
508. L. E. Post and M. Nomura, *JBC* **254**, 1064 (1979).

NOTE ADDED IN PROOF. Contrary to statements in this review article (cf. pp. 200–201), convincing evidence is now available on the existence of cruciforms (N. Panayotatos and R. D. Wells, in press (1980)). The supercoiled forms of two different DNAs (pVH51 and pBR322) contain specific cleavage sites for two endonucleases (S1 and the T7 gene 3 product) specific for single-stranded regions. Both sites are in the center of unusually long inverted repeats and coincide with the single-stranded loops of potential cruciform structures. Five other DNA molecules (G4, ϕ174, fd, M13, SV40), which lack long inverted repeats, are not cleaved specifically. The DNAs must be in a supercoiled form to be cleaved specifically; topoisomers with a larger number of supercoiled turns are cleaved more rapidly than those that are partially or totally relaxed. R-loop structures and nucleotide modification have been eliminated as possible explanations. These results indicate that superhelicity induces tertiary (cruciform) structures that expose specific bases in single-stranded conformations.

Index

A

Amino acids, DNA conformation and, 226-227
Antibodies
  complexes with nucleic acids, 113-114
  specific
    for nucleic acids, 110-111
    for nucleosides, 111-112
Anticodon loop, conformational flexibility of, 73-76

C

Chain elongation, at replication fork, 90-92
Chromatin
  active analysis of structure by nuclease digestion
    DNA sequences in repeating subunits, 11-12
    DNase II recognition of distinctive features, 17-19
    micrococcal nuclease analysis, 16-17
    preferential digestion by DNase I, 12-16
    utility in isolation, 19-20
  active, basic structure of, 42-44
  active, factors involved in nuclease hypersensitivity of, 45-46
  active, histone Hl and, 44-45
  active, proteins associated with, 20-21
    complement as deduced from nuclease studies, 21-38
    content of isolated native chromatin, 38-42
  active conformation, generation of, 46-49
  transcriptionally active ultrastructure
    electron microscopy of spread preparations and *in vivo* situation, 7-10
    nonribosomal gene chromatin, 5-7
    ribosomal gene chromatin, 3-5

D

Deoxyribonuclease II, active chromatin and, 17-19
Deoxyribonucleic acid
  effect of protein binding on conformation
    amino acids and polypeptides, 226-227
    conclusions, 237-238
    E. coli lactose repressor, 233-235
    E. coli RNA polymerase, 235-236
    general considerations, 224-226
    helix-destabilizing proteins, 227-233
    other proteins and, 236-237
  restriction fragments, preparation of large amounts, 200-203
  structure
    cruciforms, 199-200
    duplex conformations, 172-181
    long-range interactions, 194-199
    overview, 168-169
    potential recognition sites and, 169-172
    temperature-dependent conformational changes, 182-194
  structure in single-stranded viral
    biological importance, 243-244
    DNA polymerase reactions *in vitro*, 241-243
    electron microscopy, 243
    general considerations, 238
    restriction endonuclease susceptibility, 239-240
    single-strand-specific nuclease restrictions, 240-241
  secondary structure in intercistronic regions
    general considerations and definitions, 244-245
    phage lambda, 252-253
    single-stranded DNA phages, 245-252
    summary, 253
  transcription recognition sites
    bacterial promoters, 203-209
    effect of supercoiling on transcription, 209-214
    T7 late promoters, 214-217
    terminators, 217
Deoxyribonucleic acid polymerases
  general properties of, 87-90
  mechanism at replication fork
    base selection and editing, 96-102
    processive action, 94-96
    unwinding action, 92-93
  reactions *in vitro*, 241-243
Deoxyribonucleic acid replication, origins of RNA-primed

auxiliary proteins, 219–224
location of origins, 218
primase, 218–219
N^6,N^6-Dimethyladenine, antibodies specific for, 152–154
Dinitrophenol, antibodies specific for, 148–149

E

Escherichia coli
lactose repressor, DNA conformation and, 233–235
RNA polymerase, DNA conformation and, 235–236

G

Gene activity, modulation by changes in chromatin structure, 46–49

H

Histone H1, active chromatin and, 44–45

I

Immunochemical procedures
characterization of antibodies specific for nucleosides by radioimmunoassay, 127–133
immunization, 127
purification of antibodies specific for nucleosides, 133–134
synthesis and characterization of nucleoside-protein conjugates, 124–126
Initiation complex, ribosomal protein S1 and, 64–70

L

Lactose repressor, DNA conformation and, 233–235

M

N^6-Methyladenine, antibodies specific for, 154–156

5-Methylcytosine, antibodies specific for, 149–152
Micrococcal nuclease, active genes and, 16–17

N

Nuclease
active chromatin hypersensitivity to, 45–46
restrictions, single-strand specific, 240–241
Nuclease sensitivity, active chromatin and, 19–20
Nucleic acids
antibodies specific for, 110–111
complexes with antibodies, 113–114
immunochemical approaches for assessing function of modified constituents, 142–143
antibodies as site-specific probes, 143–145
antibodies specific for 7-methylguanine inhibition of mRNA translation, 145–148
immunochemical isolation, 134–135
mRNAs containing 7-methylguanine, 140–142
nucleotide sequences containing N^6-methyladenine, 135–140
immunoelectron microscopy
antibodies specific for N^6,N^6-dimethyladenine, 152–154
antibodies specific for dinitrophenol, 148–149
antibodies against N^6-methyladenine, 154–156
antibodies specific for 5-methylcytosine, 149–152
methylation of
extent of, 114–115
nucleosides methylated, 116–123
Nucleosides, antibodies specific for, 111–112
characterization of, 127–133
purification of, 133–134
Nucleoside-protein conjugates, synthesis and characterization of, 124–126

O

Oligonucleotides
immunochemical isolation of those with modified constituents, 135–142
ribosomal protein S1 and, 60–64

P

Phage(s), single-stranded DNA secondary structures, 245-252
Phage lambda, secondary structures of DNA, 252-253
Polynucleotides, ribosomal protein S1 and, 60-64
Polypeptides, DNA conformation and, 226-227
Protein(s)
  associated with active chromatin
    complement as deduced from nuclease studies, 21-38
    content of isolated native chromatin, 38-42
  DNA conformation and, 226-238
  helix-destabilizing, 227-233

R

Radioimmunoassay, for antibodies specific for nucleosides, 127-133
Recognition sites, on DNA for regulatory proteins, 169-172
Replication fork
  chain elongation at, 90-92
  mechanism of DNA polymerases at, base selection and editing, 96-102
  processive action, 94-96
  unwinding action, 92-93
Restriction endonuclease, single-stranded viral DNA and, 239-240
Ribonucleic acid
  transfer, induced conformational changes in, 71-73
    allosteric changes, 76-80
    flexibility of anticodon loop, 73-76
    kinetic data necessary to understand selection of cognate tRNA by ribosome, 81-82
    small molecules and, 80-81
Ribonucleic acid polymerase, DNA conformation and, 235-236
Ribosomal cycle, protein S1 and, 58-60
Ribosomal protein S1, complex formation with oligo- and poly-nucleotides, 60-64
  function in formation of 30S initiation complex, 64-70
  function in ribosomal cycle, 58-60
  unclarity of biochemical function, 70-71

S

Subunits, repeating, in actively transcribed DNA, 11-12
Supercoiling, transcription and, 209-215

Contents of Previous Volumes

Volume 1
"Primer" in DNA Polymerase Reactions—*F. J. Bollum*
The Biosynthesis of Ribonucleic Acid in Animal Systems—*R. M. S. Smellie*
The Role of DNA in RNA Synthesis—*Jerard Hurwitz and J. T. August*
Polynucleotide Phosphorylase—*M. Grunberg-Manago*
Messenger Ribonucleic Acid—*Fritz Lipmann*
The Recent Excitement in the Coding Problem—*F. H. C. Crick*
Some Thoughts on the Double-Stranded Model of Deoxyribonucleic Acid—*Aaron Bendich and Herbert S. Rosenkranz*
Denaturation and Renaturation of Deoxyribonucleic Acid—*J. Marmur, R. Rownd, and C. L. Schildkraut*
Some Problems Concerning the Macromolecular Structure of Ribonucleic Acids—*A. S. Spirin*
The Structure of DNA as Determined by X-Ray Scattering Techniques—*Vittoria Luzzati*
Molecular Mechanisms of Radiation Effects—*A. Wacker*

Volume 2
Nucleic Acids and Information Transfer—*Liebe F. Cavalieri and Barbara H. Rosenberg*
Nuclear Ribonucleic Acid—*Henry Harris*
Plant Virus Nucleic Acids—*Roy Markham*
The Nucleases of *Escherichia coli*—*I. R. Lehman*
Specificity of Chemical Mutagenesis—*David R. Krieg*
Column Chromatography of Oligonucleotides and Polynucleotides—*Matthys Staehelin*
Mechanism of Action and Application of Azapyrimidines—*J. Skoda*
The Function of the Pyrimidine Base in the Ribonuclease Reaction—*Herbert Witzel*
Preparation, Fractionation, and Properties of sRNA—*G. L. Brown*

Volume 3
Isolation and Fractionation of Nucleic Acids—*K. S. Kirby*
Cellular Sites of RNA Synthesis—*David M. Prescott*
Ribonucleases in Taka-Diastase: Properties, Chemical Nature, and Applications—*Fujio Egami, Kenji Takahashi, and Tsuneko Uchida*
Chemical Effects of Ionizing Radiations on Nucleic Acids and Related Compounds—*Joseph J. Weiss*
The Regulation of RNA Synthesis in Bacteria—*Frederick C. Neidhardt*
Actinomycin and Nucleic Acid Function—*E. Reich and I. H. Goldberg*
De Novo Protein in Synthesis *in Vitro*—*B. Nisman and J. Pelmont*
Free Nucleotides in Animal Tissues—*P. Mandel*

Volume 4
Fluorinated Pyrimidines—*Charles Heidelberger*
Genetic Recombination in Bacteriophage—*E. Volkin*
DNA Polymerases from Mammalian Cells—*H. M. Keir*
The Evolution of Base Sequences in Polynucleotides—*B. J. McCarthy*
Biosynthesis of Ribosomes in Bacterial Cells—*Syozo Osawa*
5-Hydroxymethylpyrimidines and Their Derivatives—*T. L. V. Ulbright*
Amino Acid Esters of RNA, Nucleotides, and Related Compounds—*H. G. Zachau and H. Feldmann*
Uptake of DNA by Living Cells—*L. Ledoux*

Volume 5
Introduction to the Biochemistry of D-Arabinosyl Nucleosides—*Seymour S. Cohen*
Effects of Some Chemical Mutagens and Carcinogens on Nucleic Acids—*P. D. Lawley*
Nucleic Acids in Chloroplasts and Metabolic DNA—*Tatsuichi Iwamura*
Enzymatic Alteration of Macromolecular Structure—*P. R. Srinivasan and Ernest Borek*
Hormones and the Synthesis and Utilization of Ribonucleic Acids—*J. R. Tata*
Nucleoside Antibiotics—*Jack J. Fox, Kyoichi A. Watanabe, and Alexander Bloch*
Recombination of DNA Molecules—*Charles A. Thomas, Jr.*
    Appendix I. Recombination of a Pool of DNA Fragments with Complementary Single-Chain Ends—*G. S. Watson, W. K. Smith, and Charles A. Thomas, Jr.*
    Appendix II. Proof that Sequences of A, C, G, and T Can Be Assembled to Produce Chains of Ultimate Length, Avoiding Repetitions Everywhere—*A. S. Fraenkel and J. Gillis*
The Chemistry of Pseudouridine—*Robert Warner Chambers*
The Biochemistry of Pseudouridine—*Eugene Goldwasser and Robert L. Heinrikson*

Volume 6
Nucleic Acids and Mutability—*Stephen Zamenhof*
Specificity in the Structure of Transfer RNA—*Kin-ichiro Miura*
Synthetic Polynucleotides—*A. M. Michelson, J. Massoulié, and W. Guschlbauer*
The DNA of Chloroplasts, Mitochondria, and Centrioles—*S. Granick and Aharon Gibor*
Behavior, Neural Function, and RNA—*H. Hydén*
The Nucleolus and the Synthesis of Ribosomes—*Robert P. Perry*
The Nature and Biosynthesis of Nuclear Ribonucleic Acids—*G. P. Georgiev*
Replication of Phage RNA—*Charles Weissmann and Severo Ochoa*

Volume 7
Autoradiographic Studies on DNA Replication in Normal and Leukemic Human Chromosomes—*Felice Gavosto*
Proteins of the Cell Nucleus—*Lubomir S. Hnilica*
The Present Status of the Genetic Code—*Carl R. Woese*
The Search for the Messenger RNA of Hemoglobin—*H. Chantrenne, A. Burny, and G. Marbaix*
Ribonucleic Acids and Information Transfer in Animal Cells—*A. A. Hadjiolov*
Transfer of Genetic Information during Embryogenesis—*Martin Nemer*
Enzymatic Reduction of Ribonucleotides—*Agne Larsson and Peter Reichard*
The Mutagenic Action of Hydroxylamine—*J. H. Phillips and D. M. Brown*
Mammalian Nucleolytic Enzymes and Their Localization—*David Shugar and Halina Sierakowska*

Volume 8
Nucleic Acids—The First Hundred Years—*J. N. Davidson*
Nucleic Acids and Protamine in Salmon Testes—*Gordon H. Dixon and Michael Smith*
Experimental Approaches to the Determination of the Nucleotide Sequences of Large Oligonucleotides and Small Nucleic Acids—*Robert W. Holley*
Alterations of DNA Base Composition in Bacteria—*G. F. Gause*
Chemistry of Guanine and Its Biologically Significant Derivatives—*Robert Shapiro*
Bacteriophage φX174 and Related Viruses—*Robert L. Sinsheimer*
The Preparation and Characterization of Large Oligonucleotides—*George W. Rushizky and Herbert A. Sober*
Purine *N*-Oxides and Cancer—*George Bosworth Brown*
The Photochemistry, Photobiology, and Repair of Polynucleotides—*R. B. Setlow*
What Really is DNA? Remarks on the Changing Aspects of a Scientific Concept—*Erwin Chargaff*
Recent Nucleic Acid Research in China—*Tien-Hsi Cheng and Roy H. Doi*

Volume 9

The Role of Conformation in Chemical Mutagenesis—*B. Singer and H. Fraenkel-Conrat*
Polarographic Techniques in Nucleic Acid Research—*E. Paleček*
RNA Polymerase and the Control of RNA Synthesis—*John P. Richardson*
Radiation-Induced Alterations in the Structure of Deoxyribonucleic Acid and Their Biological Consequences—*D. T. Kanazir*
Optical Rotatory Dispersion and Circular Dichroism of Nucleic Acids—*Jen Tsi Yang and Tatsuya Samejima*
The Specificity of Molecular Hybridization in Relation to Studies on Higher Organisms—*P. M. B. Walker*
Quantum-Mechanical Investigations of the Electronic Structure of Nucleic Acids and Their Constituents—*Bernard Pullman and Alberte Pullman*
The Chemical Modification of Nucleic Acids—*N. K. Kochetkov and E. I. Budowsky*

Volume 10

Induced Activation of Amino Acid Activating Enzymes by Amino Acids and tRNA—*Alan H. Mehler*
Transfer RNA and Cell Differentiation—*Noboru Sueoka and Tamiko Kano-Sueoka*
N^6-(Δ^2-Isopentenyl)adenosine: Chemical Reactions, Biosynthesis, Metabolism, and Significance to the Structure and Function of tRNA—*Ross H. Hall*
Nucleotide Biosynthesis from Preformed Purines in Mammalian Cells: Regulatory Mechanisms and Biological Significance—*A. W. Murray, Daphne C. Elliott, and M. R. Atkinson*
Ribosome Specificity of Protein Synthesis *in Vitro*—*Orio Ciferri and Bruno Parisi*
Synthetic Nucleotide-peptides—*Zoe A. Shabarova*
The Crystal Structures of Purines, Pyrimidines and Their Intermolecular Complexes—*Donald Voet and Alexander Rich*

Volume 11

The Induction of Interferon by Natural and Synthetic Polynucleotides—*Clarence Colby, Jr.*
Ribonucleic Acid Maturation in Animal Cells—*R. H. Burdon*
Liporibonucleoprotein as an Integral Part of Animal Cell Membranes—*V. S. Shapot and S. Ya. Davidova*
Uptake of Nonviral Nucleic Acids by Mammalian Cells—*Pushpa M. Bhargava and G. Shanmugam*
The Relaxed Control Phenomenon—*Ann M. Ryan and Ernest Borek*
Molecular Aspects of Genetic Recombination—*Cedric I. Davern*
Principles and Practices of Nucleic Acid Hybridization—*David E. Kennell*
Recent Studies Concerning the Coding Mechanism—*Thomas H. Jukes and Lila Gatlin*
The Ribosomal RNA Cistrons—*M. L. Birnstiel, M. Chipchase, and J. Speirs*
Three-Dimensional Structure of tRNA—*Friedrich Cramer*
Current Thoughts on the Replication of DNA—*Andrew Becker and Jerard Hurwitz*
Reaction of Aminoacyl-tRNA Synthetases with Heterologous tRNA's—*K. Bruce Jacobson*
On the Recognition of tRNA by Its Aminoacyl-tRNA Ligase—*Robert W. Chambers*

Volume 12

Ultraviolet Photochemistry as a Probe of Polyribonucleotide Conformation—*A. J. Lomant and Jacques R. Fresco*
Some Recent Developments in DNA Enzymology—*Mehran Goulian*
Minor Components in Transfer RNA: Their Characterization, Location, and Function—*Susumu Nishimura*
The Mechanism of Aminoacylation of Transfer RNA—*Robert B. Loftfield*
Regulation of RNA Synthesis—*Ekkehard K. F. Bautz*
The Poly(dA-dT) of Crab—*M. Laskowski, Sr.*

The Chemical Synthesis and the Biochemical Properties of Peptidyl-tRNA—*Yehuda Lapidot and Nathan de Groot*

Volume 13
Reactions of Nucleic Acids and Nucleoproteins with Formaldehyde—*M. Ya. Feldman*
Synthesis and Functions of the -C-C-A Terminus of Transfer RNA—*Murray P. Deutscher*
Mammalian RNA Polymerases—*Samson T. Jacob*
Poly(adenosine diphosphate ribose)—*Takashi Sugimura*
The Stereochemistry of Actinomycin Binding to DNA and Its Implications in Molecular Biology—*Henry M. Sobell*
Resistance Factors and Their Ecological Importance to Bacteria and to Man—*M. H. Richmond*
Lysogenic Induction—*Ernest Borek and Ann Ryan*
Recognition in Nucleic Acids and the Anticodon Families—*Jacques Ninio*
Translation and Transcription of the Tryptophan Operon—*Fumio Imamoto*
Lymphoid Cell RNA's and Immunity—*A. Arthur Gottlieb*

Volume 14
DNA Modification and Restriction—*Werner Arber*
Mechanism of Bacterial Transformation and Transfection—*Nihal K. Notani and Jane K. Setlow*
DNA Polymerases II and III of *Escherichia coli*—*Malcolm L. Gefter*
The Primary Structure of DNA—*Kenneth Murray and Robert W. Old*
RNA-Directed DNA Polymerase—Properties and Functions in Oncogenic RNA Viruses and Cells—*Maurice Green and Gray F. Gerard*

Volume 15
Information Transfer in Cells Infected by RNA Tumor Viruses and Extension to Human Neoplasia—*D. Gillespie, W. C. Saxinger, and R. C. Gallo*
Mammalian DNA Polymerases—*F. J. Bollum*
Eukaryotic RNA Polymerases and the Factors That Control Them—*B. B. Biswas, A. Ganguly, and D. Das*
Structural and Energetic Consequences of Noncomplementary Base Oppositions in Nucleic Acid Helices—*A. J. Lomant and Jacques R. Fresco*
The Chemical Effects of Nucleic Acid Alkylation and Their Relation to Mutagenesis and Carcinogenesis—*B. Singer*
Effects of the Antibiotics Netropsin and Distamycin A on the Structure and Function of Nucleic Acids—*Christoph Zimmer*

Volume 16
Initiation of Enzymic Synthesis of Deoxyribonucleic Acid by Ribonucleic Acid Primers—*Erwin Chargaff*
Transcription and Processing of Transfer RNA Precursors—*John D. Smith*
Bisulfite Modification of Nucleic Acids and Their Constituents—*Hikoya Hayatsu*
The Mechanism of the Mutagenic Action of Hydroxylamines—*E. I. Budowsky*
Diethyl Pyrocarbonate in Nucleic Acid Research—*L. Ehrenberg, I. Fedorcsák, and F. Solymosy*

Volume 17
The Enzymic Mechanism of Guanosine 5', 3'-Polyphosphate Synthesis—*Fritz Lipmann and Jose Sy*
Effects of Polyamines on the Structure and Reactivity of tRNA—*Ted T. Sakai and Seymour S. Cohen*

CONTENTS OF PREVIOUS VOLUMES 277

Information Transfer and Sperm Uptake by Mammalian Somatic Cells—*Aaron Bendich, Ellen Borenfreund, Steven S. Witkins, Delia Beju, and Paul J. Higgins*
Studies on the Ribosome and Its Components—*Pnina Spitnik-Elson and David Elson*
Classical and Postclassical Modes of Regulation of the Synthesis of Degradative Bacterial Enzymes—*Boris Magasanik*
Characteristics and Significance of the Polyadenylate Sequence in Mammalian Messenger RNA—*George Brawerman*
Polyadenylate Polymerases—*Mary Edmonds and Mary Ann Winters*
Three-Dimensional Structure of Transfer RNA—*Sung-Hou Kim*
Insights into Protein Biosynthesis and Ribosome Function through Inhibitors—*Sidney Pestka*
Interaction with Nucleic Acids of Carcinogenic and Mutagenic N-Nitroso Compounds—*W. Lijinsky*
Biochemistry and Physiology of Bacterial Ribonuclease—*Alok K. Datta and Salil K. Niyogi*

Volume 18
The Ribosome of *Escherichia coli*—*R. Brimacombe, K. H. Nierhaus, R. A. Garrett and H. G. Wittmann*
Structure and Function of 5 S and 5.8 S RNA—*Volker A. Erdmann*
High-Resolution Nuclear Magnetic Resonance Investigations of the Structure of tRNA in Solution—*David R. Kearns*
Premelting Changes in DNA Conformation—*E. Paleček*
Quantum-Mechanical Studies on the Conformation of Nucleic Acids and Their Constituents—*Bernard Pullman and Anil Saran*

Volume 19 (Symposium on mRNA: The Relation of Structure to Function)
I. The 5′-Terminal Sequence ("Cap") of mRNAs
Caps in Eukaryotic mRNAs: Mechanism of Formation of Reovirus mRNA 5′-Terminal m^7GpppGm-C—*Y. Furuichi, S. Muthukrishnan, J. Tomasz and A. J. Shatkin*
Nucleotide Methylation Patterns in Eukaryotic mRNA—*Fritz M. Rottman, Ronald C. Desrosiers and Karen Friderici*
Structural and Functional Studies on the "5′-Cap": A Survey Method of mRNA—*Harris Busch, Friedrich Hirsch, Kaushal Kumar Gupta, Manchanahalli Rao, William Spohn and Benjamin C. Wu*
Modification of the 5′-Terminals of mRNAs by Viral and Cellular Enzymes—*Bernard Moss, Scott A. Martin, Marcia J. Ensinger, Robert F. Boone and Cha-Mer Wei*
Blocked and Unblocked 5′ Termini in Vesicular Stomatitis Virus Product RNA *in Vitro:* Their Possible Role in mRNA Biosynthesis—*Richard J. Colonno, Gordon Abraham and Amiya K. Banerjee*
The Genome of Poliovirus Is an Exceptional Eukaryotic mRNA—*Yuan Fon Lee, Akio Nomoto and Eckard Wimmer*
II. Sequences and Conformations of mRNAs
Transcribed Oligonucleotide Sequences in Hela Cell hnRNA and mRNA—*Mary Edmonds, Hiroshi Nakazato, E. L. Korwek and S. Venkatesan*
Polyadenylylation of Stored mRNA in Cotton Seed Germination—*Barry Harris and Leon Dure III*
mRNAs Containing and Lacking Poly(A) Function as Separate and Distinct Classes during Embryonic Development—*Martin Nemer and Saul Surrey*
Sequence Analysis of Eukaryotic mRNA—*N. J. Proudfoot, C. C. Cheng and G. G. Brownlee*
The Structure and Function of Protamine mRNA from Developing Trout Testis—*P. L. Davies, G. H. Dixon, L. N. Ferrier, L. Gedamu and K. Iatrou*
The Primary Structure of Regions of SV40 DNA Encoding the Ends of mRNA—*Kiranur N.*

Subramanian, Prabhat K. Ghoshi, Ravi Dhar, Bayar Thimmappaya, Sayeeda B. Zain, Julian Pan and Sherman M. Weissman

Nucleotide Sequence Analysis of Coding and Noncoding Regions of Human β-Globin mRNA—Charles A. Marotta, Bernard G. Forget, Michael Cohen/Solal and Sherman M. Weissman

Determination of Globin mRNA Sequences and Their Insertion into Bacterial Plasmids—Winston Salser, Jeff Browne, Pat Clarke, Howard Heindell, Russell Higuchi, Gary Paddock, John Roberts, Gary Studnicka and Paul Zakar

The Chromosomal Arrangement of Coding Sequences in a Family of Repeated Genes—G. M. Rubin, D. J. Finnegan and D. S. Hogness

Mutation Rates in Globin Genes: The Genetic Load and Haldane's Dilemma—Winston Salser and Judith Strommer Isaacson

Heterogeneity of the 3' Portion of Sequences Related to Immunoglobulin κ-Chain mRNA—Ursula Storb

Structural Studies on Intact and Deadenylylated Rabbit Globin mRNA—John N. Vournakis, Marcia S. Flashner, MaryAnn Katopes, Gary A. Kitos, Nikos C. Vamvakopoulos, Matthew S. Sell and Regina M. Wurst

Molecular Weight Distribution of RNA Fractionated on Aqueous and 70% Formamide Sucrose Gradients—Helga Boedtker and Hans Lehrach

III. Processing of mRNAs

Bacteriophages T7 and T3 as Model Systems for RNA Synthesis and Processing—J. J. Dunn, C. W. Anderson, J. F. Atkins, D. C. Bartelt and W. C. Crockett

The Relationship between hnRNA and mRNA—Robert P. Perry, Enzo Bard, B. David Hames, Dawn E. Kelley and Ueli Schibler

A Comparison of Nuclear and Cytoplasmic Viral RNAs Synthesized Early in Productive Infection with Adenovirus 2—Heschel J. Raskas and Elizabeth A. Craig

Biogenesis of Silk Fibroin mRNA: An Example of Very Rapid Processing?—Paul M. Lizardi

Visualization of the Silk Fibroin Transcription Unit and Nascent Silk Fibroin Molecules on Polyribosomes of Bombyx mori—Steven L. McKnight, Nelda L. Sullivan and Oscar L. Miller, Jr.

Production and Fate of Balbiani Ring Products—B. Daneholt, S. T. Case, J. Hyde, L. Nelson and L. Wieslander

Distribution of hnRNA and mRNA Sequences in Nuclear Ribonucleoprotein Complexes—Alan J. Kinniburgh, Peter B. Billings, Thomas J. Quinlan and Terence E. Martin

IV. Chromatin Structure and Template Activity

The Structure of Specific Genes in Chromatin—Richard Axel

The Structure of DNA in Native Chromatin as Determined by Ethidium Bromide Binding—J. Paoletti, B. B. Magee and P. T. Magee

Cellular Skeletons and RNA Messages—Ronald Herman, Gary Zieve, Jeffrey Williams, Robert Lenk and Sheldon Penman

The Mechanism of Steroid-Hormone Regulation of Transcription of Specific Eukaryotic Genes—Bert W. O'Malley and Anthony R. Means

Nonhistone Chromosomal Proteins and Histone Gene Transcription—Gary Stein, Janet Stein, Lewis Kleinsmith, William Park, Robert Jansing and Judith Thomson

Selective Transcription of DNA Mediated by Nonhistone Proteins—Tung Y. Wang, Nina C. Kostraba and Ruth S. Newman

V. Control of Translation

Structure and Function of the RNAs of Brome Mosaic Virus—Paul Kaesberg

Effect of 5'-Terminal Structures on the Binding of Ribopolymers to Eukaryotic Ribosomes—S. Muthukrishnan, Y. Furuichi, G. W. Both and A. J. Shatkin

CONTENTS OF PREVIOUS VOLUMES 279

Translational Control in Embryonic Muscle—*Stuart M. Heywood and Doris S. Kennedy*
Protein and mRNA Synthesis in Cultured Muscle Cells—*R. G. Whalen, M. E. Buckingham and F. Gros*
VI. Summary: mRNA Structure and Function—*James E. Darnell*

Volume 20

Correlation of Biological Activities with Structural Features of Transfer RNA—*B. F. C. Clark*
Bleomycin, an Antibiotic That Removes Thymine from Double-Stranded DNA—*Werner E. G. Müller and Rudolf K. Zahn*
Mammalian Nucleolytic Enzymes—*Halina Sierakowska and David Shugar*
Transfer RNA in RNA Tumor Viruses—*Larry C. Waters and Beth C. Mullin*
Integration versus Degradation of Exogenous DNA in Plants: An Open Question—*Paul F. Lurquin*
Initiation Mechanisms of Protein Synthesis—*Marianne Grunberg-Manago and François Gros*

Volume 21

Informosomes and Their Protein Components: The Present State of Knowledge—*A. A. Preobrazhensky and A. S. Spirin*
Energetics of the Ribosome—*A. S. Spirin*
Mechanisms in Polypeptide Chain Elongation on Ribosomes—*Engin Bermek*
Synthetic Oligodeoxynucleotides for Analysis of DNA Structure and Function—*Ray Wu, Chander P. Bahl and Saran A. Narang*
The Transfer RNAs of Eukaryotic Organelles—*W. Edgar Barnett, S. D. Schwartzbach, and L. I. Hecker*
Regulation of the Biosynthesis of Aminoacid:tRNA Ligases and of tRNA—*Susan D. Morgan and Dieter Söll*

Volume 22

The -C-C-A End of tRNA and Its Role in Protein Biosynthesis—*Mathias Sprinzl and Friedrich Cramer*
The Mechanism of Action of Antitumor Platinum Compounds—*J. J. Roberts and A. J. Thomson*
DNA Glycosylases, Endonucleases for Apurinic/Apyrimidinic Sites, and Base Excision-Repair—*Thomas Lindahl*
Naturally Occurring Nucleoside and Nucleotide Antibiotics—*Robert J. Suhadolnik*
Genetically Controlled Variation in the Shapes of Enzymes—*George Johnson*
Transcription Units for mRNA Production in Eukaryotic Cells and Their DNA Viruses—*James E. Darnell, Jr.*

Volume 23

The Peptidyltransferase Center of Ribosomes—*Alexander A. Krayevsky and Marina K. Kukhanova*
Patterns of Nucleic Acid Synthesis in *Physarum polycephalum*—*Geoffrey Turnock*
Biochemical Effects of the Modification of Nucleic Acids by Certain Polycyclic Aromatic Carcinogens—*Dezider Grunberger and I. Bernard Weinstein*
Participation of Modified Nucleosides in Translation and Transcription—*B. Singer and M. Kröger*
The Accuracy of Translation—*Michael Yarus*
Structure, Function, and Evolution of Transfer RNAs (with Appendix Giving Complete Sequences of 178 tRNAs)—*Ram P. Singhal and Pamela A. M. Fallis*

RAYMOND H. FOGLER LIBRARY
DATE DUE